Research Progress and Application Prospects of Gel Electrolytes

Research Progress and Application Prospects of Gel Electrolytes

Guest Editors

Ana Isabel Ribeiro
Andrea Zille

MDPI

Basel • Beijing • Wuhan • Barcelona • Belgrade • Novi Sad • Cluj • Manchester

Guest Editors
Ana Isabel Ribeiro
Centro de Ciência e
Tecnologia Têxtil
Universidade do Minho
Guimarães
Portugal

Andrea Zille
Centro de Ciência e
Tecnologia Têxtil
Universidade do Minho
Guimarães
Portugal

Editorial Office
MDPI AG
Grosspeteranlage 5
4052 Basel, Switzerland

This is a reprint of the Special Issue, published open access by the journal *Gels* (ISSN 2310-2861), freely accessible at: https://www.mdpi.com/journal/gels/special_issues/4638YD96UE.

For citation purposes, cite each article independently as indicated on the article page online and as indicated below:

Lastname, A.A.; Lastname, B.B. Article Title. *Journal Name* **Year**, *Volume Number*, Page Range.

ISBN 978-3-7258-7300-5 (Hbk)
ISBN 978-3-7258-7301-2 (PDF)
https://doi.org/10.3390/books978-3-7258-7301-2

Contents

About the Editors

Andrea Zille

Andrea Zille is an Associate Professor at the Department of Textile Engineering and the Director of the Centre for Textile Science and Technology at the University of Minho. Throughout his career, he has been involved in European research frameworks and national projects, focusing on biotechnology and nanotechnology applied to materials. His scientific interests encompass surface functionalization of polymers, plasma modification, antimicrobial and antifouling materials, and the development of advanced soft and functional materials for textiles and related technologies. He has served as the lead author for the Intergovernmental Panel on Climate Change Refinement and, since 2021, has been listed among the world's top two percent of scientists. He also teaches international postgraduate programmes and has contributed to numerous collaborative projects that bridge fundamental research and applied material solutions.

Ana Isabel Ribeiro

Ana Isabel Ribeiro is a Researcher at the Centre for Textile Science and Technology at the University of Minho. She holds a PhD in Textile Engineering, which focused on plasma deposition of antimicrobial hybrid nanocomposites for biomedical textiles. She also holds degrees in Chemistry, Medicinal Chemistry, and Textile Engineering, reflecting a multidisciplinary background. She has contributed to European and national projects related to antibacterial coatings, smart functional textiles and, more recently, to sustainability and circular economy initiatives, including recycling of industrial textile residues and extraction of fibres from biological resources. Her scientific interests span nanotechnology, pharmacology, drug delivery, materials science, biomaterials, conductive materials and sustainability. She has authored multiple peer-reviewed publications, book chapters and conference contributions, and she coordinates collaborative research tasks with academic and industrial partners.

Preface

This Reprint compiles a selection of studies that demonstrate how fast gel electrolytes have been changing in both materials science and electrochemical engineering. By assembling this collection, we sought to offer a consistent scientific overview for researchers who need clear reference points for their work in batteries, supercapacitors, wearable electronics and next generation sensing technologies, as the field is becoming increasingly vast.

The papers collected here embody a shared drive: determining the ways in which the design, processing and integration of soft ion conducting systems could lead to the elimination of the most difficult problems, which have existed for a long time for liquid and solid electrolytes. In fact, each paper is different in terms of content and viewpoint, but the idea of the material formulation, structural control, device performance and predictive modelling have been documented by them in combination.

This Reprint serves the purpose of reaching out to scientists and engineers who seek not only foundational insights but also applied strategies for setting up safer, more resilient and more environmentally friendly electrochemical components. We believe that this collection will be useful for newcomers to the field, as well as for experts who want to deepen their understanding of structure property relationships in gels. It will also facilitate the continued interdisciplinary dialogue. Our goal is not merely to present accomplishments in the field, but also to make clear why gel electrolytes are transpiring to be the core of the next generation of energy and electronic systems.

Ana Isabel Ribeiro and Andrea Zille
Guest Editors

gels

MDPI

Editorial

Research Progress and Application Prospects of Gel Electrolytes

Andrea Zille * and Ana Isabel Ribeiro

2C2T—Centre for Textile Science and Technology, Campus Azurém, University of Minho, 4800-058 Guimarães, Portugal; a.isabel.f.ribeiro@gmail.com
* Correspondence: azille@det.uminho.pt

1. Introduction

Gel electrolytes have rapidly emerged as indispensable components in modern electrochemical systems, bridging the performance gap between liquid and solid electrolytes. These soft, ion-conducting materials combine the high ionic conductivity of liquids with the mechanical stability of solids, offering unique advantages in a single platform. In energy storage, gel polymer electrolytes (GPEs) enable safer, high-energy density batteries by mitigating the leakage, flammability, and dendrite formation issues that plague traditional liquid electrolytes. Their inherent flexibility and viscoelasticity also make them ideal for flexible and wearable electronics, where they can sustain mechanical deformation while maintaining ionic conduction. Indeed, applications now span from lithium batteries and supercapacitors to stretchable sensors and bioelectronic devices, underscoring the growing importance of gel electrolytes in delivering safe, high-performance energy storage and conversion systems.

Recognizing these trends, the first edition of this Special Issue, "Research Progress and Application Prospects of Gel Electrolytes", was conceived to highlight cutting-edge research in this vibrant field. That Special Issue attracted ten contributions encompassing a diverse range of topics, from advanced materials development to device integration. The works collectively reinforced how gel electrolytes are enabling breakthroughs in electrochemical technologies while also shedding light on remaining challenges. In the following paragraphs, we briefly summarize the key results and insights from the first edition, which form a foundation for the continued exploration in this second edition.

Realizing these patterns, the idea of the first volume of this Special Issue, "Research Progress and Application Prospects of Gel Electrolytes", was born to emphasize the forefront research of this lively sector. This edition attracted ten submissions covering a wide range of topics, starting from the development of advanced materials to device integration. The papers, as a whole, emphasized the role of gel electrolytes in enabling the breakthrough of electrochemical technologies, while at the same time, they pointed to the issues that persist. Next, we briefly identify the main findings and insights of the first volume that serve as a stepping stone for more in-depth research in this second volume.

2. Highlights from the First Edition

https://www.mdpi.com/journal/gels/special_issues/4638YD96UE (accessed on 28 November 2025).

Advances in Energy Storage: The first edition was largely dedicated to gel electrolytes for batteries and supercapacitors [contributions 1,2]. One of the primary themes was how material design can not only increase the performance of a battery but also make it safer. For instance, Mao et al. fabricated a composite gel electrolyte by incorporating a metal–organic framework (ZIF-8) in a UV-cured polymer matrix, which had an ionic conductivity of ~1.17×10^{-3} S/cm and a long electrochemical stability window of 4.7 V [contribution 3]. Consequently, this MOF-derived gel made it possible to achieve high specific capacities and enhanced rate capability in lithium metal cells, thus a better cycling performance was obtained. In their study, Yao et al. similarly prepared a UV-crosslinked PVDF-HFP/PAN hybrid gel electrolyte with LLZTO ceramic nanoparticles and achieved an ionic conductivity at room temperature of 2.8×10^{-4} S/cm [contribution 4]. It is worth noting that a $LiFePO_4$ | | Li cell with this gel gave ~164 mAh·g^{-1} at 0.1 C, and the capacity after 200 cycles was ≈89%, which represents a remarkable degree of short-term stability in solid-state cells. This indicates how far researchers have come in developing gel electrolytes that are capable of fast ion transport and stable long-term operation. Alongside these findings, scientists also face the problem of high interfacial resistance in multi-layer cell architectures. Chen et al. looked into the design of a multi-layer solid-state battery and found that the capacity in the stacked cells decreased much more rapidly compared with the single-layer cells. Three-layer stack had only 15.6% capacity retention after 100 cycles vs. single-layer cell (∼48.9% retention) [contribution 5]. This discovery exemplifies how the improvement of interfaces and architectures is indispensable as gel-based batteries are becoming thicker or of higher capacity.

Flexible Electronics and Sensors: One of the most notable changes for electronic devices due to gel electrolytes is the complete revolution of their form factors. Ribeiro et al., in their comprehensive review, examined the role of gel electrolytes in the development of textile-based power sources and wearable devices. Gels can be easily integrated into fibers, yarns, and fabric and thus not only enhance safety (no liquid leakage), but also mechanical flexibility and conformability, which are indispensable properties for powering wearable devices used in healthcare, sports, and IoT applications [contribution 6]. The review also states that gel electrolytes provide leak resistance and improved interface compatibility in such systems; however, there are still difficulties in ionic conductivity and stability over long periods under repeated deformation. The first volume of the Special Issue showed that the use of gels is spreading beyond energy storage to the field of sensing technologies. For example, Cedeño Mata et al. developed a polyvinyl alcohol-based ionic liquid gel polymer electrolyte (ILGPE) with the addition of SiO_2 nanoparticles and PVP to improve the performance of a capacitive humidity sensor [contribution 7]. The composite gel resulted in a significantly improved sensor response and hysteresis less than 0.02 over 20–90% relative humidity due to the designed pore structures and the interactions between the SiO_2/PVP fillers and the host matrix. This is a perfect example of how gel electrolytes can be reconciled with the functional needs (here water uptake dynamics) of sensor applications. These advancements in flexible and functional gels are indicative of an upcoming era in which power sources and sensors can be embedded more safely and intimately in wearable, portable, or biologically integrated devices.

Innovations in Fabrication and Design: Several works in the first edition led to the exploration of new fabrication methods for gel electrolytes and greater understanding, showing that the field is maturing and uses a variety of interdisciplinary tools. Yao et al. reported on the employment of 3D printing (direct ink writing) in the fabrication of a "ceramic-in-gel" electrolyte with a highly controlled porous microstructure [contribution 8]. The constructed gel revealed a standard sponge-like structure and reached an ionic conductivity of ~5.77×10^{-4} S/cm, which corresponds to a substantial improvement of Li-ion transport at room temperature. This innovative manufacturing method not only facilitates the design of electrolyte shapes (e.g., complex geometries or thin-film designs) for better contact in batteries and microdevices but also allows for the exploration of other possibilities in the field of electrolytes. On the other hand, data-driven methods are employed to decipher gels' complex chemistry. Xu et al. presented an interpretable machine learning framework that links PVA hydrogel formulation parameters to mechanical performance [contributions 9,10]. Using large-scale data point examination, their model speculated that PVA molecular weight was the main factor controlling the tensile properties, while the degree of hydrolysis and cross-linker content played the next most important roles in the hierarchy. Moreover, the insights gained through SHAP interpretability analysis offer a solid and understandable platform for designing the next generation of hydrogels with desirable properties. In parallel, another paper approached the topic in a different way, introducing machine learning for supercapacitors' hydrogel electrolyte optimization and revealing design factors leading to better capacitance and cyclability (e.g., polymer composition, crosslink, and density). In the end, both advanced fabrication and computational modeling methods spearhead the transformation of gel electrolytes from a trial-and-error approach toward *engineered* gels with predictable structure–property relationships.

The first volume of this Special Issue has been a splendid demonstration of gel electrolytes' plasticity and their ripple effect across numerous fields. The array of contributions ranges from high-energy batteries and flexible supercapacitors to wearable power textiles and sensors and underlines the fact that gel-based electrolytes are driving the next generation of electrochemical technologies. In addition to this, scientists have unveiled several tangible problems, e.g., how to reach a high ionic conductivity at room temperature, ensure interfacial stability in full cells, and produce at a large scale in a sustainable manner, that still need to be solved before gels can be fully applied. This set of discoveries heralds the second edition, which is intended to comprehend the progress made and further advance the field.

3. Conclusions and Outlook

The creation of gel electrolytes will expand rapidly in many innovative ways in the future. Advanced material designs that not only enhance ionic transport but also provide mechanical strength and safety are one of the main themes being discussed. New polymer matrices (e.g., bio-based or supramolecular polymers), inorganic–organic hybrid networks, and nanoengineered fillers that jointly create fast ion channels without compromising stability are among these designs. More specifically, the main design targets will be to increase Li^+ transference number and lower interfacial resistance, especially in high-current applications.

Similarly, a lot of attention is being directed toward thermal and electrochemical stability enhancement in gels in order to create batteries that are safe under more extreme conditions. Recently, it has been proven in several experiments that, for example, blend-type gel membranes can last structurally at high temperatures and for long cycles. The invention of inherently non-flammable, self-extinguishing, or thermally adaptive gel electrolytes may not only enhance the safety of EVs but also that of grid storage, thus alleviating a substantial part of the safety concern.

Moreover, fostering sustainability and retaining a good performance over long periods of time is a great challenge for gel electrolytes. As the discipline develops, the scientific community aims to replace volatile or poisonous agents with environmentally friendly ones and simplify gel electrolyte production overall. The establishment of fluorine-free gel polymer electrolytes that generate a stable, LiF-free solid–electrolyte interphase is a milestone example of this effort. By using recyclable or bio-degradable polymer gels, water-based synthesis routes, and inexpensive dopants, the research community will be in a position to guarantee that gel electrolytes are a part of the solution to the energy issue.

In addition, durability over the long term under real-world conditions (humidity, temperature fluctuations, and prolonged cycling) will be the yardstick of success. An example of research into aging mechanisms is monitoring water uptake in hydrogel electrolytes or electrode/electrolyte interphase changes over hundreds of cycles, which will aid the development of gels capable of maintaining their performance for years of operation.

Above all, the goal of this work that links different research themes is the eventual incorporation of gel electrolytes into practical devices. This second edition of our Special Issue will be enriched with papers demonstrating the use of gels in the fabrication of batteries, supercapacitors, fuel cells, sensors, and other technologies. Such papers will illuminate application prospects where gel electrolytes can be used in a lithium metal pouch cell with greatly reduced dendrite growth, or a hydrogel film is used to enable a flexible bio-sensor. We are also very interested in new ionic transport mechanisms, such as assisted ion hopping along polymer segments or deliberately phase-separated pathways that imitate biological ion channels, which, in turn, may result in electrolytes that do not compromise conductivity, selectivity, and mechanical strength. By delving into ion dynamics at a very basic level (using sophisticated spectroscopy, modeling, and theory), researchers can develop gel electrolytes that are capable of transporting ions at higher rates and with greater selectivity than ever before.

By issuing this call for papers for a second edition of the Special Issue, we are setting the bar very high. We are looking to gather studies that do not merely present new material or performance benchmarks but also shed light on the scientific understanding of gel electrolyte operation and optimization. Ideally, each paper submitted should be a small contribution to the larger puzzle: through the intelligent use of gels, how can one create safe and efficient electrochemical systems with a long duration? We hope the concerted efforts recorded in this Special Issue (2nd Edition) will represent a departure from the familiar and, thus, provide a source of new ideas for both academic and industrial researchers. The future of gel electrolytes is very promising considering the fast-paced and interdisciplinary nature of this field, which covers polymer chemistry, materials science, electrochemistry, and device engineering. We believe that persistent investigations will result in gel-derived energy and electronics devices that not only perform well but are also flexible, dependable, and eco-friendly.

In conclusion, gel electrolytes have now moved to the forefront of technological innovation in energy storage and flexible electronics. The advances in the research highlighted in the first edition have made more bold developments possible. We, as the Guest Editors of this second edition, are excited about the new discoveries and ingenious solutions that will be disclosed. Collectively, these works will facilitate the move of gel electrolytes from experimental research to practical use—thus, enabling the advent of a new era of safe, high-performing electrochemical devices.

Author Contributions: A.Z. and A.I.R. have contributed equally in all steps of the editorial conceptualization and writing. All authors have read and agreed to the published version of the manuscript.

Funding: This Editorial received no external funding.

Data Availability Statement: No new data were created or analyzed in this study.

Acknowledgments: The Guest Editors, Andrea Zille and Ana Isabel Ribeiro, express their deepest gratitude to all the authors and reviewers who had a part in this Special Issue's first edition and provided high-quality work. The authorship of the studies on gel electrolytes was considerable and very promising. Likewise, a huge thank you to the Editorial staff of *Gels* for their discreet professional support in the arrangement of both the first and the second editions. Their commitment and advice have been the driving factors behind the success of this project. The authors of the second edition are receiving a warm welcome and a huge thank you in advance for sharing their significant research with us. We hope that this collection will act as a catalyst for more breakthroughs and will lead to the establishment of new connections in the gel electrolyte community.

Conflicts of Interest: The authors declare no conflicts of interest.

List of Contributions:

1. Kuo, T.-J.; Chao, W.-T. Bidirectional Long Short-Term Memory Model of SoH Prediction for Gelled-Electrolyte Batteries under Charging Conditions. *Gels* **2023**, *9*, 989. https://doi.org/10.3390/gels9120989.
2. Choi, J.; Jung, J.C.; Jung, W. Advances in Carbon Xerogels: Structural Optimization for Enhanced EDLC Performance. *Gels* **2024**, *10*, 400. https://doi.org/10.3390/gels10060400.
3. Mao, N.; Lan, L.; Hun, Q.; Wei, J.; Liang, X.; Guo, Y. Metal–Organic Frameworks (MOF)-Derived Gel Electrolyte via UV Cross-Linking for High-Performance Lithium Metal Batteries. *Gels* **2025**, *11*, 409. https://doi.org/10.3390/gels11060409.
4. Yao, X.; Lan, L.; Hun, Q.; Lu, X.; Wei, J.; Liang, X.; Shen, P.; Long, Y.; Guo, Y. Preparation and Performance of PVDF-HFP/PAN-Based Gel Polymer Electrolytes. *Gels* **2025**, *11*, 317. https://doi.org/10.3390/gels11050317.
5. Chen, C.; Qin, W.; Hun, Q.; Wang, Y.; Liang, X.; Tan, R.; Li, J.; Guo, Y. Preparation and Failure Behavior of Gel Electrolytes for Multilayer Structure Lithium Metal Solid-State Batteries. *Gels* **2025**, *11*, 573. https://doi.org/10.3390/gels11080573.
6. Ribeiro, A.I.; Alves, C.; Fernandes, M.; Abreu, J.; Pedroso de Lima, F.; Padrão, J.; Zille, A. Gel Electrolytes in the Development of Textile-Based Power Sources. *Gels* **2025**, *11*, 392. https://doi.org/10.3390/gels11060392.
7. Cedeño Mata, M.; Orpella, A.; Dominguez-Pumar, M.; Bermejo, S. Boosting the Sensitivity and Hysteresis of a Gel Polymer Electrolyte by Embedding SiO_2 Nanoparticles and PVP for Humidity Applications. *Gels* **2024**, *10*, 50. https://doi.org/10.3390/gels10010050.
8. Yao, X.; Qin, W.; Hun, Q.; Mao, N.; Li, J.; Liang, X.; Long, Y.; Guo, Y. Fabrication and Performance Study of 3D-Printed Ceramic-in-Gel Polymer Electrolytes. *Gels* **2025**, *11*, 534. https://doi.org/10.3390/gels11070534.
9. Xu, L.; Liu, S.; Lin, A.; Su, Z.; Liang, D. Interpretable Prediction and Analysis of PVA Hydrogel Mechanical Behavior Using Machine Learning. *Gels* **2025**, *11*, 550. https://doi.org/10.3390/gels11070550.
10. Xu, L.; Liu, S.; Hu, D.; Liu, J.; Zhang, Y.; Li, Z.; Su, Z.; Liang, D. Interpretable Machine Learning Analysis of Design Factors in Hydrogel Supercapacitors. *Gels* **2025**, *11*, 464. https://doi.org/10.3390/gels11060464.

gels

MDPI

Article

Bidirectional Long Short-Term Memory Model of *SoH* Prediction for Gelled-Electrolyte Batteries under Charging Conditions

Ting-Jung Kuo [1] and Wei-Ting Chao [2,*]

[1] Department of Applied Artificial Intelligence, Ming Chuan University, Taoyuan 33348, Taiwan; tingjung@mail.mcu.edu.tw
[2] Center of Excellence for Ocean Engineering, National Taiwan Ocean University, Keelung 20224, Taiwan
* Correspondence: weiting@mail.ntou.edu.tw

Abstract: The impact of different charging currents and surrounding temperatures has always been an important aspect of battery lifetime for various electric vehicles and energy storage equipment. This paper proposes a bidirectional long short-term memory model to quantify these impacts on the aging of gel batteries and calculate their state of health. The training data set of the bidirectional long short-term memory model is collected by charging and discharging the gel battery for 300 cycles in a temperature-controlled box and an automated charge and discharge device under different operating conditions. The testing set is generated by a small energy storage device equipped with small solar panels. Data for 220 cycles at different temperatures and charging currents were collected during the experiment. The results show that the mean absolute error (MAE) and root-mean-square error (RMSE) between the training set and testing set are 0.0133 and 0.0251, respectively. In addition to the proposed model providing high accuracy, the gel battery proved to be stable and long-lasting, which makes the gel battery an ideal energy storage solution for renewable energy.

Keywords: gelled-electrolyte battery; bidirectional long short-term memory; state of health

1. Introduction

The valve-regulated lead-acid (VRLA) battery is one of the most frequently used types of rechargeable batteries. Due to low material costs, robustness, high reliability, and the ability to instantaneously discharge large currents, the VRLA is widely used in various applications, including light electric vehicles, backup power systems, and ignition systems. VRLA battery technology has undergone numerous improvements and refinements, focusing on enhancing its performance and extending its lifetime. Electrodes, separators, and electrolytes are the main components of the VRLA system. Each component is crucial to the performance and lifetime of a battery. For the electrolyte component, researchers have created technologies such as absorbent glass mats (AGM) and gelled electrolytes that use gel instead of liquid acid as the electrolyte [1]. Pyrogenic silica is a highly pure amorphous silicon dioxide capable of absorbing more than ten times its weight in acidic liquids to form a gel [2,3]. The gelled electrolytes are usually made by uniformly dispersing fumed silica in sulfuric acid [4]. The unique properties of the gel allow it to connect molecules to form a network and retain liquid, resulting in the formation of a gel structure after a certain period. Therefore, the electrolyte of the gel battery can exist in a solid state between the positive and negative plates inside the battery [5]. The advantages of gel batteries over traditional lead-acid (flooded) batteries are as follows:

- When the gel battery is charged, hydrogen gas is absorbed by the plates and converted to electrolytes. It does not require adding water or equalizing the charge, so it requires no maintenance and is very safe to use.
- The battery has a low self-discharge, instant high-power output, and long cycle life.

- The electrolyte is stable inside the battery, and the battery can be installed in any desired direction.
- At extreme temperatures, the gelled electrolyte prevents the electrolyte from evaporating or freezing and ensures high performance. If excessive gas is generated from improper charging, the safety valve automatically discharges the gas to prevent the battery from rupturing.

The heat and electrochemical polarization reaction, ohmic loss, oxygen recombination cycle, and other factors generated by a battery during the cycle will directly affect the available capacity of a battery [6]. For example, a deep-cycle discharge can cause the electrolyte to stratify quickly, decreasing the battery capacity. This is due to the irreversible degradation of the battery's grid corrosion and the high concentration of active material in the lower regions of the plates. Since the gelled electrolyte is less stratified than AGM's electrolyte, it can last longer during cycling, providing outstanding durability and enduring more cycling [7].

Compared to lithium-ion batteries, gel batteries have lower power and energy density. However, gel batteries are typically the most appropriate solution for light electric vehicles (such as golf carts, small boats, etc.), where the manufacturing cost and safety of the battery are essential factors. Additionally, the lead-acid battery system has been developed since 1859. It has a complete manufacturing process and recycling mechanism, and a lower environmental impact than lithium-ion batteries [8–10]. Therefore, it is one of the reasons why gel batteries are superior to lithium-ion batteries in terms of environmental protection.

Small solar or wind energy generators are built in urban areas and on school grounds to compensate for the energy loss caused by long-distance electricity transmission through the power grid. Renewable energy is, however, highly dependent on weather conditions due to its instability. In this regard, small battery energy storage devices are necessary. Whenever solar or wind power produces more power than the system requires, excess power is stored in the batteries When the energy produced is insufficient, the batteries power the system. Although the volume-to-capacity ratio of the gel battery is lower than that of AGM, the gel battery can provide reliable, maintenance-free power in many deep-cycle applications, such as solar energy and light service vehicles [11,12].

The above applications have heightened interest in gel batteries' charging rate and service life. Reducing the charging time is crucial to increasing people's acceptance of renewable energy. Fast charging usually involves a high current, possibly accelerating battery degradation [13–15]. Fast charging is also at odds with extending the gel battery life and lowering maintenance costs. Therefore, it is necessary to determine the aging mechanism of gel batteries and quantify the impact of different charging stresses on the battery to find the optimal charging mechanism.

An optimal battery charging mode balances charging speed and battery cycle life. Studies [16,17] have reported many charging strategies to increase charging speed, enhance charging performance, and maximize battery cycle life, such as optimizing charging strategies to suppress side reactions based on the electrochemical characteristics of the battery. These strategies can reduce battery degradation but are nearly impossible to implement in real-life applications. In the research of [18–20], optimal charging strategies are based on equivalent circuit models to adjust the charging efficiency and cycle life. These strategies are easy to implement but need more support from electrochemical mechanisms. Researchers [21,22] have proposed a multi-level constant-current charging strategy to find the best charging mode for the battery during the charging process. Researchers [23,24] have studied the impact of multi-level fast charging on the aging process of batteries. These studies compared and evaluated different charging strategies and charging stresses, including the charging current and charging cut-off voltage, but did not discuss the performance and aging mechanism of the battery under different operating temperatures.

The charging strategies mentioned previously lack physical or chemical theorems because the battery's aging mechanism under different charging current rates and operating temperatures is unclear. Establishing a model describing the relationship between the fading capacity and charging stress is very difficult. Researchers [25–27] have studied battery degradation under different operating conditions, such as the ambient temperature, state of charge (*SoC*) range, charge/discharge rate, and cut-off voltage, through which battery life can be extended. In work [28], researchers reported that to delay battery degradation, the charging process can be significantly affected by adjusting the charging current and cut-off voltage. In work [29], researchers studied the aging behavior of batteries during long-term cycling at high charging rates to explain the aging effects of batteries. In this paper, based on these reports, we propose a model for accurately estimating the state of health (*SoH*) of gel batteries under different operating conditions.

1.1. Contributions of the Paper

This paper conducted a cycle life test on commercial 12 Ah gel batteries to demonstrate the aging mechanism of the batteries under different charging currents. The bidirectional long short-term memory (BiLSTM) model was used to quantify and estimate the gel battery's *SoH* to verify its characteristics. The training set is produced by an automatic charging and discharging device in the laboratory. A working small energy storage device generates the testing set. The main contributions of this paper are summarized as follows:

- There are few studies related to gel battery performance. This study intends to conduct 300 deep cycles under different charging currents and temperatures to examine the characteristics of the gel electrolyte.
- This study develops a deep-learning model of the battery cycle life to describe the relationship between the battery's capacity fading, charging stress, and operating temperature.
- In a working environment of 25 °C, the high charging current impacts battery life but is still in a safe range. However, at a high temperature of 50 °C, in order to avoid the intense electrochemical reaction inside the battery, the charging current needs to be reduced.

1.2. Organization of the Paper

This paper is organized as follows: Section 2 introduces the design of the experiment, including the deep-cycle life testing of the battery under different charging currents and temperatures. The test results related to battery aging, including *SoC* and *SoH*, are also presented in this section. Section 3 analyzes the degradation mechanism of the battery and proposes a BiLSTM model to establish the *SoH* model. The data for the model comes from laboratories and a small energy storage device. Section 4 discusses the results in depth. Section 5 summarizes the characteristics of the gel battery and the performance of our proposed BiLSTM model.

2. Experimental Setup and Procedures

In order to verify whether the battery's *SoH* model can improve the prediction accuracy in actual cases, we propose using BiLSTM to build a deep-learning model. The model data collection was divided into a training set and testing set. Data for the training set were collected in the laboratory using automatic battery charging and discharging devices. Data for the testing set are collected when the gel battery is connected to the solar energy generation device and the microgrid. Gel battery data were collected during the charging and discharging process based on the load, voltage, current, temperature, and time.

2.1. Battery Testing Set Setup

Figure 1 illustrates a small demonstration energy storage device built on campus. When renewable energy (such as solar or wind) is abundant, the device supplies power for the campus, and the excess power is used to charge the gel batteries. The fully charged batteries are connected to the microgrid and discharged through loads (such as low-power equipment) on the microgrid. The small demonstration energy storage device has a simple air conditioner to dissipate heat. The air conditioner is only turned on when the battery's operating temperature is higher than 50 °C to keep the batteries within a safe operating temperature range as well as reduce power consumption from the air conditioner.

During the cycle process, a battery management system (BMS) monitors and records the battery's operating temperature, voltage, and current. Our main goal is to study the impact of charging strategies on the battery's cycle life under various operating conditions and to propose recommended charging strategies based on the *SoH* of the battery.

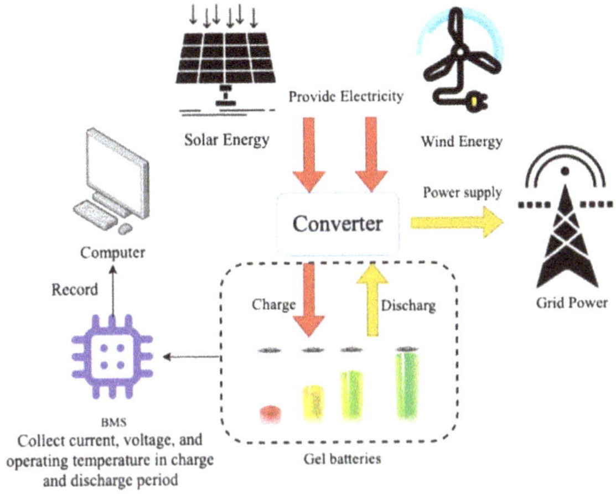

Figure 1. The architecture of a small demonstration energy storage center.

2.2. Gel Battery Training Set Setup

In order to study the aging mechanism of gel batteries and establish the *SoH* under different operating conditions, the batteries were charged under different currents and operating temperatures. The discharge process was executed at a fixed discharge current and temperature. It is used to evaluate the performance of gel batteries and determine their cycle life to determine the impact of varying charging currents and temperatures. The specifications of the tested gel batteries are listed in Table 1. The batteries were charged and discharged using a Chroma 17020 battery system from Taiwan's Chroma corporation. The experimental process is shown in Figure 2.

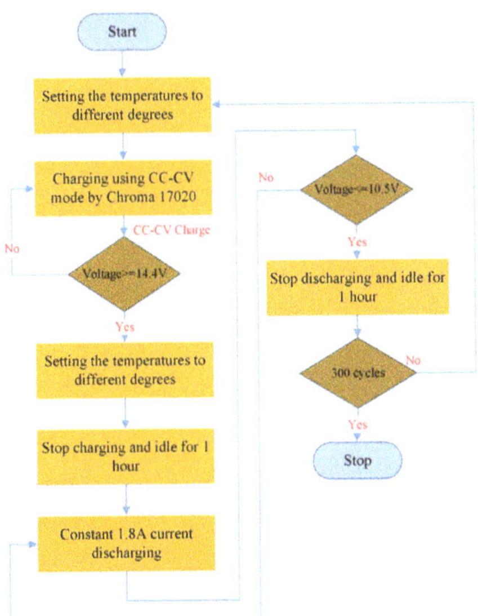

Figure 2. Experimental setup flowchart.

Table 1. Nominal specifications of the gel battery.

Item	Specification
Nominal capacity	12 Ah
Charging method	Constant current and constant voltage (CC-CV)
Max. charging current	6 A
Charging cut-off voltage	14.4 V
Max. discharging current	6 A
Discharging cut-off voltage	10.5 V

By definition, one battery cycle consists of a complete charge and discharge at the cut-off voltage. This paper used three batteries as a group to carry out corresponding experimental conditions to ensure data consistency. New batteries were charged using the constant-current and constant-voltage (CC-CV) method with different maximum currents (1.8, 3.6, and 5.4 A) until the voltage reached 14.4 V. After each complete charge, batteries were left idle for 1 h to allow the electrochemical reaction inside the battery to reach a steady state. Then, each battery was discharged at a current of 1.8 A until the voltage dropped to the 10.5 V cut-off voltage. All the tested batteries were charged and discharged 300 times in the cycle. The experimental parameter settings are listed in Table 2.

Table 2. The specifications for the operating environment of training batteries.

Battery ID	1, 2, 3	4, 5, 6	7, 8, 9	10, 11, 12	13, 14, 15	16, 17, 18
Charging current (A)	1.8	3.6	5.4	1.8	3.6	5.4
Charging temperature (°C)		25			50	
Discharging current (A)				1.8		
Discharging temperature (°C)				25		

2.3. Charge Curves Analysis of Batteries under Different Charge Current and Cycle Numbers at 25 °C

The *SoC* curves of the 1.8 A current charging behavior for 60-, 100-, 160-, 260-, and 300-cycle batteries under 25 °C are illustrated in Figure 3a. The *SoC* is defined as the state of available energy in the battery. It is a reference value based on the designed rate capacity during manufacturing. The battery is fully charged at 100% and fully discharged at 0% [30], as shown in Equation (1).

$$SoC_{(t)} = SoC_{(t0)} - \frac{1}{C_{max(t)}} \int_{t0}^{t} I_{(t)}dt \tag{1}$$

where C_{max} denotes the maximum usable capacity of the battery and I denotes the battery current at time t. Q_a denotes the available capacity of the battery through the multiple cycles.

The battery's available energy will age when it is charged and discharged repeatedly. When an aging battery is fully charged, the *SoC* cannot reach 100% capacity. Thus, Equation (2) allows the *SoC* to be modified [31].

$$SoC = \frac{Q_a}{C_{max}} \tag{2}$$

Q_a denotes the available capacity of the battery for the charging/discharging cycle.

The *SoH* refers to either the battery's capacity fade or to the power fade, which is defined as follows [32]:

$$SoH = \frac{C_{max}}{C_r} \times 100\% \tag{3}$$

where Cr is the rated capacity.

The data acquired from the experiment were used as the training set for the BiLSTM model. Hence, the relationship between the variation in the capacity, energy loss, temperature, and cycle number was further investigated.

The CC-CV mode is a charging method that switches between constant-current charging and constant-voltage charging based on the battery's current voltage. When the voltage is low, the battery is charged in the CC mode with a maximum current based on the preset. As the voltage reaches the charging cut-off voltage, the CC mode is switched to the CV mode to continue charging with a smaller current. The *SoC* curves for a battery charging in the CC mode are seen in Figure 3a when the battery starts to charge from an open-circuit voltage (OCV) to a closed-circuit voltage (CCV). The low currents used to charge the batteries indicate that they are charging in the CV mode. Compared to a brand-new battery that has not been cycled, a battery that has been cycled more times will reach the cut-off voltage more rapidly. It is evident by observing the occupied time of the CC and CV modes that as the number of cycles increases, the CC mode charging time becomes shorter, and the CV mode charging time becomes longer. As shown in Figure 3b, after 300 cycles, the *SoH* remains at 84.65%, only 13.35% below the available capacity of the brand-new battery (100%).

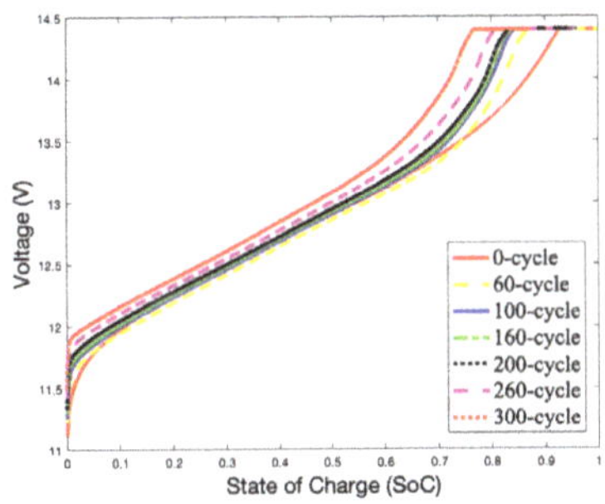

(**a**)

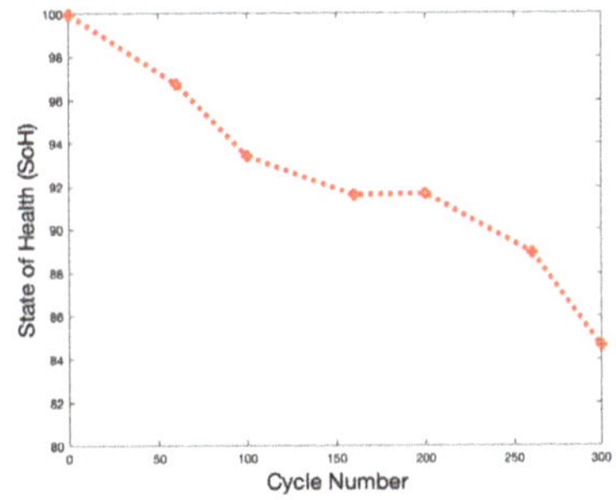

(**b**)

Figure 3. 1.8 A charge *SoC* curves and *SoH* curve in 25 °C during 300 cycles: (**a**) *SoC* curves; (**b**) *SoH* curve.

Figure 4a,b illustrate the *SoC* and *SoH* curves for gel batteries charging at 25 °C with a 3.6 A charge over 300 cycles. The battery's *SoH* is only slightly affected by the 3.6 A charging current compared to 1.8 A. After 300 cycles, the battery's capacity is 81.8%, only 18.2% lower than that of a brand-new battery.

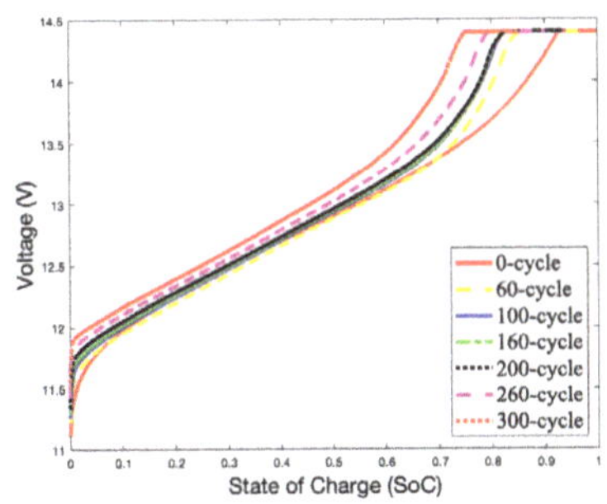

(**a**)

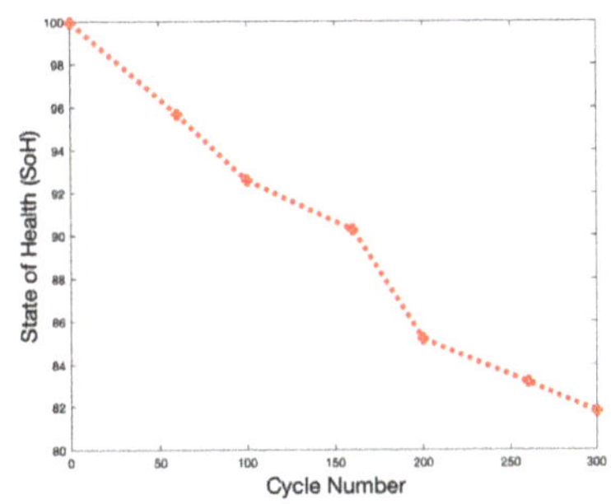

(**b**)

Figure 4. 3.6 A Charge *SoC* curves and *SoH* curve in 25 °C during 300 cycles: (**a**) *SoC* curves; (**b**) *SoH* curve.

Figure 5a,b illustrate the *SoC* and *SoH* curves charging at 25 °C with 5.4 A over 300 cycles. Compared with the 1.8 A and 3.6 A charging current, the 5.4 A charging current has a higher impact on the battery cycle life. Compared with the available power capacity of a brand-new battery, it decreases by 27.2%.

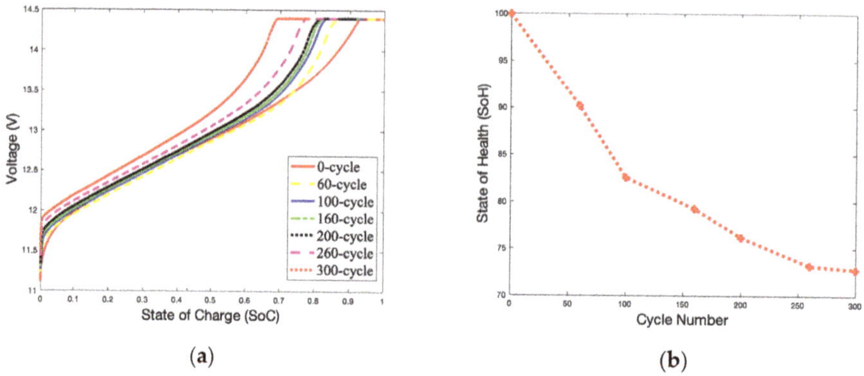

Figure 5. 5.4 A Charge *SoC* curves and *SoH* curve in 25 °C during 300 cycles: (**a**) *SoC* curves; (**b**) *SoH* curve.

2.4. Charge Curves Analysis of Batteries under Different Charge Current and Cycle Numbers at 50 °C

A temperature of 50 °C was selected to study the impact of surrounding temperatures on the battery's capacity during charging. Figure 6a shows an unstable charging *SoC* curve compared with Figure 3a. A high temperature accelerates electrochemical reactions inside the battery, and the charging curve will quickly approach the cut-off voltage. When the battery reaches 300 cycles, fluctuations in the *SoC* curve can be found in the OCV-CCV stage. Figure 6b shows that as the number of charges reaches 200 cycles, the *SoH* curve drops rapidly and declines to 73.09% at 300 cycles. As shown in Figure 3b, the *SoH* of the battery is reduced by 11.56% compared to the *SoH* of the battery with 300 cycles (84.65%).

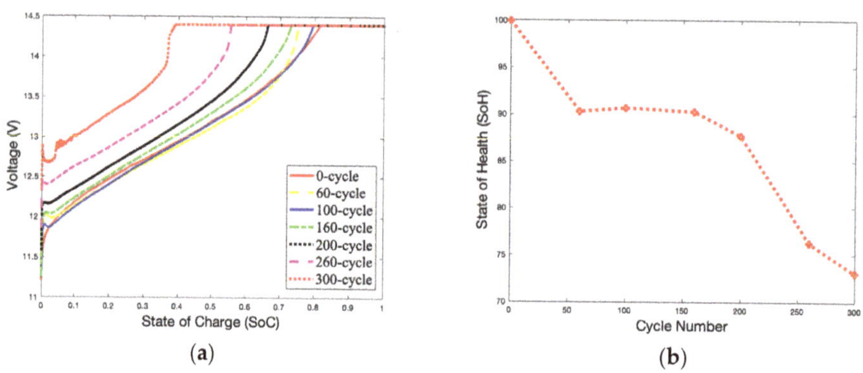

Figure 6. 1.8 A Charge *SoC* curves and *SoH* curve in 50 °C during 300 cycles: (**a**) *SoC* curves; (**b**) *SoH* curve.

Figure 7a,b illustrate the *SoC* and *SoH* curves charging at 50 °C with 3.6 A over 300 cycles. Figures 6b and 7b have similar *SoH* curves. The curves from 60 cycles to 160 cycles are relatively gentle and drop rapidly after 160 cycles. Based on 300 cycles, the *SoH* of the battery is 50.8%, which is a reduction of 22.29% as compared to Figure 6b at 73.09%.

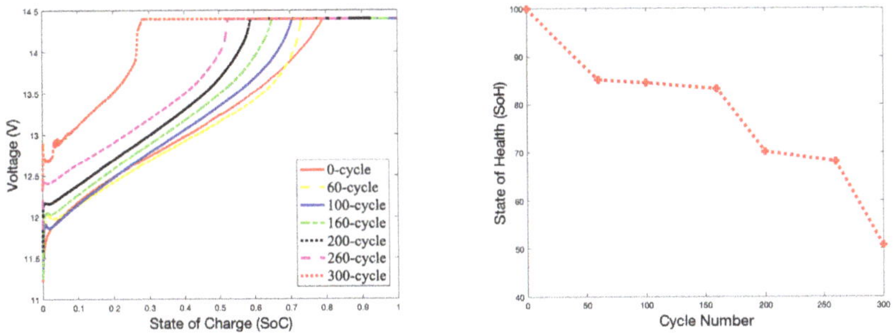

(a) (b)

Figure 7. 3.6 A Charge *SoC* curves and *SoH* curve in 50 °C during 300 cycles: (**a**) *SoC* of charge curves; (**b**) *SoH* curve.

Figure 8a,b illustrate the *SoC* and *SoH* curves charging at 50 °C with 5.4 A over 300 cycles. The battery almost performed as well as a brand-new battery (*SoH* = 100%) before 160 cycles. However, the *SoH* curve suddenly dropped rapidly after 260 cycles, and reached a failure state at 300 cycles (*SoH* = 28.61%). This may be due to side effects caused by high-current charging. The thermal mechanism of the battery during charging is very complex. It is related to the electrochemical characteristics of the battery itself, such as the reaction heat, energy loss caused by battery polarization, electrolyte decomposition, self-discharge side reactions in the battery, and Joules generated by battery resistance. Heat will cause the electrochemical reaction inside the battery to be rapid due to the characteristics of the gel electrolyte. The battery's performance was good before 160 cycled; then, the materials inside the battery began to age rapidly, causing the *SoH* of the battery to drop.

The declining trend of the *SoH* curves for the battery is partially linear, especially between 100 and 200 cycles, as show in Figures 3b and 6b. At this stage, the increase in thermal mechanisms on the electrochemical reaction rate is slightly more significant than the impact of the battery's material aging. Therefore, there is a slight increase in the battery's capacity during this period.

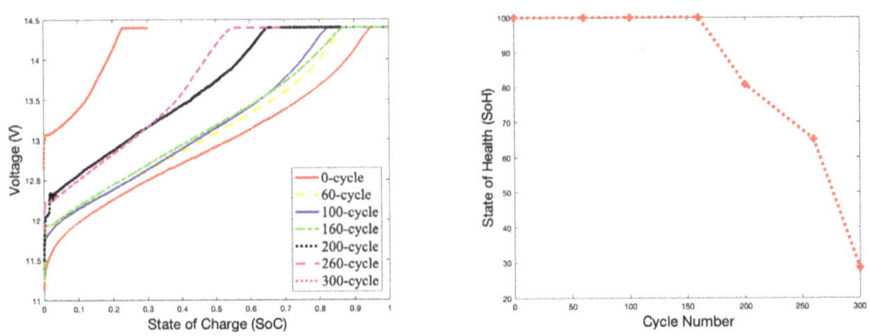

(a) (b)

Figure 8. 5.4 A Charge *SoC* curves and *SoH* curve in 50 °C during 300 cycles: (**a**) *SoC* curves; (**b**) *SoH* curve.

The results of the above experimental analysis indicate that gelled-electrolyte batteries can be charged with a higher current (preferably not exceeding 0.45 C rate). The optimal operating temperature of the gel battery is approximately 25 °C. In this temperature range, the cycle life of the gel battery can be extended.

3. Bidirectional Long Short-Term Memory (BiLSTM) Model Development

Before the BiLSTM setup process, data were divided into training sets and testing sets. The training datasets consisted of all the charging characteristics for each gel battery in the laboratory. The testing datasets consisted of data on twelve 12 Ah gel batteries connected to the 48 V microgrid in the energy storage device. While the solar panels were charging the 48 V microgrid, the battery's voltage, current, and surrounding temperature were recorded by a battery management system (BMS). The datasets collected information on the gel batteries for one year. Since rain or solar panel maintenance can affect the data, only 220 data records were usable. The acquired battery data that were used as input were battery terminal voltage (V), current (I), charge time (Ct), and operating environment temperature (T).

3.1. Long Short-Term Memory Architecture

The long short-term memory (LSTM) model is a recurrent neural network (RNN) type designed to address the vanishing gradient problem and capture long-term dependencies on sequential data. Its architecture consists of gateways, including the forget gate, input gate, and output gate [33–35], as shown in Figure 9. As seen in Figure 9, the memory cell's state is a horizontal line that runs through the LSTM unit, which acts as a memory or conveyor belt that allows information to flow through the network and preserve long-term dependencies. Its function is to act as a data bridge from data acquired in the past to currently available data [36]. When calculating the *SoH*, the capacity to recall past information makes this method especially useful in solving problems that require long sequential data or time series [37]. The LSTM cell has three gates that control the information and whether it has been updated. The operation flow of the memory cell and the function of the three gates are described as follows:

First, the initial value stored in the memory cell is called C. New data (Z_j) inputs are multiplied by the hyperbolic activation functions $g(x)$ to obtain the value $g(Zj)$. The input gate regulates the flow of new information into the memory cell. After being processed by the input gate, the new data ($C^{'}$) can be expressed as follows:

$$C' = g(Z_j)f(Z_i) \tag{4}$$

where Z_i is the parameter to control whether the gateway is open and $f(x)$ is a sigmoid activation function. When $f(Z_i) = 1$, the memory cell is updated to 0. Otherwise, it is not updated.

The forget gate is a sigmoid layer that decides what information from the memory cell state should be forgotten or retained. It takes the previous cell state (C) and the current input (Z_k) to produce a forget gate output between 0 and 1 for each component of the cell state. When $Cf(Z_k)$ is 1, then C is preserved; otherwise, it is canceled. Then, the memory cell state can be updated as follows:

$$C' = g(Z_j)f(Z_i) + Cf(Z_k) \tag{5}$$

Then, $C^{'}$ is stored in the memory cell and called C. Before being processed by the output gate, $C^{'}$ is multiplied to a hyperbolic tangent function ($h(x)$) to obtain ($h(C^{'})$). The output gate determines the next hidden state, the filtered version of the memory cell state, and is shared with the next time step. It considers the previous cell state, the current input (Z_o), and the candidate cell state ($C^{'}$) to produce a value between 0 and 1. After being processed by the output gate, the output value (y) can be expressed as follows:

$$y = h(C')f(Z_o) \tag{6}$$

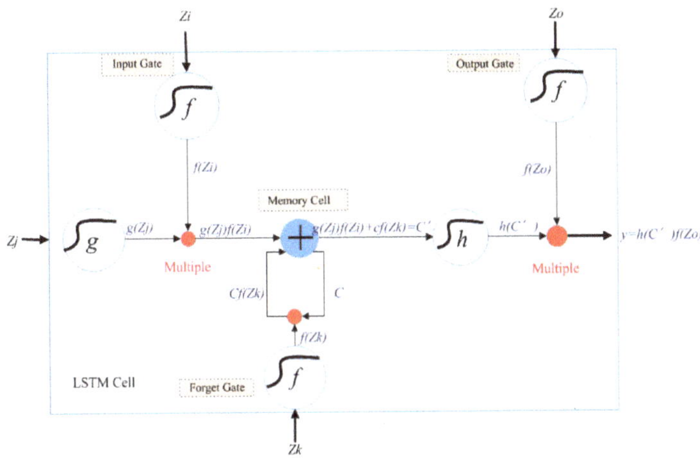

Figure 9. Structure of LSTM network (refer to [38]).

3.2. Bidirectional Long Short-Term Memory Architecture

The BiLSTM model is an extension of the LSTM architecture that captures information from past and future time steps. The BiLSTM model consists of two LSTM networks, processing the input sequence in the forward and backward direction. The forward LSTM network reads the input sequence from left to right, creating hidden states at each time step. The backward LSTM network processes the input sequence in the reverse order, reading it from right to left. It also produces hidden states but captures information on the sequence in a backward manner.

The forward and backward LSTM outputs are concatenated at each time step (t). These outputs create a combined representation that includes information from both the past and the future. The concatenated output is often passed through additional layers or used directly for downstream tasks. The BiLSTM architecture is based on research conducted in [39–41], as shown in Figure 10.

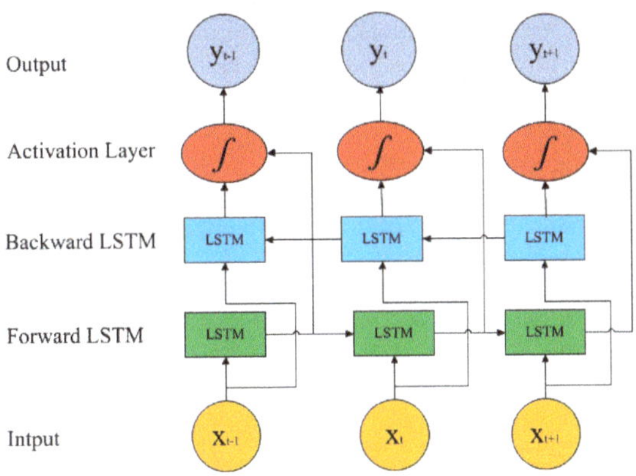

Figure 10. Bidirectional LSTM structure.

The *SoH* is a time series during the entire service life of the battery. The *SoH* value in this series is not only related to the trend change in the forward data but is also the basis for the change in the backward *SoH* data. Compared with the LSTM network, the BiLSTM network can analyze the two-way time relationship of the battery *SoH* data and extract the data's past and future features, improving prediction accuracy.

3.3. Evaluation Indicators and Parameter of The BiLSTM Model Setup

This paper utilized Matlab to construct the deep-learning architecture of the predictive model. Based on previous experiments [39,42–44], the Adam optimizer was used to train our model with the batch size and epochs set to 32 and 100, respectively. To avoid gradient explosion, the maximum gradient was set to 2; the dropout rate of each LSTM function was set to 0.3 to avoid overfitting. The hyperparameters of BiLSTM settings are listed in Table 3.

Table 3. Hyperparameters of BiLSTM.

Hyperparameters	Value
Learning Function	Adam
Max Epoch	100
Min. Batch Size	32
Dropout	0.02
Hidden Layer	10
Number of Neurons in the Hidden Layer	50

We used the following two metrics: mean absolute error (MAE) and root-mean-square error (RMSE), defined as the following:

$$MAE = \frac{1}{n}\sum_{i=1}^{n}|m_i - \hat{m}_i| \tag{7}$$

$$MSE = \sqrt{\frac{1}{n}\sum_{i=1}^{n}(m_i - \hat{m}_i)^2} \tag{8}$$

where n is the total number of data; m_i is the real SoH value; $\hat{m}_i$ is the predicted SoH value. For indicators such as mean absolute error (MAE) and RMSE, the closer they approached zero, the more accurate the prediction.

4. Results and Discussion

Matlab implemented all the analyses for the gel battery datasets with their deep-learning toolbox. Previous studies [45] used different discharge currents to estimate the *SoC* and *SoH* of the gel batteries and compared the accuracy of the LSTM model, FNN, and RNN. In this paper, we propose using the BiLSTM model to verify battery performances.

In preparing the testing set for the BiLSTM model, four gel batteries were connected in a series and sealed into a battery pack. Next, three battery packs were connected in parallel to the microgrid. Accordingly, the data set consists of the characteristics of twelve batteries. The battery pack was considered a 48 V battery to provide power to a microgrid and to be charged by solar panels to improve measurements. As part of the battery pack, there was a BMS that could record voltage, current, time, and working temperature and a fan that maintained a consistent temperature in the battery pack.

During the testing period, the batteries performed 220 cycles, which can be used in the datasets. The average temperature was recorded for each testing day. Batteries were fully discharged (42 V) and charged (57.6 V) to the default cut-off voltage as fully as possible.

The predicted (BiLSTM model) and measured (Coulomb counting) data are shown in Figure 11, where the two curves almost align. This indicates that the predicted *SoH* values are consistent with the actual values and that the BiLSTM model can accurately predict the battery's *SoH* value.

All batteries obey the laws of thermodynamics and electrochemistry, meaning they can describe the gel battery's charge and discharge characteristics and capacity fade. The BiLSTM model can predict experimental data based on the results of the battery training set, so it can be regarded as the memory model that can indirectly predict the thermodynamic behaviors of the battery.

When comparing the predicted and measured values, the error is more significant at low working temperatures, as shown in Figure 12. A list of the top 10 absolute errors in 220 cycles is also shown in Table 4. As low-temperature weather rarely occurs in Taiwan, the effect of low-temperature weather conditions was not considered during the gel battery *SoH* testing design. When prediction errors increase in low-temperature conditions, the LSTM model's computing power cannot adapt to the low-temperature conditions because the learning samples did not anticipate low-temperature samples. This can be resolved by including more diverse learning samples regarding temperature variables.

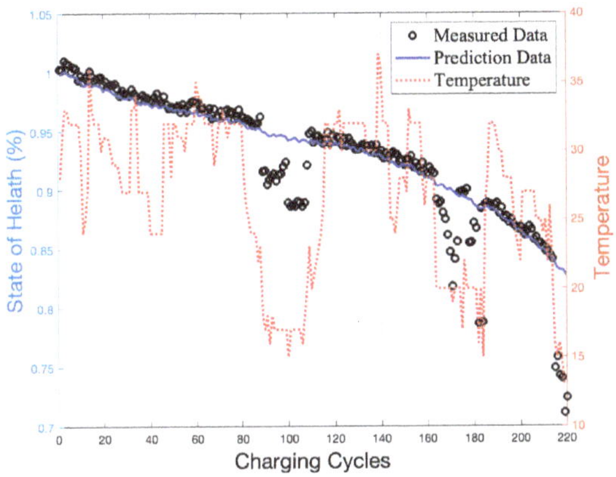

Figure 11. The differences in *SoH* between the experimental and predicted data for the tested battery.

Table 4. Top ten absolute errors in the 220 cycles.

Temperature (°C)	13	14	15	17	17	19	21	21	22	22
Cycle Number	101	121	155	150	134	122	123	133	131	125
Absolute Error	0.115	0.101	0.100	0.098	0.097	0.097	0.097	0.084	0.071	0.059

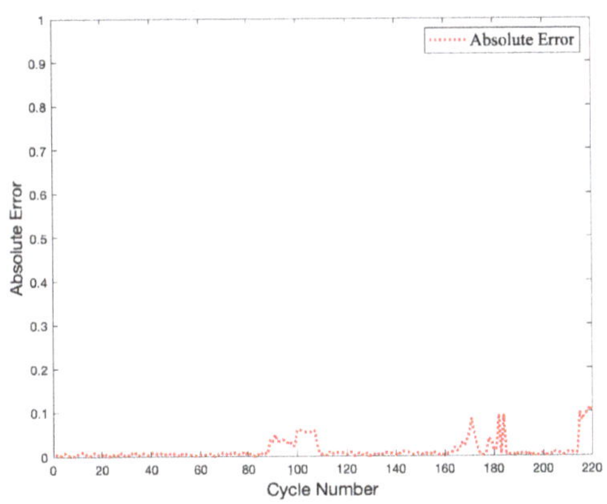

Figure 12. The absolute errors between predicted and measured results.

5. Conclusions

This paper is dedicated to the *SoH* assessment and management of gel batteries. The BiLSTM model was first used to train each battery and to estimate the *SoH* under different operation conditions. During 300 cycles, many datasets were collected using different batteries in different working environments. The datasets were applied to train the BiLSTM model. The testing dataset consists of twelve gel batteries connected to a microgrid. In the morning, the solar panels charge the batteries; in the evening, the batteries discharge the electrical appliances on the microgrid. During the operating periods of the microgrid, the charging characteristics and working temperatures of the batteries were collected by the BMS and a computer. Based on the verified results, the BiLSTM model produced an MAE of 0.0133 and an RMSE of 0.0251, indicating that the proposed model can be used for the estimation of *SoH* and the investigation of battery aging mechanisms.

The BiLSTM model structure has proven to be successful in predicting the *SoH* of the battery due to its bidirectional nature. The training process combines forward and backward data measurement, thus enabling a more robust training phase than other recurrent networks or unidirectional LSTM models. Aside from this, due to the implementation of differential electricity prices, the battery can be charged when electricity is inexpensive, and it can provide power when electricity is expensive. Using such a strategy can maximize the benefits of gel batteries. To demonstrate the effectiveness of the proposed model and optimize the model, future work will include investigating any other gel battery datasets available. In the future, the authors intend to enhance this method to reduce prediction errors, particularly for constrained input data with low temperatures.

Author Contributions: Conceptualization, T.-J.K. and W.-T.C.; methodology, T.-J.K. and W.-T.C.; validation, T.-J.K. and W.-T.C.; software, T.-J.K. and W.-T.C.; formal analysis, T.-J.K. and W.-T.C.; investigation, T.-J.K. and W.-T.C.; resources, T.-J.K.; data curation, W.-T.C.; writing—original draft preparation, T.-J.K.; writing—review and editing, W.-T.C.; visualization, T.-J.K. and W.-T.C.; supervision, T.-J.K. and W.-T.C.; project administration, T.-J.K.; funding acquisition, T.-J.K. and W.-T.C. All authors have read and agreed to the published version of the manuscript.

Funding: This research was funded by the National Science and Technology Council, Taiwan grant number NSTC 112-2222-E-130-002- and 112-2218-E-019-001-.

Institutional Review Board Statement: Not applicable.

Informed Consent Statement: Not applicable.

Data Availability Statement: Data are contained within the article.

Acknowledgments: The authors thank the anonymous reviewers for their insightful and constructive comments to improve this paper. The authors would like to express great thanks for all the support.

Conflicts of Interest: The authors declare no conflict of interest.

References

1. Chikkatti, B.S.; Sajjan, A.M.; Kalahal, P.B.; Banapurmath, N.R.; Khan, T.Y.; Khadar, S.D.A.; Shamsudeen, S.M.; Raju, A. A Novel Poly (vinyl alcohol)–tetraethylorthosilicate Hybrid Gel Electrolyte for Lead Storage Battery. *Gels* **2022**, *8*, 791. [CrossRef] [PubMed]
2. Linhares, T.; Carneiro, V.H.; Pessoa de Amorim, M.T.; Durães, L. A Comparative Thermoacoustic Insulation Study of Silica Aerogels Reinforced with Reclaimed Textile Fibres: Cotton, Polyester and Wool. *Gels* **2023**, *9*, 548. [CrossRef] [PubMed]
3. Ghica, M.E.; Mandinga, J.G.; Linhares, T.; Almeida, C.M.; Durães, L. Improvement of the Mechanical Properties of Silica Aerogels for Thermal Insulation Applications through a Combination of Aramid Nanofibres and Microfibres. *Gels* **2023**, *9*, 535. [CrossRef] [PubMed]
4. Aruchamy, K.; Ramasundaram, S.; Divya, S.; Chandran, M.; Yun, K.; Oh, T.H. Gel Polymer Electrolytes: Advancing Solid-State Batteries for High-Performance Applications. *Gels* **2023**, *9*, 585. [CrossRef] [PubMed]
5. Sultana, S.; Ahmed, K.; Jiwanti, P.K.; Wardhana, B.Y.; Shiblee, M.N.I. Ionic liquid-based gels for applications in electrochemical energy storage and conversion devices: A review of recent progress and future prospects. *Gels* **2021**, *8*, 2. [CrossRef] [PubMed]
6. Lambert, D.W.; Greenwood, P.H.; Reed, M.C. Advances in gelled-electrolyte technology for valve-regulated lead-acid batteries. *J. Power Sources* **2002**, *107*, 173–179. [CrossRef]
7. Tuphorn, H. Gelled-electrolyte batteries for electric vehicles. *J. Power Sources* **1992**, *40*, 47–61. [CrossRef]
8. Tan, S.-y.; Payne, D.J.; Hallett, J.P.; Kelsall, G.H. Developments in electrochemical processes for recycling lead–acid batteries. *Curr. Opin. Electrochem.* **2019**, *16*, 83–89. [CrossRef]
9. Joshi, B.V.; Vipin, B.; Ramkumar, J.; Amit, R. Impact of policy instruments on lead-acid battery recycling: A system dynamics approach. *Resour. Conserv. Recycl.* **2021**, *169*, 105528. [CrossRef]
10. Yanamandra, K.; Pinisetty, D.; Daoud, A.; Gupta, N. Recycling of Li-ion and lead acid batteries: A review. *J. Indian Inst. Sci.* **2022**, *102*, 281–295. [CrossRef]

11. Hollenkamp, A.; Baldsing, W.; Hamilton, J.; Rand, D. Advanced lead/acid batteries for stand-alone power-supply systems. *J. Power Sources* **1990**, *31*, 329–336. [CrossRef]
12. May, G.; Lenain, P. Gelled-electrolyte lead/acid batteries for stationary and traction applications. *J. Power Sources* **1992**, *40*, 187–193. [CrossRef]
13. Gao, Y.; Jiang, J.; Zhang, C.; Zhang, W.; Ma, Z.; Jiang, Y. Lithium-ion battery aging mechanisms and life model under different charging stresses. *J. Power Sources* **2017**, *356*, 103–114. [CrossRef]
14. Zhang, S.S. The effect of the charging protocol on the cycle life of a Li-ion battery. *J. Power Sources* **2006**, *161*, 1385–1391. [CrossRef]
15. Legrand, N.; Knosp, B.; Desprez, P.; Lapicque, F.; Raël, S. Physical characterization of the charging process of a Li-ion battery and prediction of Li plating by electrochemical modelling. *J. Power Sources* **2014**, *245*, 208–216. [CrossRef]
16. Liu, J.; Li, G.; Fathy, H.K. A computationally efficient approach for optimizing lithium-ion battery charging. *J. Dyn. Syst. Meas. Control* **2016**, *138*, 021009. [CrossRef]
17. Lu, B.; Song, Y.; Zhang, J. Selection of charge methods for lithium ion batteries by considering diffusion induced stress and charge time. *J. Power Sources* **2016**, *320*, 104–110. [CrossRef]
18. Yu, Y.; Narayan, N.; Vega-Garita, V.; Popovic-Gerber, J.; Qin, Z.; Wagemaker, M.; Bauer, P.; Zeman, M. Constructing accurate equivalent electrical circuit models of lithium iron phosphate and lead–acid battery cells for solar home system applications. *Energies* **2018**, *11*, 2305. [CrossRef]
19. Shahriari, M.; Farrokhi, M. Online state-of-health estimation of VRLA batteries using state of charge. *IEEE Trans. Ind. Electron.* **2012**, *60*, 191–202. [CrossRef]
20. Tran, N.-T.; Khan, A.B.; Choi, W. State of charge and state of health estimation of AGM VRLA batteries by employing a dual extended Kalman filter and an ARX model for online parameter estimation. *Energies* **2017**, *10*, 137. [CrossRef]
21. Lavety, S.; Keshri, R.K.; Chaudhari, M.A. Evaluation of charging strategies for valve regulated lead-acid battery. *IEEE Access* **2020**, *8*, 164747–164761. [CrossRef]
22. Taheri, A.; Zhalebaghi, M.H. A new model predictive control algorithm by reducing the computing time of cost function minimization for NPC inverter in three-phase power grids. *ISA Trans.* **2017**, *71*, 391–402. [CrossRef] [PubMed]
23. Hartani, M.A.; Hamouda, M.; Abdelkhalek, O.; Mekhilef, S. Sustainable energy assessment of multi-type energy storage system in direct-current-microgrids adopting Mamdani with Sugeno fuzzy logic-based energy management strategy. *J. Energy Storage* **2022**, *56*, 106037. [CrossRef]
24. Shuai, C.; Sun, Y.; Zhang, X.; Yang, F.; Ouyang, X.; Chen, Z. Intelligent diagnosis of abnormal charging for electric bicycles based on improved dynamic time warping. *IEEE Trans. Ind. Electron.* **2022**, *70*, 7280–7289. [CrossRef]
25. Hasanah, A.W.; Pujotomo, I.; Hariyati, R.; Qosim, M.N.; Sabaryati, J.; Viridi, S. Modeling of the Relation of SoC, SoH, DoD for VRLA Battery of Solar Power Plant in IT-PLN. In Proceedings of the International Conference on Innovation in Science and Technology (ICIST 2020), Semarang, Indonesia, 30 November 2021; pp. 24–28.
26. Yu, R.; Liu, G.; Xu, L.; Ma, Y.; Wang, H.; Hu, C. Review of Degradation Mechanism and Health Estimation Method of VRLA Battery Used for Standby Power Supply in Power System. *Coatings* **2023**, *13*, 485. [CrossRef]
27. Lu, R.; Lu, J.; Liu, P.; He, M.; Liu, J. Design of the vrla battery real-time monitoring system based on wireless communication. *Sensors* **2020**, *20*, 4350. [CrossRef]
28. Keshri, R.K.; Given, M.A.C. Multi-step Constant Current Charging Strategy for a Valve Regulated Lead-Acid Battery. In Proceedings of the 2019 IEEE Transportation Electrification Conference (ITEC-India), Bengaluru, India, 17–19 December 2019; pp. 1–5.
29. Lavety, S.; Keshri, R.; Ghosh, S.; Chaudhari, M.A. Non-linear model and parameter extraction for charge/discharge behavior of valve regulated lead-acid battery. *IEEE Trans. Energy Convers.* **2021**, *36*, 2600–2611. [CrossRef]
30. Teodorescu, R. An overview and comparison of online implementable SOC estimation methods for lithium-ion battery. *IEEE Trans. Ind. Appl.* **2017**, *54*, 1583–1591.
31. Santhanagopalan, S.; White, R.E. Online estimation of the state of charge of a lithium ion cell. *J. Power Sources* **2006**, *161*, 1346–1355. [CrossRef]
32. Zhang, M.; Yang, D.; Du, J.; Sun, H.; Li, L.; Wang, L.; Wang, K. A review of SOH prediction of Li-ion batteries based on data-driven algorithms. *Energies* **2023**, *16*, 3167. [CrossRef]
33. Park, K.; Choi, Y.; Choi, W.J.; Ryu, H.-Y.; Kim, H. LSTM-based battery remaining useful life prediction with multi-channel charging profiles. *IEEE Access* **2020**, *8*, 20786–20798. [CrossRef]
34. Ren, L.; Dong, J.; Wang, X.; Meng, Z.; Zhao, L.; Deen, M.J. A data-driven auto-CNN-LSTM prediction model for lithium-ion battery remaining useful life. *IEEE Trans. Ind. Inform.* **2020**, *17*, 3478–3487. [CrossRef]
35. Song, X.; Yang, F.; Wang, D.; Tsui, K.-L. Combined CNN-LSTM network for state-of-charge estimation of lithium-ion batteries. *IEEE Access* **2019**, *7*, 88894–88902. [CrossRef]
36. Yang, F.; Zhang, S.; Li, W.; Miao, Q. State-of-charge estimation of lithium-ion batteries using LSTM and UKF. *Energy* **2020**, *201*, 117664. [CrossRef]
37. Yayan, U.; Arslan, A.T.; Yucel, H. A novel method for SoH prediction of batteries based on stacked LSTM with quick charge data. *Appl. Artif. Intell.* **2021**, *35*, 421–439. [CrossRef]
38. Shih, S.-Y.; Sun, F.-K.; Lee, H.-y. Temporal pattern attention for multivariate time series forecasting. *Mach. Learn.* **2019**, *108*, 1421–1441. [CrossRef]

39. Sun, S.; Sun, J.; Wang, Z.; Zhou, Z.; Cai, W. Prediction of battery soh by cnn-bilstm network fused with attention mechanism. *Energies* **2022**, *15*, 4428. [CrossRef]
40. Lindemann, B.; Maschler, B.; Sahlab, N.; Weyrich, M. A survey on anomaly detection for technical systems using LSTM networks. *Comput. Ind.* **2021**, *131*, 103498. [CrossRef]
41. Tian, Y.; Wen, J.; Yang, Y.; Shi, Y.; Zeng, J. State-of-Health Prediction of Lithium-Ion Batteries Based on CNN-BiLSTM-AM. *Batteries* **2022**, *8*, 155. [CrossRef]
42. Tang, R.; Zhang, P.; Ning, S.; Zhang, Y. Prediction of Battery SOH and RUL Based on Cooperative Characteristics in Voltage-Temperature-Time Dimensions. *J. Electrochem. Soc.* **2023**, *170*, 060535. [CrossRef]
43. Zhang, W.; Peng, G.; Li, C.; Chen, Y.; Zhang, Z. A new deep learning model for fault diagnosis with good anti-noise and domain adaptation ability on raw vibration signals. *Sensors* **2017**, *17*, 425. [CrossRef] [PubMed]
44. Hu, N.; Chen, H.; Cheng, Z.; Zhang, L.; Zhang, Y. Fault diagnosis for planetary gearbox based on EMD and deep convolutional neural networks. *J. Mech. Eng.* **2019**, *55*, 9–18. [CrossRef]
45. Lai, C.-M.; Kuo, T.-J. Available capacity computation model based on long short-term memory recurrent neural network for gelled-electrolyte batteries in golf carts. *IEEE Access* **2021**, *10*, 54433–54444. [CrossRef]

gels

MDPI

Review

Advances in Carbon Xerogels: Structural Optimization for Enhanced EDLC Performance

Jongyun Choi [1], Ji Chul Jung [1,*] and Wonjong Jung [2,*]

[1] Department of Chemical Engineering, Myongji University, Yongin 17058, Republic of Korea; jongyun0122@gmail.com
[2] Department of Mechanical, Smart, and Industrial Engineering, Gachon University, Seongnam 13120, Republic of Korea
* Correspondence: jcjung@mju.ac.kr (J.C.J.); wonjongjung@gachon.ac.kr (W.J.)

Abstract: This review explores the recent progress on carbon xerogels (CXs) and highlights their development and use as efficient electrodes in organic electric double-layer capacitors (EDLCs). In addition, this work examines how the adjustment of synthesis parameters, such as pH, polymerization duration, and the reactant-to-catalyst ratio, crucially affects the structure and electrochemical properties of xerogels. The adaptability of xerogels in terms of modification of their porosity and structure plays a vital role in the improvement of EDLC applications as it directly influences the interaction between electrolyte ions and the electrode surface, which is a key factor in determining EDLC performance. The review further discusses the substantial effects of chemical activation with KOH on the improvement of the porous structure and specific surface area, which leads to notable electrochemical enhancements. This structural control facilitates improvement in ion transport and storage, which are essential for efficient EDLC charge–discharge (C–D) cycles. Compared with commercial activated carbons for EDLC electrodes, CXs attract interest for their superior surface area, lower electrical resistance, and stable performance across diverse C–D rates, which underscore their promising potential in EDLC applications. This in-depth review not only summarizes the advancements in CX research but also highlights their potential to expand and improve EDLC applications and demonstrate the critical role of their tunable porosity and structure in the evolution of next-generation energy storage systems.

Keywords: carbon xerogels; tailoring structures; EDLCs; electrochemical performance

1. Introduction

With the growing interest in the development and utilization of eco-friendly energy sources, the research on electrical energy storage devices is becoming increasingly active with the goal of ensuring the efficient use of electrical energy [1–7]. Such energy storage devices include fuel cells, lithium-ion batteries (LIBs), and electric double-layer capacitors (EDLCs), with each device finding its utilization in various fields depending on its unique advantages [8–12]. EDLCs are extensively used in industries such as electric vehicles, portable electronic devices, and renewable energy systems, owing to their rapid charge–discharge (C–D) capability, excellent durable life cycles, and high power density [13–16]. The electrode materials in EDLCs consist of carbon material as the active material, a conductive additive, and a polymer binder. Active materials, including activated carbons, carbon nanotubes, activated carbon fibers, and carbon xerogels (CXs), considerably influence the performance of EDLCs [17–26]. Specifically, due to their chemical stability, cost-effectiveness, porous structure, and high specific surface area, activated carbons are commercially preferred as the active materials for EDLCs [17,27]. However, the low electrical conductivity of activated carbons limits their use as electrode materials [27,28]. Therefore, many studies should focus on active materials superior to activated carbons for applications in EDLC electrodes.

Our research group concentrated on CXs, which, although they have similar advantages to activated carbon, possess outstanding electrical conductivity. CXs exhibit remarkable electrical conductivity due to their effective connectivity via three-dimensional (3D) networks of carbon particles [29–31]. Furthermore, physical properties, such as 3D network structure, pore size, and specific surface area of CXs, can be systematically controlled through the adjustment of synthesis parameters, which makes them highly promising as electrode materials for EDLCs [32–35]. For these reasons, numerous studies have investigated the mechanism underlying the electrochemical performances of EDLCs under the influence of various pore sizes of CXs, which occur due to synthesis conditions and the physical characteristics of CXs [36–44]. In addition, various attempts, including methods beyond structure control, such as metal doping and the application of activation processes to CXs, are being made to improve the performance of EDLC electrodes through CXs [37,43–46].

This review comprehensively introduces the methodologies used for the fabrication of CXs, with a focus on the structural control of CXs to specialize them for application in EDLC electrodes. This work further delves into CX electrochemical characteristics and their potential for industrial application in EDLC electrodes, various approaches utilized for the synthesis of CXs as electrode materials for EDLCs, and the effects of synthesis parameters on the electrochemical performance of EDLCs [47–61]. Specifically, an analysis was conducted on the effects of synthesis parameters, such as the polymerization duration, pH, and amount of catalyst on the physical properties and electrochemical performance of CXs [55–57]. Furthermore, a comparison with commercial electrode materials for EDLC electrodes was conducted to explore the commercial viability of CXs [58]. In our studies, we used an organic electrolyte and a two-electrode system, which mirror the workings of commercial EDLCs, and the results imply the great industrial relevance of our findings. This review is of considerable importance, as it compares and analyzes data collected over more than a decade of research conducted by a single research group under nearly identical conditions. This allows for a more consistent explanation of the effects of varying experimental parameters. Figure 1 provides a summary of this review.

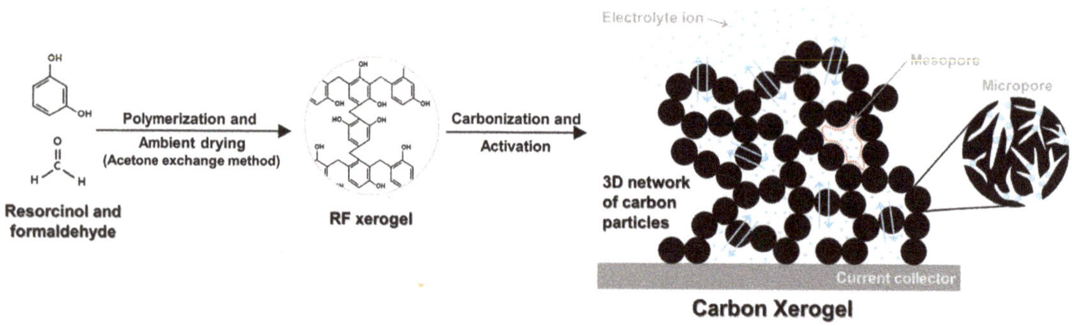

Figure 1. Schematic summary of this review.

2. CXs

CXs represent a type of carbon gel obtained through the carbonization of organic gels. In addition to xerogels, carbon gels include aerogels and cryogels, which are differentiated through the drying method of wetted organic gels [62–64]. Figure 2 illustrates the synthesis processes of various carbon gels, including carbon aerogels, xerogels, and cryogels, and Table 1 summarizes their characteristics. Aerogels can be obtained through supercritical drying to prevent the collapse of their 3D network structures caused by the strong capillary forces of water [65]. Cryogels are synthesized via freeze drying to facilitate the removal of water from the gels [66]. Notably, xerogels can be easily prepared through drying under ambient temperature and pressure conditions and can be produced by exchanging the water in the hydrogels with a substance of low surface tension before drying to prevent the collapse of their 3D network structures [62,67]. Xerogels with a stable pore structure were successfully synthesized by replacing the water in hydrogels with acetone, which has a lower surface tension, followed by drying [47,68,69]. The advantages of xerogels include a simple production method and the lack of a need for complex equipment, which makes them suitable for mass production [68]. Therefore, xerogels have a higher commercial potential compared with other carbon gels.

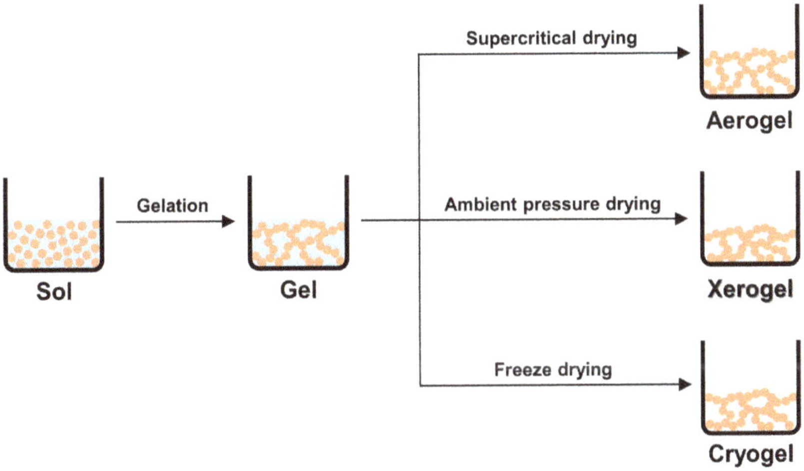

Figure 2. Synthesis processes of various carbon gels [63,64].

Table 1. Characteristics of various carbon gels [62,65,66].

Type	Drying Method	Advantages	Disadvantages
Aerogel	Supercritical drying	The structure mostly remains intact after drying.	Aerogels require supercritical drying equipment, making them expensive to produce.
Xerogel	Ambient pressure drying	The manufacturing method is very simple and highly beneficial for mass production.	Xerogels experience more structural collapse than other gels.
Cryogel	Freeze drying	After drying, the structure remains partially intact.	Cryogels necessitate freeze-drying equipment and are costly to produce.

CXs are typically synthesized via the sol–gel method, which involves the use of resorcinol and formaldehyde as reactants and sodium carbonate (Na_2CO_3) as a base catalyst to produce a resorcinol–formaldehyde (RF) gel [67,69]. The physical properties of CXs can be adjusted based on the reaction conditions during RF gel formation [55–57,69–71]. The general procedure for CX fabrication includes the following steps: (1) initiation of polymerization through the addition of resorcinol and formaldehyde to an aqueous solution containing the base catalyst Na_2CO_3, which results in sol formation; (2) aging and gelation, which leads to the formation of hydrogels with a 3D network structure; (3) implementation of the acetone exchange method, followed by drying under ambient pressure and temperature conditions to produce RF xerogels; and (4) carbonization of RF xerogels under an inert atmosphere to obtain CXs.

The physical properties of CXs, such as specific surface area, pore size, pore volume, and 3D network structure, are determined by their synthesis parameters, including polymerization duration, pH, and the ratio of resorcinol to catalyst (R/C ratio). CXs with tailored structures and physical properties for specific applications can be fabricated through precise control of the synthesis parameters [72–74]. Therefore, CXs have attracted considerable interest for various applications. For example, CXs are utilized in biomedical applications due to the porosity obtained from the sol–gel process. CXs can be engineered to be texturally suited for specific drugs, thereby optimizing effective drug delivery [62,75]. Moreover, the inherent chemical stability of carbon materials and the adjustable porous properties of CXs have been employed as supports for metal catalysts such as Pt/C and Ru/C [76]. This is particularly useful in advanced oxidation processes, such as wet air oxidation, and in producing fine chemicals. Furthermore, CXs are utilized in environmental engineering. Discharging industrial effluents containing toxic metal ions and persistent reactive dyes into the environment presents significant life-threatening effects, highlighting the urgent need for efficient solutions. CXs have shown effectiveness in adsorbing large organic molecules from wastewater, and their commercial applications are under exploration [77–79].

Most importantly, CXs show great potential uses in energy storage devices, such as fuel cells, LIBs, and EDLCs [80–82]. Research on the use of CXs in energy storage devices draws significant interest because the structural control of CXs can substantially improve the performance of these devices [81,83]. In particular, the use of CXs is gaining attention in the field of EDLCs, where the structural and physical properties of active materials considerably influence the electrochemical performance [84,85]. Their adjustable physical properties based on reaction conditions suggest the high utility of CXs as electrode materials for EDLCs. Therefore, we synthesized structurally controlled CXs by regulating the reaction conditions and used them as active materials for EDLC electrodes [52–58]. In addition, we conducted studies to understand the interrelation between the physical properties of CXs, determined by the synthesis parameters, and the electrochemical performance of EDLCs.

3. EDLC

EDLCs, known as supercapacitors, derive their name from their working principle. EDLCs store energy through electrolyte ion adsorption on the electrode surface, which forms a Helmholtz double layer and utilizes electrostatic attraction [86,87]. Importantly, EDLCs store electrical energy via the adsorption–desorption process of electrolyte ions on the electrode surface. Based on this operation principle, EDLCs offer several advantages, including long-term durability, rapid C–D rates, and high power densities [88]. In contrast, LIBs store electrical energy through redox reactions at the electrodes. Specifically, compared to LIBs, which maintain 80% of their initial capacity after 3000 cycles, EDLCs exhibit significantly long-term durability, maintaining over 80% of their capacity even after 10,000 cycles [89–91]. This difference makes EDLCs and LIBs complementary to each other, as both types of devices bring unique benefits and offset each other's limitations [88]. These complementary characteristics enable the combined application of both types of energy storage devices across various industries [92].

Combining EDLCs with LIBs creates a powerful energy management system capable of handling sudden increases in electricity demand. This effectiveness is due to the ability of EDLCs to charge and discharge rapidly. For instance, EDLCs help stabilize the power grid by instantly supplying power, which is particularly useful for balancing the unpredictable nature of renewable energy sources like solar and wind [13]. Additionally, EDLCs significantly enhance the efficiency and overall performance of electric vehicles and hybrid electric vehicles by improving energy use during crucial actions such as regenerative braking and acceleration [16]. Moreover, by adeptly managing high-power demands, EDLCs help extend the lifespan of these vehicles' batteries. Consequently, EDLCs make substantial contributions to various fields, including renewable energy systems, uninterruptible power supplies, and electric vehicles [93,94]. The pivotal role of energy storage technologies, particularly EDLCs, in balancing energy supply and demand and enhancing the reliability of the electricity supply cannot be overstated.

The electrochemical performance of EDLCs is considerably affected by several key components, including the electrolyte, conducting additive, binder, and active material [95–101].

Electrolytes can be divided into aqueous and organic types, with the latter being preferred in the industry due to their higher electrical conductivity and stability at elevated voltages [102,103]. Conducting additives, such as carbon black, increase the electrical conductivity of the electrodes [98]. Binders, such as polytetrafluoroethylene and polyvinylidene fluoride (PVDF), bond the conductive additive and active material to the current collector [99]. The active material, which provides the necessary space for electrolyte ion adsorption, greatly influences the EDLC performance [100,101]. To enhance the performance of EDLCs, researchers have utilized carbon materials, such as activated carbons, carbon nanotubes, activated carbon fibers, and CXs, with extensive studies being conducted on active materials [101,104,105]. Table 2 provides a summary of the results of relevant studies that employed a two-electrode system analogous to commercial EDLCs.

Table 2. Overview of EDLC electrode materials and their electrochemical performances.

Electrode Materials	Electrolyte	Specific Capacitance (F/g)	Current Density (A/g)	Reference
Activated carbon	6 M KOH in H_2O	122	1.0	[106]
		262	0.05	[107]
		198	1.0	[108]
		148	0.5	[109]
		254	1.0	[110]
		93	0.1	[111]
		155	0.2	[112]
	1 M Na_2SO_4 in H_2O	84	1.0	[113]
	1 M $LiClO_4$ in PC	192	1.0	[114]
	1 M $(C_2H_5)_4NBF_4$ in PC	82	0.1	[115]
	1 M $SBPBF_4$ in PC	82	1.0	[116]
	1 M $TEABF_4$ in AN	115	0.2	[117]
	1 M $TEABF_4$ in EC/DMC	88	1.0	[118]
	1 M $TEABF_4$ in PC	41	1.0	[119]
	1 M Et_4NBF_4 in PC	299	1.0	[120]
	1 M Et_4NBF_4 in ACN	100	0.1	[121]
Carbon nanotube	1 M Na_2SO_4 in H_2O	150	0.15	[122]
	1.96 M $TEMABF_4$ in PC	44	2.0	[123]
	0.6 M TEABF4 in CPAME	96	0.5	[124]
Activated carbon fiber	6 M KOH in H_2O	210	0.04	[125]
	1 M H_2SO_4 in H_2O	248	1.0	[126]
	1 M $TEABF_4$ in AN	112	1.0	[127]
Carbon aerogel	6 M KOH in H_2O	384	0.01	[128]
		112	1.0	[129]
	1 M Et_4NBF_4 in AN	149	0.5	[130]
Carbon xerogel	6 M KOH in H_2O	112	1.0	[45]
		201	1.0	[131]
		266	1.0	[132]
	1 M Na_2SO_4 in H_2O	124	1.0	[133]
	1 M H_2SO_4 in H_2O	141	0.2	[134]

The physical properties of electrode materials, including their specific surface areas, pore sizes, and pore volumes, are pivotal determinants of EDLC performance [97,98]. This highlights the importance of understanding the relationship between the physical properties of electrode materials and the electrochemical performance of EDLCs. Notably, EDLCs exhibit a markedly improved electrochemical performance when their pore size closely matches that of the electrolyte ions [135,136]. In addition, a high specific surface area is crucial for the enhanced performance of EDLCs.

According to the International Union of Pure and Applied Chemistry (IUPAC) standards, electrode material pores can be categorized based on their diameters into micropores (widths not exceeding approximately 2 nm), mesopores (between 2 and 50 nm), and macropores (exceeding 50 nm) [137]. Among these pores, the micropores and mesopores within active materials considerably affect EDLC performance. Electrolyte ion adsorption is highly influenced by the specific surface area of active materials, which is largely determined by the volume of micropores and mesopores. In addition, mesopores act as conduits for ion movement, which affects the electrochemical performance of EDLCs [138]. However, EDLC electrochemical performance is not entirely dependent on the specific surface area or pore size [139]. Various factors, such as electrical conductivity and the composition of the 3D network structure, exert a complex influence [88]. Consequently, research and development should focus on active materials for EDLCs with precisely controlled properties, including electrical conductivity, pore structure, and specific surface area.

CXs can have a tailored pore volume and size through the adjustment of synthesis parameters. Moreover, the 3D carbon particle network of CXs can improve electrical conductivity. This property has prompted numerous researchers to investigate CXs as EDLC electrode materials. To enhance EDLC performance with CXs, researchers have proposed strategies that not only control physical properties but also include metal doping [45,46,140]. Doping CXs with metals such as Co, Ni, Cu, Mn, Fe, and Zn considerably increases the specific capacitance of EDLCs, and such an improvement is attributed to the presence of metal oxides and the resulting Faradaic redox reactions [48–51]. However, redox reactions can reduce some key benefits of EDLCs, including durable life cycles and power density. Therefore, optimization of the CX structure has been found to be more effective than metal doping for achieving high-performance EDLCs.

Coin-type EDLCs were assembled to analyze the electrochemical performance of CXs, with the fabrication process detailed in Figure 3. The electrodes were produced using various carbon materials (CXs, activated CXs (ACXs), and activated carbons) as the active material, Super-P as the conducting additive, and PVDF as the binder. A slurry was prepared by mixing the active material, conductive agent, and binder at a mass ratio of 8:1:1 with 1-methyl-2-pyrrolidone as the solvent. Then, the slurry was coated onto an Al current collector. The organic electrolyte 1 M tetraethylammonium tetrafluoroborate dissolved in acetonitrile (1 M $TEABF_4$/ACN) was utilized in the coin-type EDLCs. The assembled EDLCs were evaluated for their electrochemical performance through cyclic voltammetry (CV), galvanostatic charge–discharge (GCD), electrochemical impedance spectroscopy (EIS), and four-point probe (FPP) measurements.

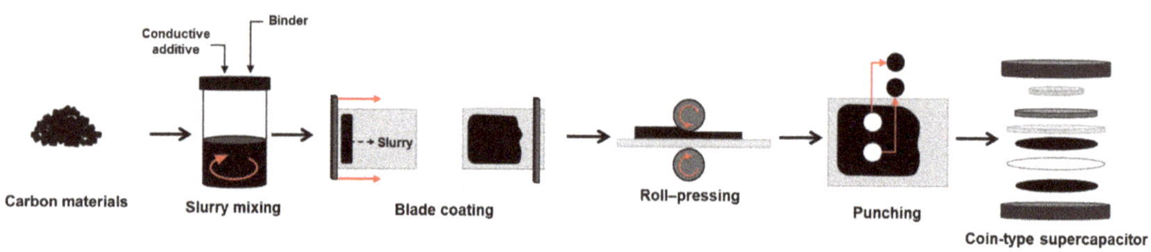

Figure 3. Scheme of the coin-cell EDLC fabrication process [22,111,115].

4. Tailoring Structure of Carbon Xerogels by Synthesis Parameters

Xerogels were produced by subjecting RF gels to the drying–wetting method under ambient conditions. During the drying–wetting process, the high surface tension of water (71.99 dyne/cm) posed a risk of pore structure collapse [141]. To mitigate this issue, our research group replaced the water in the RF gels with acetone, which has a lower surface tension (20.66 dyne/cm), before proceeding to the drying process [47]. As a result, the integrity of the pore structure was successfully preserved, which enabled the production of CXs with minimal structural collapse. The produced CXs were then utilized in our studies, which focused on designing CXs with precisely controlled structures through the manipulation of synthesis parameters. The synthesis process involved adjusting variables, including the polymerization duration, pH, and R/C ratio. Figure 4 concisely illustrates the structural alterations resulting from these varied conditions.

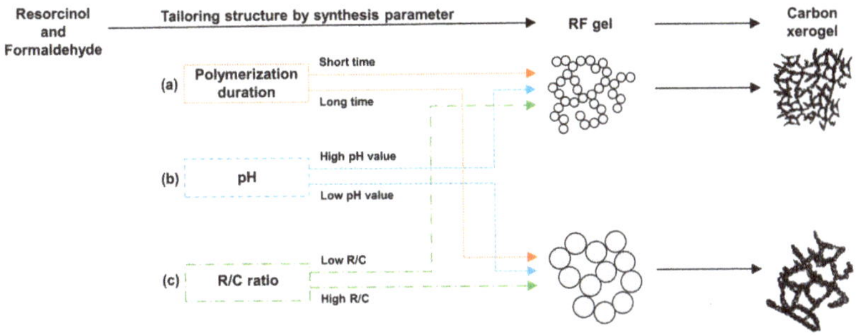

Figure 4. Structural changes of CXs based on synthesis parameters: (**a**) polymerization duration, (**b**) pH, and (**c**) R/C ratio [55–57,142,143].

4.1. Polymerization Duration

As the physical properties of CXs varied depending on the RF gels prior to carbonization, understanding of the synthesis process of RF gels was essential. We optimized the polymerization duration to control the physical properties of the CXs, which led to an in-depth understanding of the RF gels' synthesis process [55]. Adjusting the polymerization duration substantially affected the physical properties of the CXs, including their pore structure and specific surface area, which are vital for the electrochemical performance of EDLCs. Therefore, a systematic analysis is needed to determine the mechanism underlying the influence of polymerization duration on the physical properties of CXs and its consequent effects on the electrochemical performance of EDLCs. Figure 4a illustrates the variations in the CX structure over polymerization durations ranging from 1 h to 20 h.

Mesopore formation in CXs was evident across all polymerization durations (Figure 5a,b). The presence of these mesopores suggests the formation of a 3D network structure within CXs. Furthermore, the data presented in Table 3 show that the specific surface area, pore size, and pore volume increased with prolonged polymerization. This information is crucial for understanding the mechanism underlying the influence of polymerization duration on the structural characteristics of CXs. As the polymerization time increases, larger particle sizes could be generated [142]. As a result, these larger particles caused the creation of large pores and high pore volumes during the gelation process. Conversely, a short polymerization duration resulted in small molecular weights per particle, which led to the formation of smaller particles. These smaller particles formed a tightly interconnected 3D network structure within the CXs, which was characterized by small pore sizes and volumes. Therefore, the adjustment of polymerization duration plays a key role in the customizing of the structure of CXs, which makes it a critical parameter in the synthesis of CXs optimized for EDLC electrode materials.

Table 3. Physical properties of carbon xerogels (CX_polymerization duration) [55].

Samples	S_{BET} (m^2/g) [a]	D_{avg} (nm) [b]	Total Pore Volume (cm^3/g)
CX_1	645	12.2	0.7
CX_2	828	12.2	1.0
CX_5	796	16.1	1.2
CX_10	853	16.1	1.3
CX_20	998	21.3	2.1

[a] Specific surface area was calculated by the Brunauer–Emmet–Teller (BET) plot. [b] Average pore diameter was determined by the Barrett–Joyner–Halenda (BJH) method with the desorption branch.

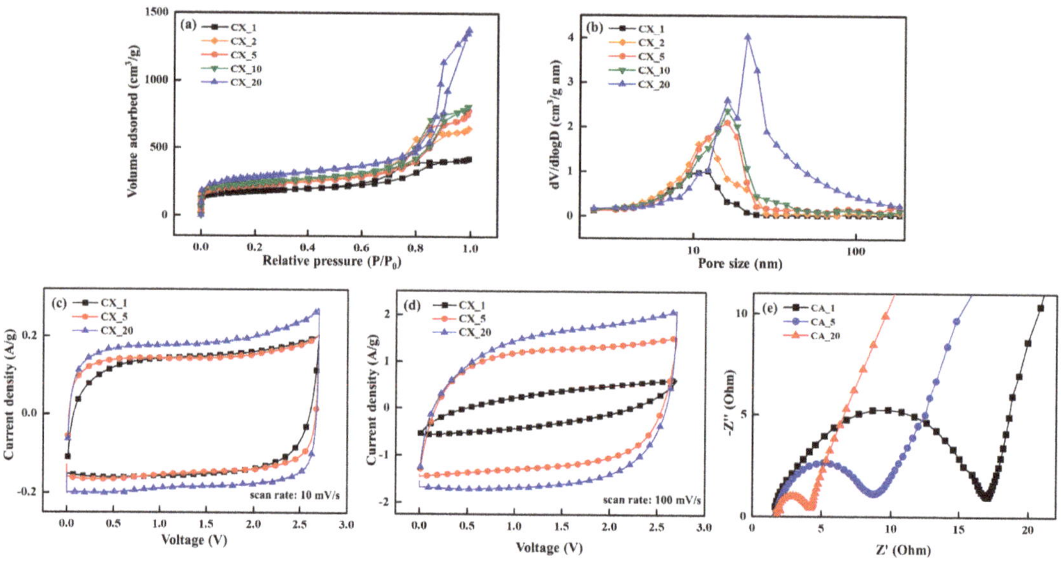

Figure 5. (**a**) N_2 adsorption–desorption isotherm, (**b**) pore size distribution, (**c**) cyclic voltammograms at a scan rate of 10 mV/s, (**d**) cyclic voltammograms at a scan rate of 100 mV/s, and (**e**) Nyquist plots of carbon xerogels (CX_polymerization druation) [55].

CV and EIS analyses were conducted on the CXs (CX_1, CX_5, and CX_20), and the CV curves and Nyquist plots are illustrated in Figure 5c–e. Under optimal operational conditions, EDLCs maintain the magnitude of the current and linearly change in response to variations in voltage, irrespective of the flow direction [144]. Thus, an ideal CV curve should have a shape closely resembling a symmetrical rectangle [145]. At low scan rates under all polymerization durations, the CV curve maintained its rectangular shape. However, at a high scan rate of 100 mV/s, the shape of the CV curve transitioned from rectangular to being rugby-ball-shaped. This transformation indicates a decline in the electrochemical performance of the CXs. Notably, CX_1, which was synthesized within a short polymerization period, exhibited a more distorted rugby-ball-shaped CV curve at higher scan rates compared with other CXs. This condition prevented CX_1 from maintaining its capacitance at fast scan rates. This result can be attributed to the restricted mobility of electrolyte ions, as supported by the Nyquist plots obtained from EIS analysis.

The Nyquist plots were divided into three distinct sections: the bulk-solution resistance, which results from the mobility of electrolyte ions outside the electrode; the charge transfer resistance (R_{ct}), which includes the inherent electronic resistance of the electrode material and the resistance against ion movement within it; and Warburg resistance, which arises from the diffusion of electrolyte ions and the internal resistance within the cell [23–26]. The semicircles in the Nyquist plots correspond to the R_{ct} at the electrode, with small semicircle diameters signifying a low R_{ct} [146]. Thus, the R_{ct} contributes to determining the intrinsic resistance of EDLC electrode materials.

CX_1 exhibited the lowest capacitance retention property and the highest R_{ct}. Notably, the R_{ct} decreased with the extended polymerization duration (Figure 5e). This change in R_{ct} is closely linked to the structural transformations in CXs. Prolonged polymerization led to the formation of large particles, which in turn led to CXs with abundant pores. These pores facilitated electrolyte ion movement and acted as pathways for improved ion mobility within the CXs. As a result, enhanced electrolyte ion mobility was observed, reducing ion resistance. This outcome is consistent with the results obtained from other electrochemical analyses.

Thus, the control over the polymerization duration was a critical factor in the customization of CXs as active materials for EDLCs. The duration of polymerization profoundly influenced the growth of polymer particles. Notably, a short polymerization duration led to the formation of small particles. These small particles faced challenges in the formation of CXs with well-developed pores. The low porosity of CXs hindered the mobility of electrolyte ions, resulting in an increased R_{ct}, which negatively influenced the overall electrochemical performance of the EDLCs. Therefore, methodical control of the polymerization time is crucial for the structural modification of CXs to optimize their effectiveness as electrode materials for EDLCs.

4.2. pH of Polymerization Solution

Figure 4b demonstrates the structural modifications that occurred in response to pH levels during the synthesis process of RF gels. Previous research has shown that pH levels significantly influence the formation of initial particle nuclei [142,146,147]. At high pH levels, numerous nuclei formed at the reaction initiation, leading to the formation of small particles due to spatial constraints and limited reactants. Through gelation, the small particles synthesized at high pH levels amalgamated to form a dense structure, which hindered pore formation. As indicated in Table 4 and Figure 6a,b, the CXs synthesized under high pH levels presented poorly developed pore structures and low specific surface areas. By contrast, synthesis at low pH levels resulted in the formation of fewer nuclei, which allowed for the growth of relatively large particles. These larger particles combined and formed a 3D network structure, resulting in the production of CXs with well-developed pores and high specific surface areas.

Table 4. Physical properties and specific capacitances (calculated from GCD measurements) of carbon xerogels (CX_pH) [56].

Samples	S_{BET} (m^2/g) [a]	D_{avg} (nm) [b]	V_{total} (cm^3/g) [c]	Specific Capacitance (F/g)		
				1 A/g	3 A/g	5 A/g
CX_9.5	758.8	4.7	0.89	9.8	5.2	3.1
CX_10	613.0	2.9	0.45	8.0	4.4	3.2
CX_11	267.6	1.8	0.12	2.8	1.0	0.8
CX_12	36.8	3.1	0.03	0.9	-	-

[a] Specific surface area was calculated by the BET plot. [b] Average pore diameter was determined by the BJH method with the desorption branch. [c] Total pore volume.

The electrochemical performance of EDLCs was substantially influenced by the specific surface area and structural characteristics of the active materials. For the CXs synthesized at relatively low pH levels (9.5 and 10), rectangular-shaped CV curves with a wide area were observed at various scan rates (Figure 6c–e). Although the geometry of the CV curve transformed from a rectangular to a rugby-ball shape with the increase in scan rates, the exceptional electrical conductivity of the 3D network structure ensured the preservation of a relatively rectangular shape. In summary, the CXs synthesized at low pH levels showed superior electrochemical performances, as evidenced by the CV curves.

Comparable outcomes were elucidated through GCD analysis. The specific capacitance (C_g) derived from the GCD analysis was calculated using Equation (1), with the results tabulated in Table 4.

$$C_g = \frac{I \cdot \Delta t}{m \cdot \Delta V} \quad (1)$$

where C_g denotes the specific capacitance, m represents the mass of the active electrode material, ΔV indicates the voltage variation during the discharge phase, I is the current observed during discharge, and Δt signifies the discharge duration.

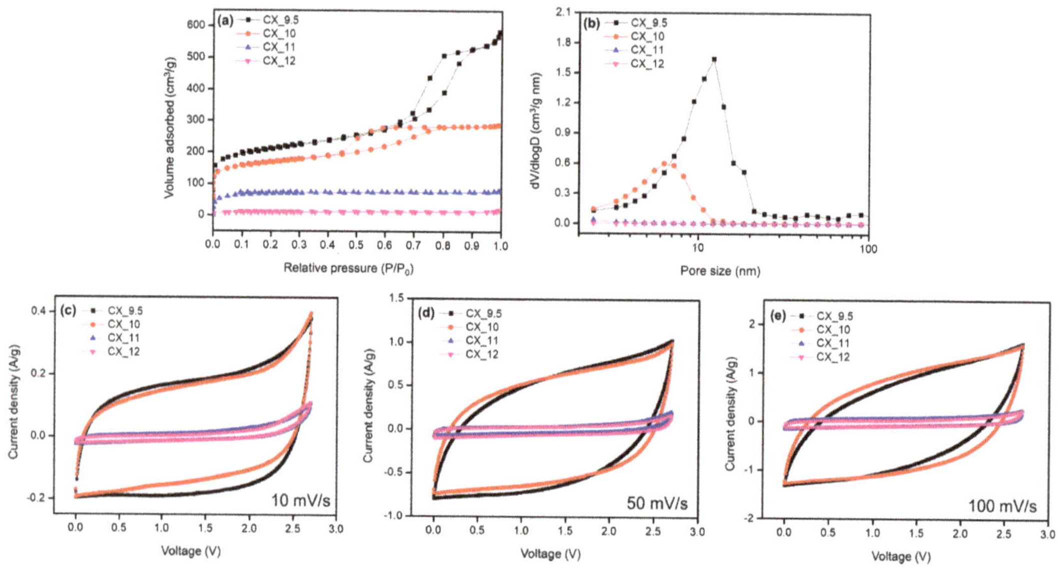

Figure 6. (**a**) N_2 adsorption–desorption isotherm, (**b**) pore size distribution, and (**c**) cyclic voltammograms at a scan rate of 10 mV/s, (**d**) 50 mV/s, and (**e**) 100 mV/s of carbon xerogels (CX_pH) [56].

CXs synthesized at low pH levels demonstrated superior specific capacitance across all current densities. These CXs possessed high specific surface areas, which contributed to their elevated specific capacitances at low C–D rates (low current density). Furthermore, the electrical conductivity facilitated by the 3D network connections and pore structures helped in maintaining a high specific capacitance at high C–D rates (high current density). By contrast, the CXs synthesized at high pH levels exhibited markedly low specific capacitances at low current densities. This finding was a consequence of their reduced specific surface areas and poorly developed pore structures.

During the synthesis of RF gels for CX production, pH played a crucial role in the initial nucleation. This nucleation process influenced the subsequent polymer growth and the formation of 3D network structures and pore structure development. Variations in the physical characteristics of CXs due to pH changes showed a direct correlation with the electrochemical performance of the EDLCs. Overall, a relatively low pH benefited particle growth, positively affecting the development of 3D network structures. Low pH conditions induced physical properties in the CXs, such as an increased specific surface area and an improved pore structure, which contributed to the high performance of the EDLCs. Thus, meticulous pH control during the polymerization process is crucial for optimizing the physical properties of CXs, thereby enhancing the overall efficiency and performance of EDLCs.

4.3. Ratio of Resorcinol to Catalyst

Figure 4c provides an overview of the mechanism of the CX structure underlying its influence on the molar R/C ratio. The R/C ratio was closely associated with the number of nuclei generated during the initial step of RF gel synthesis [41,57]. The nuclei formed during this initial step substantially influenced the physical properties of the CXs. Specifically, at a low R/C ratio, which indicates a high amount of catalyst, numerous nuclei formed, and the catalyst induced a vigorous reaction. Otherwise, at a high R/C ratio, which reflects a low amount of catalyst, a less vigorous reaction occurred due to the reduced catalyst amount.

As elucidated in Figure 7 and Table 5, an increase in the R/C ratio resulted in elevated specific surface areas, pore dimensions, and pore volumes. To facilitate a more intuitive understanding, scanning electron microscope (SEM) images of CXs according to the R/C ratio are displayed in Figure 8. As the R/C ratio increased, the proportion of catalyst per resorcinol decreased, resulting in fewer reaction initiation sites. The decrease in the number of reaction initiation sites implied the reduced formation of nuclei. This effect provided more space for particle growth, resulting in the formation of larger particles. Consequently, these larger particles formed a 3D network, leading to the development of relatively larger pores.

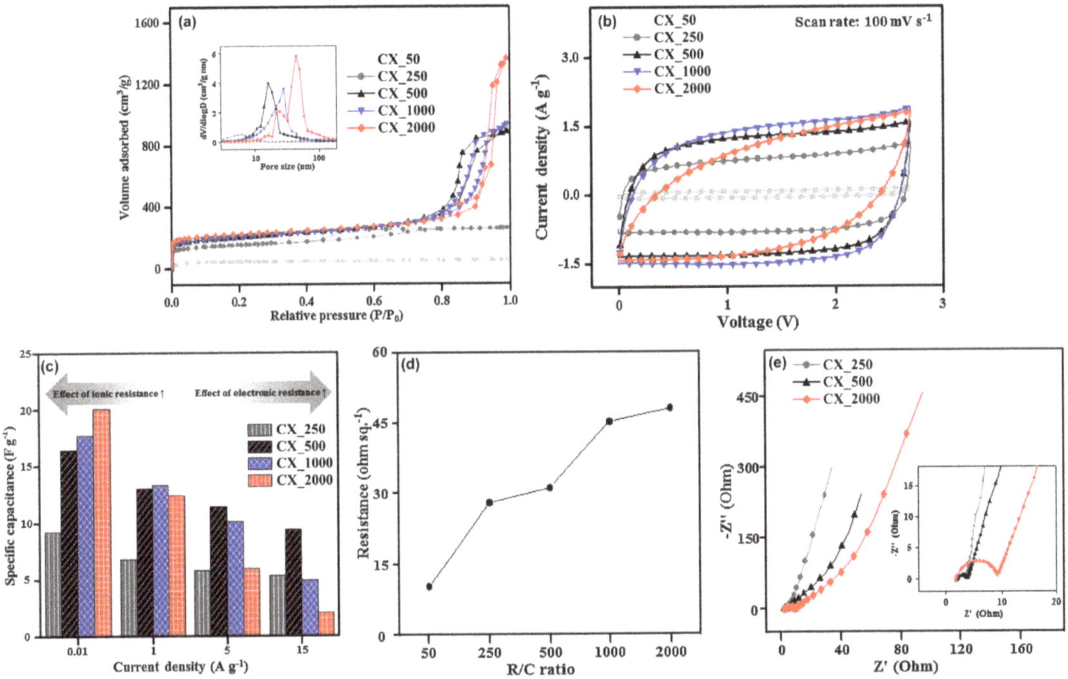

Figure 7. (**a**) N_2 adsorption–desorption isotherm, (**b**) cyclic voltammograms at a scan rate of 100 mV/s, (**c**) specific capacitance to current density, (**d**) sheet resistance, and (**e**) Nyquist plots of carbon xerogels (CX_R/C ratio) [57].

Table 5. Physical properties of carbon xerogels (CX_R/C ratio) [57].

Samples	S_{BET} (m^2/g) [a]	D_{avg} (nm) [b]	Total Pore Volume (cm^3/g)
CX_50	162.5	2.0	0.09
CX_250	564.4	2.8	0.43
CX_500	730.6	7.5	1.39
CX_1000	752.6	7.8	1.46
CX_2000	847.3	10.1	2.11

[a] Specific surface area was calculated by the BET plot. [b] Average pore diameter was determined by the BJH method with the desorption branch.

By contrast, CXs synthesized with a low R/C ratio suffered from limitations in pore structure development due to the large number of particles generated by the high catalyst quantity. Specifically, CX_50, which had a notably low R/C ratio, demonstrated a substantially reduced specific surface area and pore volume compared with other CXs. This finding suggests that excessively low R/C ratios resulted in the formation of particles that were too small to maintain a stable and effective 3D network structure of CXs for EDLCs.

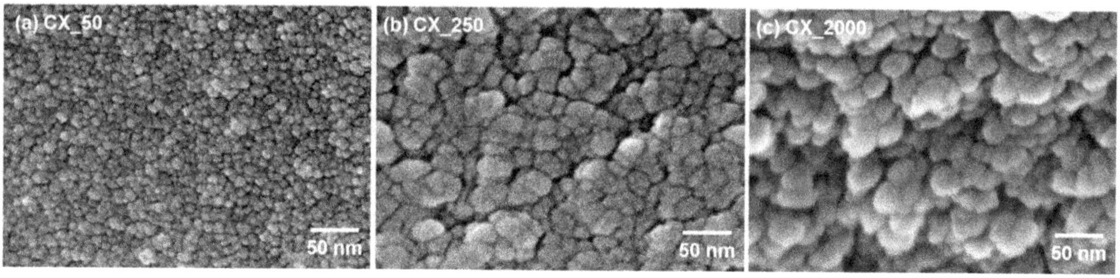

Figure 8. SEM image of carbon xerogels synthesized using different R/C ratios: (**a**) CX_50, (**b**) CX_250, and (**c**) CX_2000 [57].

Synthesized with a high R/C ratio, CX_2000 was expected to exhibit a superior electrochemical performance given its high specific surface area and abundant pores. However, the results from CV, GCD, and EIS analyses indicated unexpected deviations (Figure 7b, c, and e, respectively). Different from other CXs, CX_2000 displayed a rugby-ball-shaped CV curve at a scan rate of 100 mV/s, diverging from the expected rectangular shape. This finding suggests a decrease in the electrochemical efficiency. Furthermore, the GCD analysis highlighted that despite its notable specific capacitance at low current densities due to its high specific surface area, CX_2000 failed to maintain this capacitance at high current densities. In contrast, although CXs synthesized with a low R/C ratio exhibited a low specific capacitance at low current densities, they effectively maintained their specific capacitance at high current densities.

The electrochemical performance of a CX is intricately related to its physical and resistance properties. CXs fabricated at a high R/C ratio exhibited a high specific surface area and superior porosity. At a high R/C ratio, the development of large particles led to an improved pore structure but adversely affected the 3D network connectivity among particles. As a result, the reduced connectivity among particles increased the electronic resistance of the electrode material. Conversely, at a low R/C ratio, despite the less developed pore structure, leading to increased ionic resistance, the well-connected 3D network of particles considerably reduced the electronic resistance of the electrode material.

The structural modifications of the CXs substantially influenced the resistance properties of the EDLCs. This relationship was demonstrated by the analysis findings of the Nyquist plots and FPP measurements (Figure 7d,e). The Nyquist plots showed that the decrease in the R/C ratio led to a smaller semicircle size, which directly signified a low R_{ct} in the EDLC electrodes. The R_{ct} encompassed the ionic resistance, which involved the obstruction of electrolyte ion movement within the electrode materials, and the intrinsic electronic resistance of electrode materials themselves [58,145]. Despite the less developed pore structure of CXs fabricated with a low R/C ratio, which presents a disadvantage for ion transport, a low R_{ct} was observed. This result implies that the relatively high ionic resistance was compensated for by the low electronic resistance resulting from the well-connected 3D network structure. FPP measurements, which were used in the direct assessment of the electronic resistance of CXs, revealed that the electronic resistance increased with the increase in the R/C ratio. Thus, the CXs with a large pore size exhibited a high electronic resistance due to their poorly interconnected 3D carbon particle networks, which is consistent with and reinforces the conclusions drawn from other electrochemical analyses.

In conclusion, various studies have confirmed that the R/C ratio plays a crucial role in determining the physical properties and electrochemical performance of CXs [41,67]. Specifically, the importance of the R_{ct}, which includes ionic and electronic resistances, for the electrochemical efficacy of EDLC cells should be emphasized. An exhaustive examination explored how the structural characteristics of CXs influence these resistance properties. In the synthesis of CXs for EDLC electrode materials, the R/C ratio emerged as a fundamental factor. This finding necessitates meticulous optimization to achieve an improved electrochemical performance of EDLCs.

5. ACXs for EDLCs

Activated carbons are used as electrode materials in commercial EDLCs because of their porous properties and extensive specific surface areas. These properties are advantageous as they provide a substantial area for electrolyte ion adsorption, which improves the capacitance of EDLCs. Therefore, CXs with large specific surface areas may exhibit superior electrochemical performances as electrode materials for EDLCs.

For the attainment of a high specific surface area in CXs, an additional activation process at high temperatures was conducted. This process produced ACXs using various acids, bases, or gases as activating agents [52–54,59,148]. The activation methods included chemical activation, which uses chemical activators (H_3PO_4, NaOH, and KOH), and physical activation, which employs gases (CO_2 and steam). The preparation of commercial activated carbons for EDLCs mainly involves activation with steam or KOH, with KOH activation being particularly effective in increasing the specific surface area. Therefore, employing KOH activation in CXs to synthesize ACXs with a specific surface area and understanding the influence of physical property changes on electrochemical performance is essential.

During the activation reaction, micropores developed within the CXs, which led to an increase in the specific surface area. In addition, the activation reaction temperature contributed to determining the specific surface area and pore structure. Consequently, systematic research was conducted to explore the effect of the activation temperature on the electrochemical performance of ACXs [52,53]. Figure 9 illustrates the results of the physical and electrochemical analyses of ACXs. After the activation under various temperature conditions, the N_2 adsorption–desorption isotherms exhibited hysteresis loops. This observation suggests that preservation of the 3D network structure and mesopores in the CXs occurred post-activation. Moreover, as the activation reaction progressed, the micropore development improved, increasing the specific surface area. Higher activation temperature was associated with an increase in the specific surface area and a decrease in the average pore size. This phenomenon was attributed to the vigorous progression of the reaction at high activation temperatures. Table 6 summarizes the specific surface areas and average pore sizes of the ACXs.

The increase in the specific surface area achieved through activation improved the adsorption space for electrolyte ions, which enhanced the capacitance across all C–D rates. The benefits of CX-based materials, such as the superior electrical conductivity provided by their 3D network structure and the abundance of pores, contributed to the improved performance of EDLCs. Notably, ACXs performed excellently under rapid C–D rates. As the activation temperature increased, capacitance showed an enhancement even at high C–D rates. However, a decline in capacitance was observed in the ACXs activated at excessively high temperatures (900 °C). Similarly, the CV curves for ACXs activated at high temperatures changed into a rugby-ball shape with the increase in scan rates. This finding suggests that the harsh activation temperatures caused the collapse of pores in the ACXs and 3D network structures, which resulted in elevated resistance and diminished capacitance. In conclusion, activation at an optimal temperature is necessary for the synthesis of ACXs for EDLC applications. Moreover, EDLCs that utilize ACXs as electrode materials effectively retained their initial capacitance after 1000 C–D cycles, which indicates their substantial potential for use in commercial EDLC products.

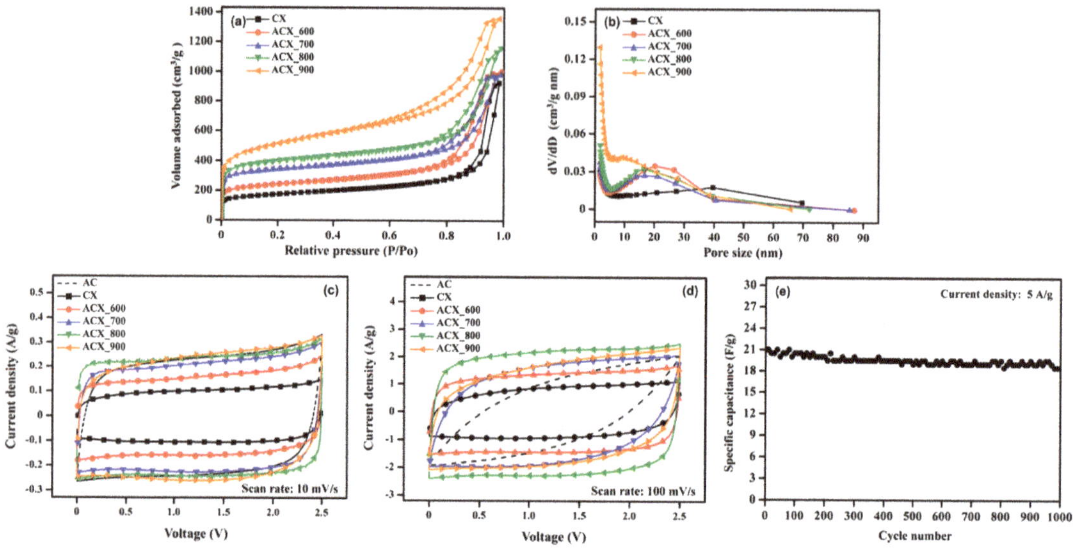

Figure 9. (**a**) N_2 adsorption–desorption isotherm, (**b**) pore size distribution of activated carbon xerogels (ACX_activation temperature), (**c**) cyclic voltammograms at a scan rate of 10 mV/s and (**d**) 100 mV/s of activated carbon (AC) and activated carbon xerogels (ACX_activation temperature), and (**e**) cycle stability of ACX_800 [53].

Table 6. Physical properties and specific capacitances (calculated from GCD measurements) of carbon xerogel (CX) and activated carbon xerogels (ACX_activation temperature) [53].

Samples	S_{BET} (m^2/g) [a]	D_{avg} (nm) [b]	Specific Capacitance (F/g)		
			1 A/g	3 A/g	5 A/g
CX	628	17.7	8.7	6.6	5.0
ACX_600	895	14.4	14.5	13.8	12.0
ACX_700	1291	12.3	20.7	16.8	14.0
ACX_800	1493	12.4	23.3	23.1	21.0
ACX_900	1860	7.53	21.2	16.5	14.0

[a] Specific surface area was calculated by the BET plot. [b] Average pore diameter was determined by the BJH method with the desorption branch.

6. Cost-Efficient Production of CXs for EDLCs

Various synthesis parameters and activation methods were investigated to synthesize CXs for EDLC electrode materials. Notably, ACXs exhibited superior electrochemical performance, which was attributed to their high specific surface area, abundant porosity, and 3D network structure. This observation underscores ACXs' considerable potential as active materials for EDLC electrodes. Nevertheless, price competitiveness is crucial for the commercial success of ACXs, prompting research and development aimed at achieving cost-efficient production [149]. An understanding of the conversion mechanism of ACXs from their precursors has led us to propose two strategies for the cost-efficient manufacture of ACXs [54,60]. Figure 10 illustrates a schematic that aids in the comprehension of the proposed methods.

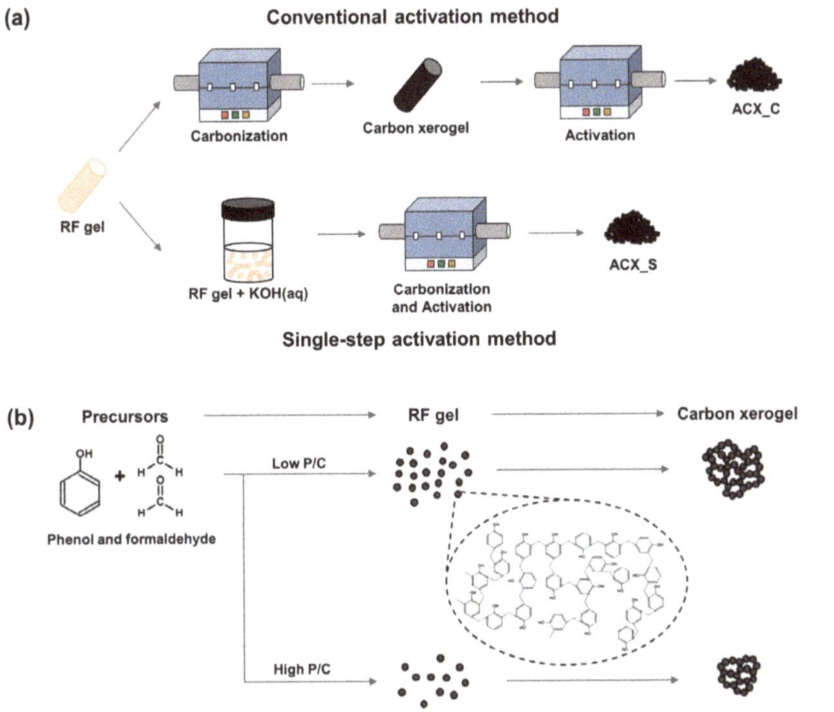

Figure 10. Preparation procedure for (**a**) activated carbon xerogels single-step activation method and (**b**) carbon xerogels using phenol precursor [54,60].

The first method involves a single-step activation process, which transforms RF gels directly into ACXs to provide an efficient solution [54]. CXs must be activated for the utilization of high-performance EDLCs. Additional procedures during the manufacturing stages contribute significantly to the reduced economic efficiency. The single-step activation process considerably expands the economic benefits by simplifying the process and thereby increasing the overall manufacturing efficiency. As a result, the single-step activation approach is considered a pivotal strategy for substantially boosting the commercial viability of ACXs.

The second method involves the use of phenol as a more cost-effective precursor compared to resorcinol [60]. Reducing the raw material costs directly affects the economic efficiency of the final products. Therefore, the utilization of inexpensive raw materials lowers the overall manufacturing costs and ultimately promotes the market competitiveness of the final product. Furthermore, phenol precursors may require unique manufacturing parameters that are distinct from those for resorcinol. One such parameter is the ratio of phenol to catalyst, which can lead to superior pore structures or improved electrochemical performance. Consequently, the application of phenol precursors is considered a critical strategy to improve the economic efficiency of the process and widen commercialization.

7. Comparison of CXs and Commercial Electrode Materials

Studies on the effect of structural modifications in CXs due to variations in the synthesis environment has shown the high potential of CXs as electrode materials for EDLCs. Note that rather than simply using CXs, utilizing ACXs can maximize the performance of EDLCs. Additionally, for practical applications, a comprehensive comparative analysis must be conducted between CXs and commercial activated carbon, which is commonly used as an electrode material for EDLCs, to ensure their applicability and effectiveness. Hence, using activation methods, we successfully synthesized an ACX with a superior performance and conducted a comparative analysis using commercial activated carbons (YP50f, MSP20, and CEP21KS) [58]. Figure 11 illustrates the physical attributes and electrochemical performance of this ACX compared with commercial activated carbons.

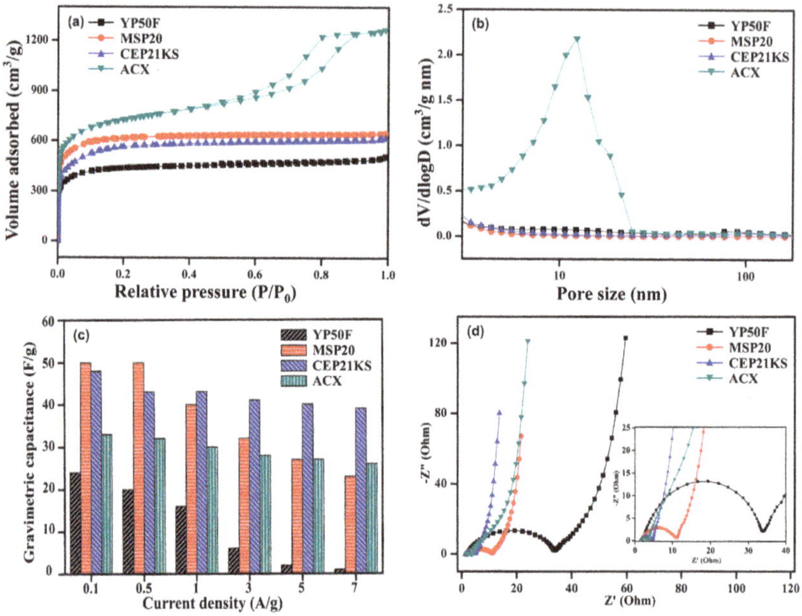

Figure 11. (**a**) N_2 adsorption–desorption isotherm, (**b**) pore size distribution, (**c**) specific capacitance to current density, and (**d**) Nyquist plots of activated carbon materials [58].

YP50F (Kuraray Co., Tokyo, Japan), MSP20 (Kansai Coke & Chemicals Co., Hyogo, Japan), and CEP21KS (Power Carbon Technology Co., Gumi, Republic of Korea) are high-quality activated carbons commercially available as electrode materials for EDLCs. Table 7 provides detailed properties of these commercial activated carbons, along with ACX. These carbon materials vary based on their precursor materials and activation method. YP50F, which is derived from coconut shells via physical activation, has a relatively low specific surface area and contains a high amount of impurities [150]. These properties suggest YP50F exhibits low capacitance due to its low specific surface area and high electrical resistance due to the presence of impurities. Empirical evaluations confirmed its inferior performance compared with other carbon materials. By contrast, MSP20 and CEP21KS, which were synthesized through chemical activation using KOH, substantially enhanced the electrochemical performance of EDLCs by providing a high specific surface area. As a result, MSP20 and CEP21KS demonstrated superior electrochemical properties compared with YP50F.

The ACX was synthesized via KOH activation and was characterized by its abundant micropores and mesopores, along with a 3D network structure that increased electrical conductivity. These properties provided the ACX with a high capacitance and low resistance, which ensured that its electrochemical performance was comparable to and potentially exceeded commercial standards. Notably, the ACX demonstrated exceptional capacitance retention under C–D rates, a feature closely related to its resistance properties. FPP measurements, which aim to determine the essential electronic resistance of electrode materials, showed that CEP21KS exhibited the lowest electronic resistance. Nonetheless, according to the Nyquist plots, the ACX possessed the smallest R_{ct}, which suggests that the R_{ct} encompassed both electronic and ionic resistances. The low electronic resistance of the

ACX was attributed to its highly interconnected 3D network structure, and the extensive porosity facilitated a low ionic resistance due to the abundance of ion pathways.

Table 7. Physical properties and specific capacitances (calculated from GCD measurements) of commercial activated carbons (YP50F, MSP20, and CEP21KS) and ACX [58].

	YP50F	MSP20	CEP21KS	ACX
Maker	Kuraray Co., Tokyo, Japan	Kansai Coke & Chemicals Co., Hyogo, Japan	Power Carbon Technology Co., Gumi, Republic of Korea	-
Activation method	Physical activation by steam	Chemical activation by KOH	Chemical activation by KOH	Chemical activation by KOH
Raw material	Coconut shell (hard carbon)	Phenolic resin (hard carbon)	Cokes (soft carbon)	Phenolic resin (hard carbon)
S_{BET} (m^2/g) [a]	1590	2330	2090	2690
D_{avg} (nm) [b]	1.8	1.8	1.9	2.9
V_{micro} (cm^3/g) [c]	0.7	1.0	1.0	0.6
V_{meso} (cm^3/g) [d]	0.2	0.1	0.1	1.1
V_{total}(cm^3/g) [e]	0.8	1.0	1.1	2.0
R_{sheet} (Ohm/sq.) [f]	1504.0	442.9	281.0	377.8

[a] Specific surface area was calculated by the BET plot. [b] Average pore diameter was determined by the BJH method with the desorption branch. [c] Micropore volume. [d] Mesopore volume. [e] Total pore volume. [f] Sheet resistance: confirmed by four-point probe measurement.

The ACX demonstrated considerable potential for improving the electrochemical performance of EDLCs, underscoring the importance of structural control and activation in optimizing their performance. In addition, we confirmed the industrial value of ACXs, which were specifically optimized for use as electrode materials of EDLCs.

8. Conclusions

This review comprehensively examined an array of research focused on the application of CXs, derived from resorcinol and formaldehyde, as electrode materials for EDLCs. It investigated the mechanism underlying the effects of adjusted synthesis parameters (such as the polymerization time, pH, and R/C ratio) on the structure of CXs. Detailed analyses of how these structural changes on the electrochemical performance of CXs were provided, highlighting the importance of appropriate synthesis conditions. Specifically, CXs comprising large particles exhibited low ionic resistance due to their large pore diameters and high pore volumes, whereas CXs made of smaller particles formed smaller pores but attained low electronic resistances through a highly developed 3D network. These observations emphasize the importance of precise synthesis conditions for CX preparation. The high specific surface area of electrode materials substantially contributes to the performance of EDLCs. This study explored how the use of an ACX, produced through the activation of CXs with KOH, can improve EDLC performance. The superior electrochemical performance of ACXs compared with CXs was elaborated and the influence of ACX activation temperature on EDLC performance was carefully examined. Through the design of the ACX structure for use as EDLC electrode materials and their comparison with commercial activated carbons (YP50F, MSP20, and CEP21KS), this review assessed the commercial viability of ACXs. ACXs exhibited a higher specific surface area and a more abundant pore structure than commercial activated carbons. Notably, ACXs exhibited the lowest charge transfer resistance (R_{ct}), which is a critical factor directly associated with EDLC efficiency. The low R_{ct} of ACXs was attributed to their high porosity and superior 3D network connectivity, resulting in low electronic resistances. These results reveal the superior functionality of ACXs as electrode materials for EDLCs and underscore their extensive potential for commercial applications. In conclusion, tailoring the structure of ACXs is pivotal for the improved performance of EDLCs, with an emphasis on their potential as next-generation materials for efficient energy storage systems. The development and optimization of ACXs for EDLC applications were evaluated to accelerate the advancement of energy storage technology and open up new possibilities for technological innovation.

Author Contributions: Conceptualization, J.C.; investigation, J.C.; writing—original draft preparation, J.C.; writing—review and editing, J.C.J. and W.J.; visualization, J.C.; supervision, J.C.J. and W.J.; funding acquisition, J.C.J. and W.J. All authors have read and agreed to the published version of the manuscript.

Funding: This work was supported by the National Research Foundation of Korea (NRF) grant funded by the Korea government (MSIT) (2021R1F1A1046272) of J.C. Jung and the Gachon University research fund of 2022 (GCU-202301140001) of W. Jung.

Institutional Review Board Statement: Not applicable.

Informed Consent Statement: Not applicable.

Data Availability Statement: Not applicable.

Conflicts of Interest: The authors declare no conflict of interest.

References

1. Mahmood, D.; Javaid, N.; Ahmed, G.; Khan, S.; Monteiro, V. A review on optimization strategies integrating renewable energy sources focusing uncertainty factor–Paving path to eco-friendly smart cities. *Sustain. Comput.* **2021**, *30*, 100559. [CrossRef]
2. Riti, J.S.; Shu, Y. Renewable energy, energy efficiency, and eco-friendly environment (R-E^5) in Nigeria. *Energy Sustain. Soc.* **2016**, *6*, 13. [CrossRef]
3. Ahmad, F.; Khalid, M.; Panigrahi, B.K. Development in energy storage system for electric transportation: A comprehensive review. *J. Energy Storage* **2021**, *43*, 103153. [CrossRef]
4. Aneke, M.; Wang, M. Energy storage technologies and real life applications—A state of the art review. *Appl. Energy* **2016**, *179*, 350–377. [CrossRef]
5. Olabi, A.G.; Mahmoud, M.; Soudan, B.; Wilberforce, T.; Ramadan, M. Geothermal based hybrid energy systems, toward eco-friendly energy approaches. *Renew. Energy* **2019**, *147*, 2003–2012. [CrossRef]
6. Amin, M.R.; Ruhul, M.; Bhuiyan, A. Overcome the future environmental challenges through sustainable and renewable energy resources. *Micro Nano Lett.* **2022**, *17*, 402–416.
7. Sharma, S.; Basu, S.; Shetti, N.P.; Kamali, M.; Walvekar, P.; Aminabhavi, T.M. Waste-to-energy nexus: A sustainable development. *Environ. Pollut.* **2020**, *267*, 115501. [CrossRef] [PubMed]
8. Kouchachvili, L.; Yaïci, W.; Entchev, E. Hybrid battery/supercapacitor energy storage system for the electric vehicles. *J. Power Sources* **2017**, *374*, 237–248. [CrossRef]
9. Hasan, M.K.; Mahmud, M.; Ahasan Habib, A.K.M.; Motakabber, S.M.A.; Islam, S. Review of electric vehicle energy storage and management system: Standards, issues, and challenges. *J. Energy Storage* **2021**, *41*, 102940. [CrossRef]
10. Hannan, M.A.; Hoque, M.M.; Mohamed, A.; Ayob, A. Review of energy storage systems for electric vehicle applications: Issues and challenges. *Renew. Sustain. Energy Rev.* **2016**, *69*, 771–789. [CrossRef]
11. Carrette, L.; Friedrich, K.A.; Stimming, U. Fuel Cells: Principles, Types, Fuels, and Applications. *ChemPhysChem* **2000**, *1*, 162–193. [CrossRef]
12. Dicks, A.L. The role of carbon in fuel cells. *J. Power Sources* **2006**, *156*, 128–141. [CrossRef]
13. Zhang, L.; Hu, X.; Wang, Z.; Sun, F.; Dorrell, D.G. A review of supercapacitor modeling, estimation, and applications: A control/management perspective. *Renew. Sustain. Energy Rev.* **2017**, *81*, 1868–1878. [CrossRef]
14. Lin, Z.; Goikolea, E.; Balducci, A.; Naoi, K.; Taberna, P.L.; Salanne, M.; Yushin, G.; Simon, P. Materials for supercapacitors: When Li-ion battery power is not enough. *Mater. Today* **2018**, *21*, 419–436. [CrossRef]
15. Lim, J.M.; Jang, Y.S.; Van T. Nguyen, H.; Kim, J.S.; Yoon, Y.; Park, B.J.; Seo, D.H.; Lee, K.; Han, Z.; Ostrikov, K.; et al. Advances in high-voltage supercapacitors for energy storage systems: Materials and electrolyte tailoring to implementation. *Nanoscale Adv.* **2023**, *5*, 615–626. [CrossRef]
16. Libich, J.; Máca, J.; Vondrák, J.; Čech, O.; Sedlaříková, M. Supercapacitors: Properties and applications. *J. Energy Storage* **2018**, *17*, 224–227. [CrossRef]
17. Schütter, C.; Pohlmann, S.; Balducci, A. Industrial Requirements of Materials for Electrical Double Layer Capacitors: Impact on Current and Future Applications. *Adv. Energy Mater.* **2019**, *9*, 1900334. [CrossRef]
18. Lee, H.; Kim, H.; Kang, S.; Park, S.; An, K.; Kim, B. Effects of pore structures on electrochemical behaviors of polyacrylonitrile (PAN)-based activated carbon nanofibers. *J. Ind. Eng. Chem.* **2014**, *21*, 736–740. [CrossRef]
19. Liu, X.; Li, S.; Mi, R.; Mei, J.; Liu, L.; Cao, L.; Lau, W.; Liu, H. Porous structure design of carbon xerogels for advanced supercapacitor. *Appl. Energy* **2015**, *153*, 32–40. [CrossRef]
20. Jiang, J.M.; Zhang, Y.D.; Nie, P.; Xu, G.Y.; Shi, M.Y.; Wang, J.; Wu, Y.T.; Fu, R.R.; Dou, H.G.; Zhang, X. Progress of Nanostructured Electrode Materials for Supercapacitors. *Adv. Sustain. Syst.* **2017**, *2*, 1700110. [CrossRef]

21. Yoo, J.; Yang, I.; Kim, M.; Jung, J.C. Effects of Molecular Weight of Polyvinylidene Fluoride Binder on Electrochemical Performances of Organic Electric Double-Layer Capacitors. *Polymer* **2019**, *43*, 838–845.
22. Yang, I.; Yoo, J.; Lee, J.; Kim, M.; Chul, J. High-Performance Organic Electric Double-Layer Capacitors Using Allergen-Derived Activated Carbons. *J. Electrochem. Soc.* **2019**, *116*, A3950–A3958. [CrossRef]
23. Jung, M.; Yang, I.; Choi, D.; Lee, J.; Jung, J.C. Activated carbons derived from polyethylene terephthalate for coin-cell supercapacitor electrodes. *Korean J. Chem. Eng.* **2023**, *40*, 2442–2454. [CrossRef] [PubMed]
24. Yang, I.; Mok, J.H.; Jung, M.; Yoo, J.; Kim, M.; Choi, D.; Jung, J.C. Polyethylene-Derived Activated Carbon Materials for Commercially Available Supercapacitor in an Organic Electrolyte System. *Macromol. Rapid Commun.* **2022**, *43*, 2200006. [CrossRef] [PubMed]
25. Jung, M.; Yang, I.; Yoo, J.; Kim, M.; Chul, J. Low-Cost Activated Carbon Electrodes from Waste Maple Leaves for Organic Electric Double-Layer Capacitors. *J. Electrochem. Soc.* **2021**, *168*, 080532. [CrossRef]
26. Yang, I.; Jung, M.; Kim, M.; Choi, D.; Jung, J.C. Physical and chemical activation mechanisms of carbon materials based on the microdomain model. *J. Mater. Chem. A* **2021**, *9*, 9815–9825. [CrossRef]
27. Davies, A.; Yu, A. Material advancements in supercapacitors: From activated carbon to carbon nanotube and graphene. *Can. J. Chem. Eng.* **2011**, *89*, 1342–1357. [CrossRef]
28. Yang, I.; Yoo, J.; Kwon, D.; Choi, D.; Kim, M.; Jung, J.C. Improvement of a commercial activated carbon for organic electric double-layer capacitors using a consecutive doping method. *Carbon* **2020**, *160*, 45–53. [CrossRef]
29. Ibarra Torres, C.E.; Serrano Quezada, T.E.; Kharissova, O.V.; Kharisov, B.I.; De La Fuente, M.I.G. Carbon-based aerogels and xerogels: Synthesis, properties, oil sorption capacities, and DFT simulations. *J. Environ. Chem. Eng.* **2020**, *9*, 104886. [CrossRef]
30. Wu, X.; Zhang, X.; Wang, X.; Zhang, C.; Zhu, Q.; Du, A.; Zhang, Z.; Shen, J. Aqueous-based, high-density nanoporous carbon xerogels with high specific surface area for supercapacitors. *J. Porous Mater.* **2021**, *29*, 87–95. [CrossRef]
31. Shen, J.; Hou, J.; Guo, Y.; Xue, H.; Wu, G.; Zhou, B. Microstructure Control of RF and Carbon Aerogels Prepared by Sol-Gel Process. *J. Sol-Gel Sci. Technol.* **2005**, *36*, 131–136. [CrossRef]
32. Cuadrado-Collados, C.; Farrando-Pérez, J.; Martínez-Escandell, M.; Ramírez-Montoya, L.A.; Menéndez, J.A.; Arenillas, A.; Montes-Morán, M.A.; Silvestre-Albero, J. Well-defined meso/macroporous materials as a host structure for methane hydrate formation: Organic versus carbon xerogels. *Chem. Eng. J.* **2020**, *402*, 126276. [CrossRef]
33. Rey-Raap, N.; Angel Menéndez, J.; Arenillas, A. Simultaneous adjustment of the main chemical variables to fine-tune the porosity of carbon xerogels. *Carbon* **2014**, *78*, 490–499. [CrossRef]
34. Calvo, E.G.; Menéndez, J.A.; Arenillas, A. Influence of alkaline compounds on the porosity of resorcinol-formaldehyde xerogels. *J. Non-Cryst. Solids* **2016**, *452*, 286–290. [CrossRef]
35. Rey-Raap, N.; Angel Menéndez, J.; Arenillas, A. RF xerogels with tailored porosity over the entire nanoscale. *Microporous Mesoporous Mater.* **2014**, *195*, 266–275. [CrossRef]
36. Lin, C.; Ritter, J.A.; Popov, B.N. Correlation of Double-Layer Capacitance with the Pore Structure of Sol-Gel Derived Carbon Xerogels. *J. Electrochem. Soc.* **1999**, *146*, 3639–3643. [CrossRef]
37. Zeller, M.; Lorrmann, V.; Reichenauer, G.; Wiener, M.; Pflaum, J. Relationship Between Structural Properties and Electrochemical Characteristics of Monolithic Carbon Xerogel-Based Electrochemical Double-Layer Electrodes in Aqueous and Organic Electrolytes. *Adv. Energy Mater.* **2012**, *2*, 598–605. [CrossRef]
38. Sillars, F.B.; Fletcher, S.I.; Mirzaeian, M.; Hall, P.J. Effect of activated carbon xerogel pore size on the capacitance performance of ionic liquid electrolytes. *Energy Environ. Sci.* **2010**, *4*, 695–706. [CrossRef]
39. Anneser, K.; Reichstein, J.; Braxmeier, S.; Reichenauer, G. Carbon xerogel based electric double layer capacitors with polymer gel electrolytes—Improving the performance by adjusting the type of electrolyte and its processing. *Electrochim. Acta* **2018**, *278*, 196–203. [CrossRef]
40. Fernández, P.S.; Arenillas, A.; Calvo, E.G.; Menéndez, J.A.; Martins, M.E. Carbon xerogels as electrochemical supercapacitors. Relation between impedance physicochemical parameters and electrochemical behaviour. *Int. J. Hydrogen Energy* **2012**, *37*, 10249–10255. [CrossRef]
41. Mezzavilla, S.; Zanella, C.; Aravind, P.R.; Della Volpe, C.; Sorarù, G.D. Carbon xerogels as electrodes for supercapacitors. The influence of the catalyst concentration on the microstructure and on the electrochemical properties. *J. Mater. Sci.* **2012**, *47*, 7175–7180. [CrossRef]
42. Wang, S.; Miao, J.; Liu, M.; Zhang, L.; Liu, Z. Hierarchical porous N-doped carbon xerogels for high performance CO_2 capture and supercapacitor. *Colloid Surf. A Physicochem. Eng. Asp.* **2021**, *616*, 126285. [CrossRef]
43. Castro, C.D.; Izquierdo, M.T.; Diossa, G.; Zapata-Benabithe, Z.; Quintana, G.C. Synthesis of bio-based xerogels from lignin precipitated from the black liquor of the paper industry for supercapacitors electrodes. *Biomass Bioenergy* **2021**, *155*, 106296. [CrossRef]
44. Sun, L.; Gong, Y.; Li, D.; Pan, C. Biomass-derived porous carbon materials: Synthesis, designing, and applications for supercapacitors. *Green Chem.* **2022**, *24*, 3864–3894. [CrossRef]
45. Wasfey, M.A.; Abdelwahab, A.; Carrasco-Marín, F.; Pérez-Cadenas, A.F.; Abdullah, H.H.; Yahia, I.S.; Farghali, A.A. Nickel Cobaltite Functionalized Silver Doped Carbon Xerogels as Efficient Electrode Materials for High Performance Symmetric Supercapacitor. *Materials* **2020**, *13*, 4906. [CrossRef]

46. Zapata-Benabithe, Z.; De Vicente, J.; Carrasco-Marín, F.; Moreno-Castilla, C. Synthesis, surface characteristics, and electrochemical capacitance of Cu-doped carbon xerogel microspheres. *Carbon* **2012**, *55*, 260–268. [CrossRef]
47. Lee, Y.J.; Jung, J.C.; Yi, J.; Baeck, S.; Yoon, J.R.; Song, I.K. Preparation of carbon aerogel in ambient conditions for electrical double-layer capacitor. *Curr. Appl. Phys.* **2009**, *10*, 682–685. [CrossRef]
48. Lee, Y.J.; Jung, J.C.; Park, S.; Seo, J.G.; Baeck, S.; Yoon, J.R.; Yi, J.; Song, I.K. Preparation and characterization of metal-doped carbon aerogel for supercapacitor. *Curr. Appl. Phys.* **2010**, *10*, 947–951. [CrossRef]
49. Lee, Y.J.; Jung, J.C.; Park, S.; Seo, J.G.; Baeck, S.; Yoon, J.R.; Yi, J.; Song, I.K. Effect of preparation method on electrochemical property of Mn-doped carbon aerogel for supercapacitor. *Curr. Appl. Phys.* **2011**, *11*, 1–5. [CrossRef]
50. Lee, Y.J.; Jung, J.C.; Park, S.; Seo, J.G.; Baeck, S.; Yoon, J.R.; Yi, J.; Song, I.K. Preparation and performance of cobalt-doped carbon aerogel for supercapacitor. *Korean J. Chem. Eng.* **2011**, *28*, 492–496. [CrossRef]
51. Lee, Y.J.; Jung, J.C.; Park, S.; Seo, J.G.; Baeck, S.; Yoon, J.R.; Yi, J.; Song, I.K. Nano-Sized Ni-Doped Carbon Aerogel for Supercapacitor. *J. Nanosci. Nanotechnol.* **2011**, *11*, 6528–6532. [CrossRef] [PubMed]
52. Kwon, S.H.; Lee, E.; Kim, B.; Kim, S.; Lee, B.; Kim, M.; Jung, J.C. Activated carbon aerogel as electrode material for coin-type EDLC cell in organic electrolyte. *Curr. Appl. Phys.* **2014**, *14*, 603–607. [CrossRef]
53. Kwon, S.H.; Lee, E.; Kim, B.; Kim, S.; Lee, B.; Kim, M.; Jung, J.C. Preparation of activated carbon aerogel and its application to electrode material for electric double layer capacitor in organic electrolyte: Effect of activation temperature. *Korean J. Chem. Eng.* **2015**, *32*, 248–254. [CrossRef]
54. Kwon, S.H.; Kim, B.; Kim, S.; Lee, B.; Kim, M.; Jung, J.C. Preparation of Nano-Porous Activated Carbon Aerogel Using a Single-Step Activation Method for Use as High-Power EDLC Electrode in Organic Electrolyte. *J. Nanosci. Nanotechnol.* **2016**, *16*, 4598–4604. [CrossRef] [PubMed]
55. Yang, I.; Kim, S.; Kwon, S.H.; Lee, J.H.; Kim, M.; Jung, J.C. Pore size-controlled carbon aerogels for EDLC electrodes in organic electrolytes. *Curr. Appl. Phys.* **2016**, *16*, 665–672. [CrossRef]
56. Jung, J.C.; Jung, W. pH-Controlled Synthesis of Carbon Xerogels for Coin-Type Organic Supercapacitor Electrodes. *Korean J. Mater. Res.* **2023**, *33*, 430–438. [CrossRef]
57. Yang, I.; Kim, S.; Kwon, S.H.; Kim, M.; Jung, J.C. Relationships between pore size and charge transfer resistance of carbon aerogels for organic electric double-layer capacitor electrodes. *Electrochim. Acta* **2017**, *223*, 21–30. [CrossRef]
58. Yang, I.; Kwon, D.; Kim, M.; Jung, J.C. A comparative study of activated carbon aerogel and commercial activated carbons as electrode materials for organic electric double-layer capacitors. *Carbon* **2018**, *132*, 503–511. [CrossRef]
59. Inn, S.H.; Jung, J.C.; Kim, M. Pore Structure and Electrochemical Properties of Carbon Aerogels as an EDLC-Electrode with Different Preparation Conditions. *Korean J. Mater. Res.* **2018**, *28*, 50–61. [CrossRef]
60. Yoo, J.; Yang, I.; Kwon, D.; Jung, M.; Kim, M.; Jung, J.C. Low-Cost Carbon Xerogels Derived from Phenol–Formaldehyde Resin for Organic Electric Double-Layer Capacitors. *Energy Technol.* **2021**, *9*, 2000918. [CrossRef]
61. Seo, H.I.; Kim, S.; Jung, J.C.; Kim, M. Preparation and Characterization of Carbon Aerogel Activated with KOH and CO_2: Effect of Pore Size Distribution on Electrochemical Properties as EDLC Electrodes. *Polymer* **2016**, *40*, 557–586.
62. Abdul Khalil, H.P.S.; Yahya, E.B.; Tajarudin, H.A.; Balakrishnan, V.; Nasution, H. Insights into the Role of Biopolymer-Based Xerogels in Biomedical Applications. *Gels* **2022**, *8*, 334. [CrossRef] [PubMed]
63. Pascuta, M.S.; Varvara, R.; Teleky, B.; Szabo, K.; Plamada, D.; Nemeş, S.; Mitrea, L.; Martău, G.A.; Ciont, C.; Călinoiu, L.F.; et al. Polysaccharide-Based Edible Gels as Functional Ingredients: Characterization, Applicability, and Human Health Benefits. *Gels* **2022**, *8*, 524. [CrossRef] [PubMed]
64. Zhao, S.; Malfait, W.J.; Guerrero-alburquerque, N.; Koebel, M.M.; Nyström, G. Biopolymer Aerogels and Foams: Chemistry, Properties, and Applications. *Angew. Chem. Int. Ed.* **2018**, *57*, 7580–7608. [CrossRef] [PubMed]
65. Lee, J.; Park, S. Recent advances in preparations and applications of carbon aerogels: A review. *Carbon* **2020**, *163*, 1–18. [CrossRef]
66. Gun'ko, V.M.; Savina, I.N.; Mikhalovsky, S.V. Cryogels: Morphological, structural and adsorption characterisation. *Adv. Colloid Interface Sci.* **2013**, *187*, 1–46. [CrossRef] [PubMed]
67. Samancı, M.; Daş, E.; Yurtcan, A.B. Effect of solvent exchange on the properties of carbon xerogel and carbon xerogel/polypyrrole composites for supercapacitors. *Carbon Lett.* **2021**, *31*, 1287–1308. [CrossRef]
68. Rojas-Cervantes, M.L. Some strategies to lower the production cost of carbon gels. *J. Mater. Sci.* **2014**, *50*, 1017–1040. [CrossRef]
69. Figueiredo, J.L.; Pereira, M.F.R. Synthesis and functionalization of carbon xerogels to be used as supports for fuel cell catalysts. *J. Energy Chem.* **2013**, *22*, 195–201. [CrossRef]
70. Job, N.; Pirard, R.; Marien, J.; Pirard, J. Porous carbon xerogels with texture tailored by pH control during sol–gel process. *Carbon* **2004**, *42*, 619–628. [CrossRef]
71. Zubizarreta, L.; Arenillas, A.; Domínguez, A.; Menéndez, J.A.; Pis, J.J. Development of microporous carbon xerogels by controlling synthesis conditions. *J. Non-Cryst. Solids* **2008**, *354*, 817–825. [CrossRef]
72. Rocha, R.P.; Restivo, J.; Sousa, J.P.S.; Órfão, J.J.M.; Pereira, M.F.R.; Figueiredo, J.L. Nitrogen-doped carbon xerogels as catalysts for advanced oxidation processes. *Catal. Today* **2014**, *241*, 73–79. [CrossRef]
73. Kakunuri, M.; Sharma, C.S. Resorcinol-formaldehyde derived carbon xerogels: A promising anode material for lithium-ion battery. *J. Mater. Res.* **2017**, *33*, 1074–1087. [CrossRef]
74. Canal-Rodríguez, M.; Arenillas, A.; Menéndez, J.A.; Beneroso, D.; Rey-Raap, N. Carbon xerogels graphitized by microwave heating as anode materials in lithium-ion batteries. *Carbon* **2018**, *137*, 384–394. [CrossRef]

75. Bailón-García, E.; Maldonado-Hódar, F.J.; Carrasco-Marín, F.; Pérez-Cadenas, A.F.; Bosi, S.; Prato, M. The use of functionalized carbon xerogels in cells growth. *Mater. Sci. Eng. C* **2019**, *100*, 598–607. [CrossRef] [PubMed]
76. Figueiredo, J.L.; Pereira, M.F.R.; Serp, P.; Kalck, P.; Samant, P.V.; Fernandes, J.B. Development of carbon nanotube and carbon xerogel supported catalysts for the electro-oxidation of methanol in fuel cells. *Carbon* **2006**, *44*, 2516–2522. [CrossRef]
77. Pillai, A.; Kandasubramanian, B. Carbon Xerogels for Effluent Treatment. *J. Chem. Eng. Data* **2020**, *65*, 2255–2270. [CrossRef]
78. Álvarez, S.; Ribeiro, R.S.; Gomes, H.T.; Sotelo, J.L.; García, J. Synthesis of carbon xerogels and their application in adsorption studies of caffeine and diclofenac as emerging contaminants. *Chem. Eng. Res. Des.* **2015**, *95*, 229–238. [CrossRef]
79. Rashed, Y.; Messele, S.A.; Zeng, H.; El-Din, G.M. Mesoporous carbon xerogel material for the adsorption of model naphthenic acids: Structure effect and kinetics modelling. *Environ. Technol.* **2020**, *41*, 3534–3543. [CrossRef]
80. Job, N.; Lambert, S.D.; Zubiaur, A.; Cao, C.; Pirard, J. Design of Pt/Carbon Xerogel Catalysts for PEM Fuel Cells. *Catalysts* **2015**, *5*, 40–57. [CrossRef]
81. Chavhan, M.P.; Ganguly, S. Activated Xerogel Nanoporous-materials For Energy Storage Applications. *Mater. Today Proc.* **2018**, *5*, 9754–9759. [CrossRef]
82. Kakunuri, M.; Vennamalla, S.; Sharma, C.S. Synthesis of carbon xerogel nanoparticles by inverse emulsion polymerization of resorcinol–formaldehyde and their use as anode materials for lithium-ion battery. *RSC Adv.* **2015**, *5*, 4747–4753. [CrossRef]
83. Li, F.; Xie, L.; Sun, G.; Kong, Q.; Su, F.; Cao, Y.; Wei, J.; Ahmad, A.; Guo, X.; Chen, C. Resorcinol-formaldehyde based carbon aerogel: Preparation, structure and applications in energy storage devices. *Microporous Mesoporous Mater.* **2019**, *279*, 293–315. [CrossRef]
84. Huang, M.; Yoo, S.J.; Lee, J.; Yoon, T. Electrochemical properties of an activated carbon xerogel monolith from resorcinol-formaldehyde for supercapacitor electrode applications. *RSC Adv.* **2021**, *11*, 33192–33201. [CrossRef]
85. Chen, L.; Deng, J.; Yuan, Y.; Hong, S.; Yan, B.; He, S.; Lian, H. Hierarchical porous graphitized carbon xerogel for high performance supercapacitor. *Diam. Relat. Mater.* **2022**, *121*, 108781. [CrossRef]
86. Da Silva, L.M.; Cesar, R.; Moreira, C.M.R.; Santos, J.H.M.; De Souza, L.G.; Pires, B.M.; Vicentini, R.; Nunes, W.; Zanin, H. Reviewing the fundamentals of supercapacitors and the difficulties involving the analysis of the electrochemical findings obtained for porous electrode materials. *Energy Storage Mater.* **2020**, *27*, 555–590. [CrossRef]
87. Burt, R.; Birkett, G.; Zhao, X.S. A review of molecular modelling of electric double layer capacitors. *Phys. Chem. Chem. Phys.* **2014**, *16*, 6519–6538. [CrossRef]
88. Patel, K.K.; Singhal, T.; Pandey, V.; Sumangala, T.P.; Sreekanth, M.S. Evolution and recent developments of high performance electrode material for supercapacitors: A review. *J. Energy Storage* **2021**, *44*, 103366. [CrossRef]
89. Kim, D.; Jin, X.; Cho, Y.; Lim, J.; Yan, B.; Ko, D.; Kim, D.K.; Piao, Y. Facile preparation of N-doped porous carbon nanosheets derived from potassium citrate/melamine for high-performance supercapacitors. *J. Electroanal. Chem.* **2021**, *892*, 115302. [CrossRef]
90. Fan, X.; Zhang, W.; Xu, Y.; Zheng, J.; Li, Y.; Fan, X.; Zhang, F.; Ji, J.; Peng, W. Porous structure engineering of N-doped carbons for enhanced mass transfer towards High-Performance supercapacitors and Li-Ion batteries. *J. Colloid Interface Sci.* **2022**, *624*, 51–59. [CrossRef]
91. Lomas, T.; Poochai, C.; Sukjit, P.; Wasapinyokul, K.; Mensing, J.P.; Tuantranont, A.; Sriprachuabwong, C. Electrolytic exfoliation of few-layer graphene/sodium dodecylbenzenesulfonate for coin- and cylindrical-cell supercapacitor electrodes. *Diam. Relat. Mater.* **2023**, *136*, 110020. [CrossRef]
92. Dutta, A.; Mitra, S.; Basak, M.; Banerjee, T. A comprehensive review on batteries and supercapacitors: Development and challenges since their inception. *Energy Storage* **2022**, *5*, e339. [CrossRef]
93. Sharma, P.; Kumar, V. Current Technology of Supercapacitors: A Review. *J. Electron. Mater.* **2020**, *49*, 3520–3532. [CrossRef]
94. Dar, M.A.; Majid, S.R.; Satgunam, M.; Siva, C.; Ansari, S.; Arularasan, P.; Rafi Ahamed, S. Advancements in Supercapacitor electrodes and perspectives for future energy storage technologies. *Int. J. Hydrogen Energy* **2024**, *70*, 10–28. [CrossRef]
95. Liu, X.; Wen, Y.; Chen, X.; Tang, T.; Mijowska, E. Co-etching effect to convert waste polyethylene terephthalate into hierarchical porous carbon toward excellent capacitive energy storage. *Sci. Total Environ.* **2020**, *723*, 138055. [CrossRef] [PubMed]
96. Portet, C.; Yushin, G.; Gogotsi, Y. Effect of Carbon Particle Size on Electrochemical Performance of EDLC. *J. Electrochem. Soc.* **2008**, *155*, A531–A536. [CrossRef]
97. Liu, D.; Xu, B.; Zhu, J.; Tang, S.; Xu, F.; Li, S.; Jia, B.; Chen, G. Preparation of Highly Porous Graphitic Activated Carbon as Electrode Materials for Supercapacitors by Hydrothermal Pretreatment-Assisted Chemical Activation. *ACS Omega* **2020**, *5*, 11058–11067. [CrossRef] [PubMed]
98. Ayaganov, Z.; Pavlenko, V.; Haque, S.F.B.; Tanybayeva, A.; Ferraris, J.; Zakhidov, A.; Mansurov, Z.; Bakenov, Z.; Ng, A. A comprehensive study on effect of carbon nanomaterials as conductive additives in EDLCs. *J. Energy Storage* **2024**, *78*, 110035. [CrossRef]
99. Rajeevan, S.; John, S.; George, S.C. The effect of poly(vinylidene fluoride) binder on the electrochemical performance of graphitic electrodes. *J. Energy Storage* **2021**, *39*, 102654. [CrossRef]
100. Pan, Z.; Yang, J.; Kong, J.; Loh, X.J.; Wang, J.; Liu, Z. "Porous and Yet Dense" Electrodes for High-Volumetric-Performance Electrochemical Capacitors: Principles, Advances, and Challenges. *Adv. Sci.* **2021**, *9*, 2103953. [CrossRef]
101. Forouzandeh, P.; Kumaravel, V.; Pillai, S.C. Electrode Materials for Supercapacitors: A Review of Recent Advances. *Catalysts* **2020**, *10*, 969. [CrossRef]

102. Saha, M.; Kumar, A.; Kanaoujiya, R.; Behera, K.; Trivedi, S. A Comprehensive Review of Novel Emerging Electrolytes for Supercapacitors: Aqueous and Organic Electrolytes Versus Ionic Liquid-Based Electrolytes. *Energy Fuels* **2024**, *38*, 8528–8552. [CrossRef]
103. Lewandowski, A.; Olejniczak, A.; Galinski, M.; Stepniak, I. Performance of carbon–carbon supercapacitors based on organic, aqueous and ionic liquid electrolytes. *J. Power Sources* **2010**, *195*, 5814–5819. [CrossRef]
104. Borchardt, L.; Zhu, Q.; Casco, M.E.; Berger, R.; Zhuang, X.; Kaskel, S.; Feng, X.; Xu, Q. Toward a molecular design of porous carbon materials. *Mater. Today* **2017**, *20*, 592–610. [CrossRef]
105. Lee, Y.J.; Park, H.W.; Park, S.; Song, I.K. Electrochemical properties of Mn-doped activated carbon aerogel as electrode material for supercapacitor. *Curr. Appl. Phys.* **2012**, *12*, 233–237. [CrossRef]
106. Li, X.; Jiang, Y.; Wang, P.; Mo, Y.; Lai, W.; Li, Z.; Yu, R.; Du, Y.; Zhang, X.; Chen, Y. Effect of the oxygen functional groups of activated carbon on its electrochemical performance for supercapacitors. *New Carbon Mater.* **2020**, *35*, 232–243. [CrossRef]
107. Xu, B.; Wu, F.; Chen, S.; Zhou, Z.; Cao, G.; Yang, Y. High-capacitance carbon electrode prepared by PVDC carbonization for aqueous EDLCs. *Electrochim. Acta* **2009**, *54*, 2185–2189. [CrossRef]
108. Wei, H.; Wang, H.; Li, A.; Li, H.; Cui, D.; Dong, M.; Lin, J.; Fan, J.; Zhang, J.; Hou, H.; et al. Advanced porous hierarchical activated carbon derived from agricultural wastes toward high performance supercapacitors. *J. Alloys Compd.* **2019**, *820*, 153111. [CrossRef]
109. Chen, H.; Guo, Y.; Wang, F.; Wang, G.; Qi, P.; Guo, X.; Dai, B.; Yu, F. An activated carbon derived from tobacco waste for use as a supercapacitor electrode material. *New Carbon Mater.* **2017**, *32*, 592–599. [CrossRef]
110. Taslim, R.; Apriwandi, A.; Taer, E. Novel Moringa oleifera Leaves 3D Porous Carbon-Based Electrode Material as a High-Performance EDLC Supercapacitor. *ACS Omega* **2022**, *7*, 36489–36502. [CrossRef]
111. Jiang, C.; Yakaboylu, G.A.; Yumak, T.; Zondlo, J.W.; Sabolsky, E.M.; Wang, J. Activated carbons prepared by indirect and direct CO_2 activation of lignocellulosic biomass for supercapacitor electrodes. *Renew. Energy* **2020**, *155*, 38–52. [CrossRef]
112. Kim, K.; Song, Y.; Ahn, H. N-doped mesoporous activated carbon derived from protein-rich biomass for energy storage applications. *Korean J. Chem. Eng.* **2023**, *40*, 1071–1076. [CrossRef]
113. Adan-Mas, A.; Alcaraz, L.; Arévalo-Cid, P.; López-Gómez, F.A.; Montemor, F. Coffee-derived activated carbon from second biowaste for supercapacitor applications. *Waste Manag.* **2021**, *120*, 280–289. [CrossRef] [PubMed]
114. Hsieh, T.; Wang, H.; Yu, G.; Huang, G.; Lin, J. Meso-pore dominant activated carbon from spent coffee grounds for high-performance electrochemical capacitors in organic electrolyte. *J. Environ. Chem. Eng.* **2021**, *9*, 106418. [CrossRef]
115. Kim, J.; Kim, S.; Kim, B.; Lee, H. Effects of Oxygen-Containing Functional Groups on the Electrochemical Performance of Activated Carbon for EDLCs. *Nanomaterials* **2023**, *13*, 262. [CrossRef] [PubMed]
116. Kim, J.; Kim, Y.; Kang, S.; Lee, H.; Kim, B. Preparation and Characterization of Asphalt Pitch-Derived Activated Carbons with Enhanced Electrochemical Performance as EDLC Electrode Materials. *Minerals* **2023**, *13*, 802. [CrossRef]
117. Zhou, J.; Zhang, Z.; Li, Z.; Zhu, T.; Zhuo, S. One-step and template-free preparation of hierarchical porous carbons with high capacitive performance. *RSC Adv.* **2015**, *5*, 46947–46954. [CrossRef]
118. Yue, Z.; Dunya, H.; Ashuri, M.; Kucuk, K.; Aryal, S.; Antonov, S.; Alabbad, B.; Segre, C.U.; Mandal, B.K. Synthesis of a Very High Specific Surface Area Active Carbon and Its Electrical Double-Layer Capacitor Properties in Organic Electrolytes. *ChemEngineering* **2020**, *4*, 43. [CrossRef]
119. Li, Q.; Jiang, Y.; Jiang, Z.; Zhu, J.; Gan, X.; Qin, F.; Tang, T.; Luo, W.; Guo, N.; Liu, Z.; et al. Ultrafast pore-tailoring of dense microporous carbon for high volumetric performance supercapacitors in organic electrolyte. *Carbon* **2022**, *191*, 19–27. [CrossRef]
120. Jiang, Y.; Chen, J.; Zeng, Q.; Zou, Z.; Li, J.; Zeng, L.; Sun, W.; Ming Li, C. Facile method to produce sub-1 nm pore-rich carbon from biomass wastes for high performance supercapacitors. *J. Colloid Interface Sci.* **2021**, *612*, 213–222. [CrossRef]
121. Gurten Inal, I.I.; Aktas, Z. Enhancing the performance of activated carbon based scalable supercapacitors by heat treatment. *Appl. Surf. Sci.* **2020**, *514*, 145895. [CrossRef]
122. Niu, Z.; Zhang, Y.; Zhang, Y.; Lu, X.; Liu, J. Enhanced electrochemical performance of three-dimensional graphene/carbon nanotube composite for supercapacitor application. *J. Alloys Compd.* **2020**, *820*, 153114. [CrossRef]
123. Honda, Y.; Takeshige, M.; Shiozaki, H.; Kitamura, T.; Yoshikawa, K.; Chakrabarti, S.; Suekane, O.; Pan, L.; Nakayama, Y.; Yamagata, M.; et al. Vertically aligned double-walled carbon nanotube electrode prepared by transfer methodology for electric double layer capacitor. *J. Power Sources* **2008**, *185*, 1580–1584. [CrossRef]
124. Yoo, Y.; Park, J.; Kim, W. Understanding and controlling the rest potential of carbon nanotube-based supercapacitors for energy density enhancement. *Appl. Surf. Sci.* **2018**, *433*, 765–771. [CrossRef]
125. Abedi, Z.; Leistenschneider, D.W.; Chen, W.; Ivey, D.G. Superior Performance of Electrochemical Double Layer Supercapacitor Made with Asphaltene Derived Activated Carbon Fibers. *Energy Technol.* **2020**, *8*, 2000588. [CrossRef]
126. Fan, L.; Sun, P.; Yang, L.; Xu, Z.; Han, J. Facile and scalable synthesis of nitrogen-doped ordered mesoporous carbon for high performance supercapacitors. *Korean J. Chem. Eng.* **2020**, *37*, 166–175. [CrossRef]
127. Vijayakumar, M.; Santhosh, R.; Adduru, J.; Rao, T.N.; Karthik, M. Activated carbon fibres as high performance supercapacitor electrodes with commercial level mass loading. *Carbon* **2018**, *140*, 465–476. [CrossRef]
128. Xu, F.; Xu, J.; Xu, H.; Lu, Y.; Yang, H.; Tang, Z.; Lu, Z.; Fu, R.; Wu, D. Fabrication of novel powdery carbon aerogels with high surface areas for superior energy storage. *Energy Storage Mater.* **2016**, *7*, 8–16. [CrossRef]
129. Lee, E.J.; Lee, Y.J.; Kim, J.K.; Lee, M.; Yi, J.; Yoon, J.R.; Song, J.C.; Song, I.K. Oxygen group-containing activated carbon aerogel as an electrode material for supercapacitor. *Mater. Res. Bull.* **2015**, *70*, 209–214. [CrossRef]

130. Liu, N.; Shen, J.; Liu, D. Activated high specific surface area carbon aerogels for EDLCs. *Microporous Mesoporous Mater.* **2013**, *167*, 176–181. [CrossRef]
131. Lv, Y.; Ding, L.; Wu, X.; Guo, N.; Guo, J.; Hou, S.; Tong, F.; Jia, D.; Zhang, H. Coal-based 3D hierarchical porous carbon aerogels for high performance and super-long life supercapacitors. *Sci. Rep.* **2020**, *10*, 7022. [CrossRef] [PubMed]
132. Ma, Y.; Wu, D.; Wang, T.; Jia, D. Nitrogen, Phosphorus Co-doped Carbon Obtained from Amino Acid Based Resin Xerogel as Efficient Electrode for Supercapacitor. *ACS Appl. Energy Mater.* **2019**, *3*, 957–969. [CrossRef]
133. Calvo, E.G.; Lufrano, F.; Staiti, P.; Brigandì, A.; Arenillas, A.; Menéndez, J.A. Optimizing the electrochemical performance of aqueous symmetric supercapacitors based on an activated carbon xerogel. *J. Power Sources* **2013**, *241*, 776–782. [CrossRef]
134. Canal-Rodríguez, M.; Menéndez, J.A.; Arenillas, A. Performance of carbon xerogel-graphene hybrids as electrodes in aqueous supercapacitors. *Electrochim. Acta* **2018**, *276*, 28–36. [CrossRef]
135. Jiang, D.; Jin, Z.; Henderson, D.; Wu, J. Solvent Effect on the Pore-Size Dependence of an Organic Electrolyte Supercapacitor. *J. Phys. Chem. Lett.* **2012**, *3*, 1727–1731. [CrossRef] [PubMed]
136. Largeot, C.; Portet, C.; Chmiola, J.; Taberna, P.; Gogotsi, Y.; Simon, P. Relation between the Ion Size and Pore Size for an Electric Double-Layer Capacitor. *J. Am. Chem. Soc.* **2008**, *130*, 2073–2731. [CrossRef] [PubMed]
137. Sing, K.S.W.; Everett, D.H.; Haul, R.A.W.; Moscou, L.; Pierotti, R.A.; Rouquérol, J.; Siemieniewska, T. Reporting Physisorption Data for Gas/Solid Systems with Special Reference to the Determination of Surface Area and Porosity (Recommendations 1984). *Pure Appl. Chem.* **1985**, *57*, 603–619. [CrossRef]
138. Li, W.; Chen, D.; Li, Z.; Shi, Y.; Wan, Y.; Wang, G.; Jiang, Z.; Zhao, D. Nitrogen-containing carbon spheres with very large uniform mesopores: The superior electrode materials for EDLC in organic electrolyte. *Carbon* **2007**, *45*, 1757–1763. [CrossRef]
139. Jiang, H.; Lee, P.S.; Li, C. 3D carbon based nanostructures for advanced supercapacitors. *Energy Environ. Sci.* **2013**, *6*, 41–53. [CrossRef]
140. Liu, Z.; Li, J.; Yang, Y.; Mi, J.H.; Tan, X.L. Synthesis, characterisation and magnetic examination of Fe, Co and Ni doped carbon xerogels. *Mater. Res. Innov.* **2012**, *16*, 362–367. [CrossRef]
141. Vargaftik, N.B.; Volkov, B.N.; Voljak, L.D. International Tables of the Surface Tension of Water. *J. Chem. Eng. Data* **1983**, *12*, 817–820. [CrossRef]
142. Sharma, C.S.; Kulkarni, M.M.; Sharma, A.; Madou, M. Synthesis of carbon xerogel particles and fractal-like structures. *Chem. Eng. Sci.* **2009**, *64*, 1536–1543. [CrossRef]
143. Elkhatat, A.M.; Al-muhtaseb, S.A. Advances in Tailoring Resorcinol-Formaldehyde Organic and Carbon Gels. *Adv. Mater.* **2011**, *23*, 2887–2903. [CrossRef] [PubMed]
144. Allagui, A.; Freeborn, T.J.; Elwakil, A.S.; Maundy, B.J. Reevaluation of Performance of Electric Double-layer Capacitors from Constant-current Charge/Discharge and Cyclic Voltammetry. *Sci. Rep.* **2016**, *6*, 38568. [CrossRef] [PubMed]
145. Pell, W.G.; Conway, B.E. Voltammetry at a de Levie brush electrode as a model for electrochemical supercapacitor behaviour. *J. Electroanal. Chem.* **2001**, *500*, 121–133. [CrossRef]
146. Mei, B.; Munteshari, O.; Lau, J.; Dunn, B.; Pilon, L. Physical Interpretations of Nyquist Plots for EDLC Electrodes and Devices. *J. Phys. Chem. C* **2018**, *122*, 194–206. [CrossRef]
147. Zanto, E.J.; Al-Muhtaseb, S.A.; Ritter, J.A. Sol−Gel-Derived Carbon Aerogels and Xerogels: Design of Experiments Approach to Materials Synthesis. *Ind. Eng. Chem. Res.* **2002**, *41*, 3151–3162. [CrossRef]
148. De Moraes, N.P.; Boldrin, F.H.C.; Campos, T.M.B.; Thim, G.P.; Lianqing, Y.; De Vasconcelos Lanza, M.R.; Rodrigues, L.A. Black-wattle tannin/kraft lignin H_3PO_4-activated carbon xerogels as excellent and sustainable adsorbents. *Int. J. Biol. Macromol.* **2022**, *227*, 58–70. [CrossRef] [PubMed]
149. Calvo, E.G.; Ania, C.O.; Zubizarreta, L.; Menéndez, J.A.; Arenillas, A. Exploring New Routes in the Synthesis of Carbon Xerogels for Their Application in Electric Double-Layer Capacitors. *Energy Fuels* **2010**, *24*, 3334–3339. [CrossRef]
150. Kim, J.; Park, I.; Seo, J.; Lee, J. A Development of High Power Activated Carbon Using the KOH Activation of Soft Carbon Series Cokes. *Trans. Electr. Electron. Mater.* **2014**, *15*, 81–86. [CrossRef]

gels

MDPI

Article

Metal–Organic Frameworks (MOF)-Derived Gel Electrolyte via UV Cross-Linking for High-Performance Lithium Metal Batteries

Naiyao Mao [1,†], Lingxiao Lan [1,†], Qiankun Hun [1], Jianghua Wei [1,*], Xinghua Liang [1,*] and Yifeng Guo [2]

[1] Guangxi Key Laboratory of Automobile Components and Vehicle Technology, Guangxi University of Science & Technology, Liuzhou 545006, China; maonaiyao@gmail.com (N.M.); llx2685062@163.com (L.L.); 13383371921@163.com (Q.H.)

[2] School of Mechanical Engineering, Chengdu University, Chengdu 610106, China; guobujia2000@163.com

* Correspondence: 15314157652@163.com (J.W.); lxh304@126.com (X.L.)

† These authors contributed equally to this work.

Abstract: Gel electrolytes (GEs) play a pivotal role in the advancement of lithium metal batteries by offering high energy density and enhanced rate capability. Nevertheless, their real-world application is hampered by relatively low ionic conductivity and significant interfacial resistance at room temperatures. In this work, we developed a gel electrolyte membrane (GEM) by embedding Zeolitic Imidazolate Framework-8 (ZIF-8) metal–organic frameworks (MOFs) material into a poly(vinylidene fluoride-co-hexafluoropropylene) (PVDF-HFP) matrix through UV curing. The composite membrane, with 4 wt% ZIF-8, exhibited an ionic conductivity of 1.17×10^{-3} S/cm, an electrochemical stability window of 4.7 V, and a lithium-ion transference number of 0.7. The test results indicate that the electrochemical performance of LFP//GEM//Li battery has an initial specific capacity of 168 mAh g^{-1} at 0.1 C rate. At 1 C, the discharge capacity was 88 mAh g^{-1}, and at 2 C, it was 68 mAh g^{-1}. Enhanced ionic transport, improved electrochemical stability, and optimized lithium-ion migration collectively contributed to superior rate performance and prolonged cycle life. This study offers novel insights and methodological advances for next-generation lithium metal batteries technologies.

Keywords: gel electrolyte; Metal–Organic Frameworks; PVDF-HFP; UV curing technology; lithium metal batteries

1. Introduction

Recent advancements in electric vehicle (EV) technology and energy storage infrastructure for power grids have driven the need for improved electrochemical performance in lithium metal batteries (LMBs). At the same time, market demands have become increasingly rigorous, particularly regarding energy density and safety, which are now key considerations in modern battery design [1,2]. Among the emerging solutions, LMBs utilizing gel electrolytes have attracted significant attention due to their potential to surpass the energy density limits of conventional liquid electrolyte systems [3]. However, several challenges persist. These include high interfacial energy barriers that impede lithium-ion transport [4], poor rate performance, and limited cycle life. Additionally, gel electrolytes typically exhibit lower ionic conductivity at room temperature compared to their liquid counterparts [5,6]. Addressing these limitations, the development of advanced gel electrolyte membranes that offer both high ionic conductivity and stable interfaces has become a critical focus in the field of electrochemical energy storage [7].

The application of polymer electrolyte materials in lithium metal batteries has made significant progress [8–10]. Poly(vinylidene fluoride) (PVDF)-based polymers, as high molecular weight materials, have been widely employed in the early development of gel electrolyte membranes due to their highly stable chemical properties and outstanding mechanical durability [11,12]. In recent years, researchers have synthesized PVDF-HFP copolymers by copolymerizing vinylidene fluoride (PVDF) with hexafluoropropylene (HFP) [13], which increased the proportion of amorphous regions in the copolymer, markedly enhancing ion transport properties [14,15]. Liang et al. further improved the system by incorporating garnet-type $Li_7La_3Zr_2O_{12}$ (LLZO) nanoparticles into the PVDF-HFP matrix. The plasticizing effect of LLZO particles endowed the composite with excellent electrochemical performance and cycling stability [16]. Sohn et al. fabricated porous polyethylene separators coated with a PVDF-HFP/PMMA composite via a simple dipping process under 40% relative humidity, achieving a maximum ionic conductivity of 1.69 mS/cm [17].

Metal–organic frameworks (MOFs) are a class of porous materials with tunable pore sizes and exceptionally high specific surface areas, offering a novel approach to the design of lithium metal batteries [18,19]. These materials feature well-ordered channel structures and high porosity [20], which provide continuous pathways and abundant active sites for ion transport. Moreover, the organic ligands and metal nodes in MOFs can be flexibly modified through chemical functionalization to optimize their electrochemical performance [21]. The high thermal stability of MOFs [22], along with their composite structures with polymers, enhances the mechanical strength of electrolytes, suppresses lithium dendrite growth, and accommodates volume changes during charge–discharge cycles [23]. Wang et al. constructed a three-dimensional MOF crosslinked network asymmetric solid polymer electrolyte (SPE) via in situ polymerization. By selectively sieving anions through the MOF layers, the lithium-ion transference number was elevated to 0.72. The nanostructured MOF layers serve as a shield, effectively suppressing Li dendrite growth and ensuring uniform Li^+ transport [24]. Sun et al. proposed a novel design for a polyethylene oxide (PEO)-based SPE by incorporating activated Copper 1,3,5-phenyltricarboxylic acid (HKUST-1) MOFs into the PEO electrolyte. This approach not only improved the lithium transference number and mechanical properties but also enhanced the overall electrochemical performance [25]. Dong et al. combined experimental and theoretical methods to systematically study the effects of pore size and metal vacancies in porous metal oxides (MOFs) on ion transport characteristics and electrochemical stability. It demonstrated that a metal–organic framework electrolyte comprising non-electroactive metal coordination nodes demonstrates at least 20% wider voltage tolerance range relative to redox-active metal node analogues [26].

UV crosslinking offers notable advantages in the fabrication of battery electrolyte membranes. Unlike conventional thermal crosslinking methods, UV crosslinking does not require high temperatures. This reduces the risk of thermal damage to the membrane and helps maintain the uniformity of its internal structure. Furthermore, UV crosslinking significantly enhances production efficiency, as it typically requires much shorter reaction times compared to thermal methods [27]. For example, Kim et al. employed UV photopolymerization to synthesize a cross-linked polymer matrix composed of poly(ethylene glycol) diacrylate (PEGDA) and dipentaerythritol hexaacrylate (DPHA). When applied in lithium-ion batteries, the resulting membrane achieved a high lithium-ion transference number of 0.65 [28].

In this study, a rigid flexible strategy is used for the design of gel electrolytes. The metal–organic framework (MOF) is used as a support to realize the high mechanical strength of the electrolytes. The flexible polymer PVDF-HFP enables the electrolyte to adapt to the volume change in the internal materials of the battery during service. PVDF-HFP and Zeolitic Imidazolate Framework-8 (ZIF-8) were utilized and processed via a rapid UV-curing technique to fabricate the poly(vinylidene fluoride-co-hexafluoropropylene)/ZIF-8 composite (abbreviated as PHZ) with excellent performance. The incorporation of ZIF-8 enabled the formation of ion channels and modulated the mobility of polymer chains, synergistically enhancing Li^+ transport. Consequently, this method promotes the advancement of lithium metal batteries with excellent electrochemical performance and enhanced safety.

2. Results and Discussion

As shown in Figure 1a–c,g–i, by analyzing the cross-sectional scanning electron microscopy (SEM) images of the electrolyte membranes, it is evident that, compared to the PVDF-HFP electrolyte membrane without MOFs, the PVDF-HFP matrix with added ZIF-8 forms a well-connected porous network structure. In the subsequent conductivity test, the conductivity of PHZ film was one unit higher than that of PVDF-HFP film. It indicates that the incorporation of MOFs into PVDF-HFP facilitates the construction of effective ion transport channels, thereby enhancing lithium-ion conduction [29].

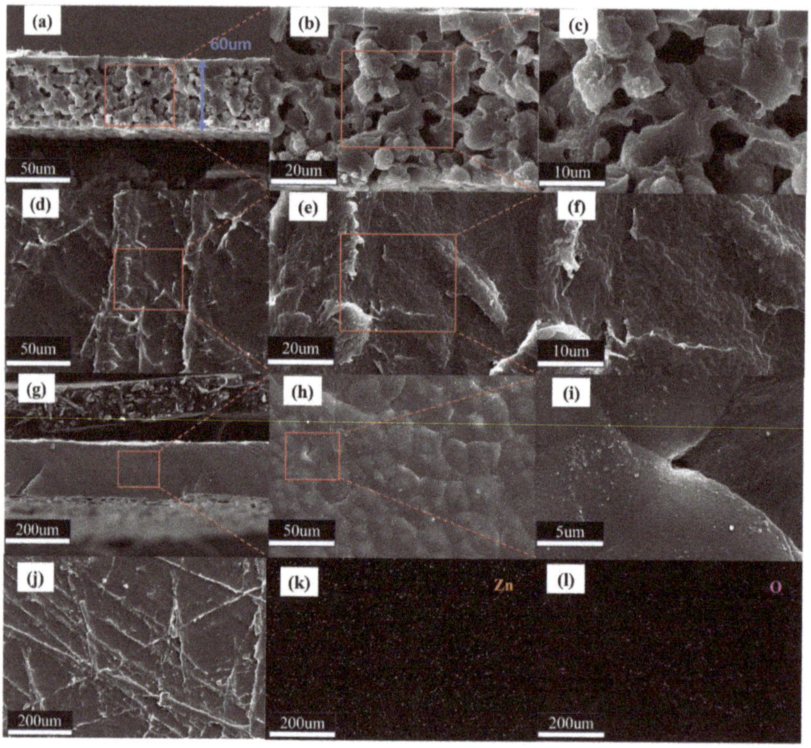

Figure 1. (**a–c**) Sectional view SEM images of PHZ electrolyte membranes. (**d–f**) Surface SEM images of PHZ electrolyte membranes. (**g–i**) Cross-sectional and surface SEM images of PVDF-HFP electrolyte membranes. (**j–l**) Elemental distribution of Zn and O in PHZ electrolyte membranes.

Meanwhile, in contrast to membranes without MOFs, the surface SEM images of PHZ electrolyte membranes exhibit a smooth morphology and seamless integration between ZIF-8 and PVDF-HFP, demonstrating strong mutual integration between the two components. Furthermore, as shown in Figure 1j,k,l the uniform distribution of Zn elements further confirms that the adopted preparation method ensures a homogeneous dispersion of ZIF-8 within the composite material.

The XRD patterns of the PHZ electrolyte membrane are shown in Figure 2a. The XRD spectrum of PVDF-HFP primarily exhibits a broad peak around $2\theta \approx 18^\circ$ to 20°, indicating its low crystallinity and high amorphous content [30]. After the incorporation of ZIF-8, the peak disappears and the amorphous region increases, suggesting that ZIF-8 hinders the orderly arrangement of PVDF-HFP molecular chains, suppressing nucleation and crystal growth, ultimately leading to a decrease in crystallinity [31]. A higher proportion of the amorphous region reduces the entanglement of molecular chains and increases the inter-chain voids (free volume), providing space for segmental motion and Li^+ migration, thereby enhancing lithium-ion transport [32].

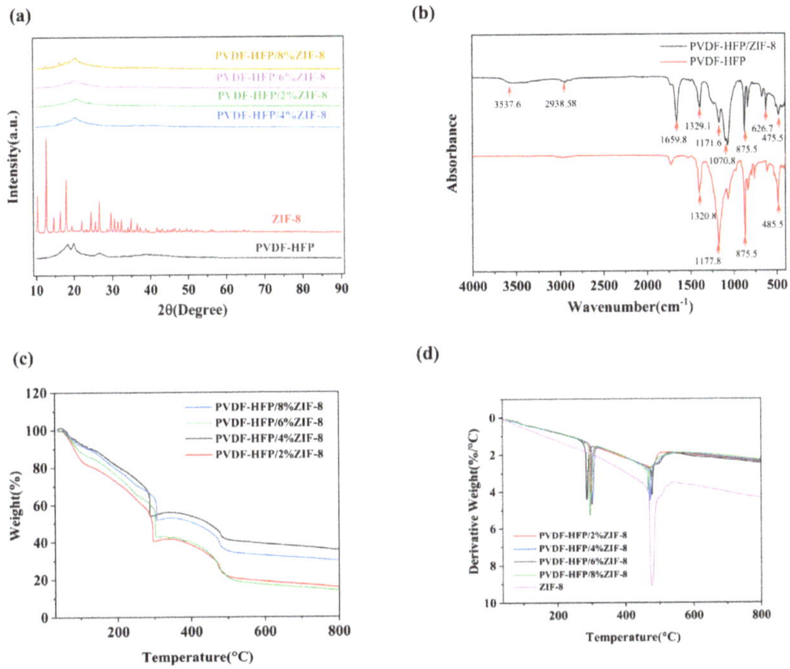

Figure 2. (**a**) XRD patterns of ZIF-8, PVDF-HFP, and their composite. (**b**) FTIR spectra of PHZ and its synthetic materials. (**c**) Thermogravimetric (TG) curves of PHZ materials with different proportions of ZIF-8 added. (**d**) Differential Scanning Calorimeter (DSC) curves of PHZ materials with different ratios of ZIF-8 added.

In order to further investigate the interaction between polymers and MOFs, FTIR spectroscopy analysis was performed on PHZ electrolyte membranes and PVDF-HFP. The FTIR spectrum provides the following insights: the peak at 3537 cm^{-1} corresponds to the coordination of the C-O-C bond, which is typically attributed to the interaction between Li^+ and ether oxygen. The peaks at 2938 cm^{-1} (C–H stretching vibration), 1329 cm^{-1} (C-F stretching vibration), and 626 cm^{-1} (CF_2 out-of-plane bending vibration) confirm the presence of the PVDF-HFP backbone. The peak at 875 cm^{-1} suggests that the HFP copolymer effectively reduces crystallinity. The peak at 1659 cm^{-1} corresponds to the C=O stretching vibration in PVDF. Additionally, the peak at 475 cm^{-1} is attributed to the vibration between the metal and organic ligand, representing the Zn–N stretching vibration in ZIF-8, confirming its successful incorporation into the polymer matrix. The peaks at 1171 cm^{-1} (C–F stretching vibration) and 1070 cm^{-1} indicate the presence of ClO_4^- in the polymer [33].

Thermogravimetric analysis (TGA) reveals that below 300 °C, the composite material undergoes mass loss due to the desorption of adsorbed water, residual solvents, or low-molecular-weight volatile components, as well as the decomposition of lithium salt. The intermediate stable stage (350–450 °C) indicates that the main structure (PVDF-HFP crystalline region and ZIF-8) maintains good thermal stability within this temperature range. As the temperature increases, the PVDF-HFP polymer backbone undergoes substantial breakdown, with its typical degradation process accompanied by the release of gases such as HF, leading to a sharp mass loss. Simultaneously, the organic ligands in ZIF-8 also enter their decomposition phase, further accelerating the overall weight reduction. The addition of lithium salt causes the decomposition temperature of ZIF-8 to shift earlier [34]. After decomposition, ZIF-8 may form inorganic products such as ZnO, which are relatively stable [35]. DSC testing also showed that PHZ material exhibited significant decomposition reactions at 300 °C, accompanied by endothermic peaks, corresponding to the weight loss changes in PHZ at different ratios in TG testing. A clear endothermic peak also appeared at around 500 °C, corresponding to the change in TG weight loss of ZIF-8 in PHZmaterial.

Polymer gel electrolyte membranes with different proportions of ZIF-8 were systematically tested using coin-type symmetric cells and an electrochemical workstation. As shown in Figure 3c, the results indicate that when 4% ZIF-8 is added to the electrolyte membrane, its ionic conductivity reaches the highest value of 1.17×10^{-3} S/cm. However, with a further increase in ZIF-8 content, the ionic conductivity gradually decreases. This phenomenon can be attributed to the intensified trends of particle agglomeration and interfacial blocking effects when the ZIF-8 loading surpasses 4%, resulting in the obstruction of ion transport pathways and subsequent impairment of ionic transport kinetics. Compared to pure PVDF-HFP electrolyte membranes (1×10^{-4} S/cm), the conductivity of PHZ increases by an order of magnitude.

Additionally, Figure 3b shows the linear sweep voltammetry (LSV) curves of PHZ electrolyte membranes with different ZIF-8 addition ratios. When the amount of ZIF added is 4%, the electrochemical window of the electrolyte membrane expands to 4.7 V, which is significantly better than that of PHZ membranes with other addition ratios. In PVDF-HFP-based electrolyte membranes, excessive addition of MOF may lead to insufficient binder and affect the mechanical strength of the membrane. If it is too low, it cannot effectively regulate the pore structure. Therefore, an appropriate ratio of MOF to PVDF HFP can expand the range of battery LSV [36]. The porous network structure of ZIF-8 forms additional ion-conduction pathways within the polymer matrix, thereby accelerating ion migration. Given that the PHZ electrolyte membrane achieved the highest conductivity when 4% of zif-8 was added, we conducted lithium migration tests on this specific proportion of PHZ, as illustrated in Figure 3d. The synergistic effect between ZIF-8 and the local segmental motion of the amorphous region in PVDF-HFP enhances lithium-ion transference [37,38], achieving a lithium-ion transference number (TLi^{+}) of 0.7.

To further evaluate the electrochemical performance of the PHZ solid-state battery, rate performance and cycling tests were conducted at room temperature (25 °C) by an electrochemical workstation (Figure 4a, b). The initial discharge capacity is 168 mAh g^{-1}. At a discharge rate of 0.2 C, the discharge capacity was 141 mAh g^{-1}, while at 0.5 C, it was 105 mAh g^{-1}. At 1 C, the discharge capacity was 88 mAh g^{-1}, and it was 68 mAh g^{-1} at 2 C. When the discharge rate returned to 0.1 C, the battery still maintained a discharge capacity of 168 mAh g^{-1}. In contrast, the solid-state battery with the PVDF-HFP electrolyte membrane (without ZIF-8) exhibited an initial discharge capacity of only 149 mAh g^{-1} at 0.1 C. At 0.2 C, the discharge capacity was 133 mAh g^{-1}; at 0.5 C, it was 77 mAh g^{-1}; at 1 C, it was 48 mAh g^{-1}; and at 2 C, it was 28 mAh g^{-1}. Comparatively, the porous structure of PHZ provides more space for lithium-ion insertion and storage, allowing more lithium ions to participate in the reaction under the same active material content. This enhances the discharge-specific capacity and enables efficient ion transport channels, leading to excellent high-rate performance. The nano confinement effect of MOFs can suppress the possible chain breakage or crystallization of PVDF-HFP in long-term cycling, thereby promoting the integrity and stability of PHZ structures [39].

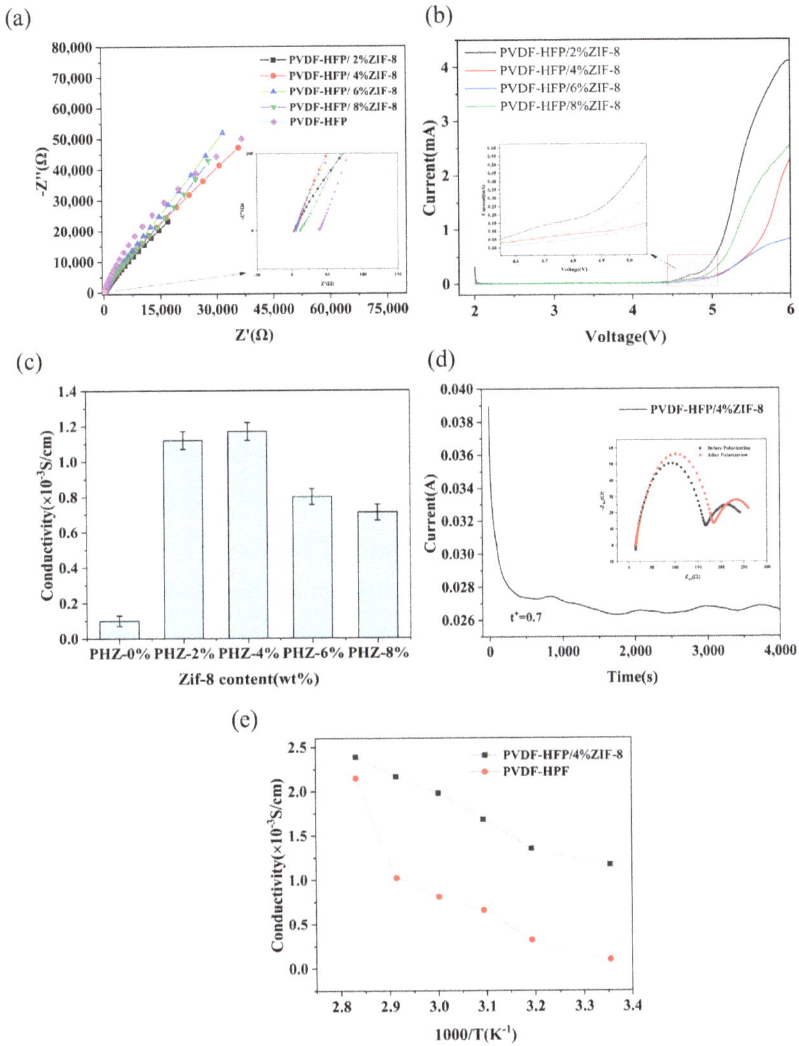

Figure 3. (**a**) Conductivity diagram of PHZ composite electrolyte membranes with different mass fractions of ZIF-8. (**b**) Electrochemical window test of PHZ electrolyte membranes. (**c**) Bar chart comparing conductivity with different mass fractions of ZIF-8. (**d**) Test chart of ion migration number of PHZ electrolyte membranes. (**e**) Comparison of conductivity between PHZ electrolyte membranes and PVDF-HFP electrolyte membrane at different temperatures.

In the cycling test, the PHZ solid-state battery exhibited an initial discharge capacity of 169 mAh g^{-1} at a 0.1 C discharge rate. After 100 cycles, the remaining discharge capacity was 162 mAh g^{-1}, with a capacity retention rate of 95.8% and a Coulombic efficiency of

99.5% (Figure 4c). The polarization remained around 100 mV(Figure 4d), with negligible change after cycling, demonstrating the outstanding cycling stability of PHZ.

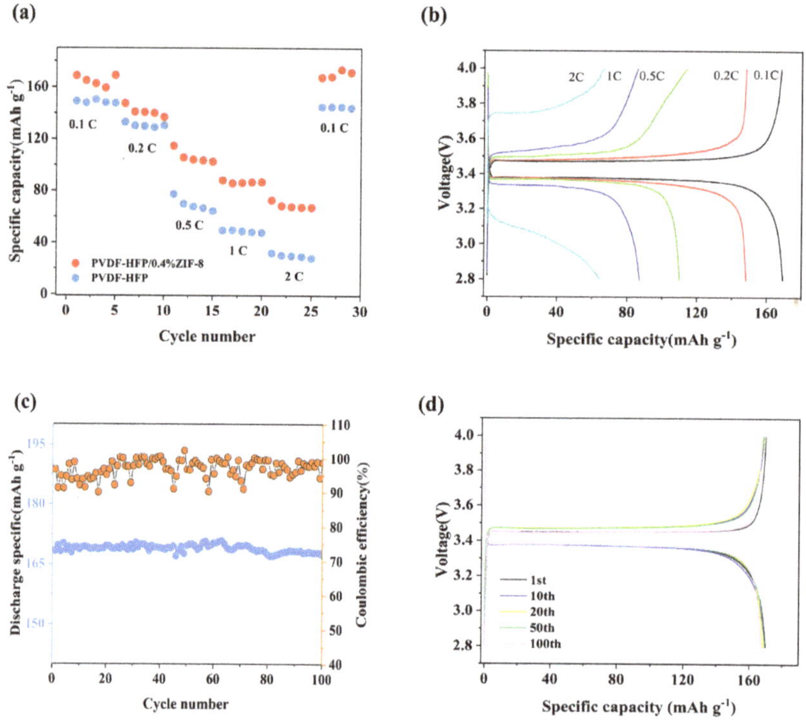

Figure 4. (**a**) Rate performance of PVDF-HFP and PHZ batteries at 25 °C. (**b**) Charge–discharge curves of the PHZ battery at 0.1 C, 0.2 C, 0.5 C, 1 C, and 2 C. (**c**) Cycling performance of the PHZ battery over 100 cycles at 25 °C. (**d**) Charge–discharge curves of the PHZ battery over 100 cycles.

3. Conclusions

In summary, we successfully fabricated the PHZ polymer electrolyte membrane by incorporating the metal–organic framework material ZIF-8 into PVDF-HFP and employing UV light-curing rapid prototyping for full polymerization. The porous network structure of ZIF-8 formed additional ion transport channels within the polymer matrix, accelerating lithium-ion migration. The uniform Li^+ flux distribution regulated by ZIF-8 channels effectively reduced dendrite growth risks and improved battery cycling stability. The PHZ polymer electrolyte membrane with 4 wt% ZIF-8 exhibited an ionic conductivity of 1.17×10^{-3} S/cm, an electrochemical stability window of 4.7 V, and a lithium transference number of 0.7. The solid-state battery assembled with PHZ polymer electrolyte demonstrated an initial discharge capacity of 168 mAh g^{-1} at a 0.1 C discharge rate, with a capacity retention rate of 95.8% and a Coulombic efficiency of 99.5% after 100 cycles. at 1 C, it was 48 mAh g^{-1}; and at 2 C, it was 28 mAh g^{-1}. These results indicate excellent rate and cycling performance. Therefore, the PHZ electrolyte material offers a novel approach and method for future lithium metal batteries research and development.

4. Materials and Methods

4.1. Materials

The raw materials include lithium perchlorate ($LiClO_4$, 99%, Aladdin, Shanghai, China), poly(vinylidene fluoride-co-hexafluoropropylene) (PVDF-HFP, Mw = 600,000, Macklin, Shanghai, China), N,N-dimethylformamide (DMF, 99.8%, Aladdin, Shanghai, China), polyvinylidene fluoride (PVDF, Arkema, France); 1-methyl-2-pyrrolidone (NMP, Aladdin, Shanghai, China); conductive carbon black (99.9%, Kejing, Shenzhen, China); lithium iron phosphate ($LiFePO_4$, LFP, 99.9%, Macklin, Shanghai, China); zeolitic imidazolate framework-8 (ZIF-8, 99%, Tanyu New Materials, Guangdong, China); 2-hydroxy-2-methylpropiophenone (1173, Chembridge, Beijing, China); polyurethane acrylate (PUA, RYOJI, Frankfurt, Germany); ethoxylated trimethylolpropane triacrylate (ETPTA, Mn = 428, Macklin, Shanghai, China).

4.2. Preparation of Electrolyte Membrane

A total of 8 g of N,N-dimethylformamide (DMF) was measured into a beaker, and the zeolitic imidazolate framework-8 (ZIF-8) MOFs were dispersed into the DMF. The mixture was sonicated for 30 min to ensure complete dispersion. Subsequently, poly(vinylidene fluoride-co-hexafluoropropylene) (PVDF-HFP) and a specific amount of lithium perchlorate ($LiClO_4$) were dissolved/dispersed into the ZIF-8/DMF solution. The solution was stirred using a magnetic stirrer at 45 °C with a stirring speed of 450 r/min for 4 h to ensure thorough mixing. After stirring, ethoxylated trimethylolpropane triacrylate (ETPTA) and polyurethane acrylate (PUA) were added to the solution, followed by an additional 15 min of stirring. Finally, 2-hydroxy-2-methylpropiophenone (HMPP) was introduced, and the solution was stirred for another 15 min. The resulting mixture was transferred to an ultrasonic device and sonicated for 30 min to remove air bubbles. The prepared polymer composite consisted of PVDF-HFP and ZIF-8, with the mass fraction of ZIF-8 in the composite set at 0 wt%, 2 wt%, 4 wt%, 6 wt%, and 8 wt%. These samples are hereafter referred to as PHZ-0%, PHZ-2%, PHZ-4%, PHZ-6%, and PHZ-8%, respectively. The production process is shown in Figure 5.

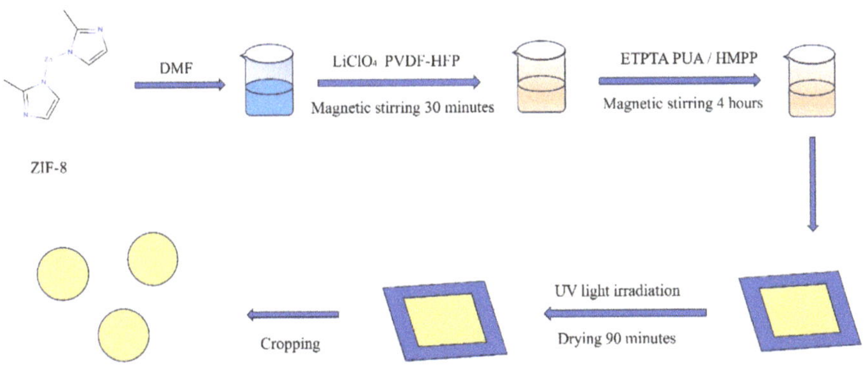

Figure 5. Preparation process of PHZ polymer.

The prepared material was poured onto a polytetrafluoroethylene (PTFE) plate and spread into a thin film using a scraper. The film was then placed in a UV curing machine for photopolymerization and subsequently vacuum-dried at 60 °C for 90 min (The light source is an ultraviolet lamp, with a peak wavelength of 365 nm). Finally, the electrolyte membrane was peeled off and cut into small circular discs with a diameter of 19 mm. The discs were stored in a glovebox (<0.01 PPM H_2O and O_2) for future use.

4.3. Battery Assembly

The lithium metal battery was assembled using an electrolyte membrane, a $LiFePO_4$ (LFP) cathode, and a lithium (Li) anode. The cathode was fabricated by coating an aluminum foil with a slurry mixture containing 80 wt% lithium iron phosphate (LFP), 10 wt% polyvinylidene fluoride (PVDF) binder, and 10 wt% carbon black additive. This cathode configuration achieved an active material mass loading of about 1.8 mg cm^{-2}. After cell assembly, the battery was stored in an argon-filled glovebox for 12 h, followed by a 1 h drying process in a vacuum oven maintained at 60 °C.

4.4. Material Characterization

The thermal decomposition curve was obtained using a thermogravimetric (TG) analyzer (Manufacturer: METTLER TOLEDO, Greifensee, Switzerland, Model: DSC 3+). For this analysis, PHZ samples were weighed (mg) and subjected to thermal analysis in the temperature range of 30–800 °C at a heating rate of 10 °C/min. The thermal analysis was conducted under a nitrogen (N_2) atmosphere with a flow rate of 100 mL/min. Fourier-transform infrared (FTIR) spectra were recorded using an FTIR spectrometer (Manufacturer: Thermo, Waltham, MA, USA, Model: Nicolet iS5) operating in attenuated total reflectance (ATR) mode. Infrared measurements were performed by mixing a small amount of potassium bromide (KBr) to absorb polymer peaks in the wavenumber range of 4000–400 cm^{-1}, with a resolution of 4 cm^{-1}. Phase analysis of the materials was conducted using an X-ray diffractometer (Manufacturer: Dandong Haoyuan, Dandong, China, Model: DX-2700, Cu-Kα, 40 kV × 30 mA). The scanning range was set from 5° to 80° (2θ), with a scanning rate of 0.5° and a step size of 0.1° min^{-1}. The overall morphology of the composite materials was observed using a scanning electron microscope (SEM) (Manufacturer: Hitachi, Tokyo, Japan, Model: HITACHI SU8010). The cross-section of the materials was obtained via liquid nitrogen freeze-fracture to examine the internal pore distribution and the embedding state of ZIF-8 particles.

4.5. Electrochemical Testing

Ionic conductivity can reflect whether polymer gel electrolytes can meet the standards of solid-state batteries. A coin cell symmetric configuration (stainless steel (SS)/CSE/SS) was assembled and tested on an electrochemical workstation (DH 7000, Jiangsu, China) using electrochemical impedance spectroscopy (EIS) in the frequency range of 0.1 Hz to 1 MHz with an amplitude of 10 mV. The ionic conductivity (σ) can be calculated using the following equation:

$$\sigma = \frac{L}{RS}. \tag{1}$$

In this equation, L denotes the electrolyte membrane thickness (cm), R is the bulk resistance (Ω) obtained from EIS fitting, and S represents the effective contact area (cm^2) between the electrode and the electrolyte membrane.

The lithium-ion transference number (${t_{Li}}^+$) is a key parameter for evaluating the contribution of lithium ions to the total ionic conductivity of the electrolyte and serves as an important criterion for determining the performance of lithium-ion batteries. The ${t_{Li}}^+$ of the composite gel electrolyte (CGE) is measured using a symmetric cell (Li/CGE/Li) by combining electrochemical impedance spectroscopy (EIS) and direct current (DC) polarization on an electrochemical workstation. First, the initial impedance value (R_0) is obtained through EIS testing. A small DC voltage is then applied, and the current is allowed to reach a steady-state value (Is). After removing the DC voltage, the impedance spectrum is measured again to obtain the interfacial resistance after polarization (R_s).

The lithium-ion transference number ($t_{Li}{}^{+}$) is calculated using the following equation:

$$t_{Li^+} = \frac{I_s(\Delta V - R_0 I_0)}{I_0(\Delta V - R_S I_S)}. \quad (2)$$

In Equation (2), R_0 and R_s represent the impedance values before and after DC polarization, respectively; ΔV is the applied voltage across the battery (10 mV); and I_0 and I_s denote the initial and steady-state currents, respectively [40].

To assess the electrochemical stability range of the CGE, a Li/CGE/WE cell was built and subjected to linear sweep voltammetry (LSV) over a voltage span of 2.0–6.0 V (vs. Li/Li^+) at a scan rate of 0.5 mV/s.

The rate performance and cycling stability of the battery were evaluated using a battery testing system (Neware, Dongguan, China). Coin cells were assembled with $LiFePO_4$ as the cathode and lithium metal as the anode, and their cycling and rate performance were measured. The tests were conducted within a voltage range of 2.8–4.0 V.

Author Contributions: Conceptualization, N.M. and X.L.; methodology, Q.H.; software, L.L.; validation, X.L., N.M. and Q.H.; formal analysis, N.M.; investigation, J.W.; resources, X.L.; data curation, N.M.; writing—original draft preparation, N.M.; writing—review and editing, X.L.; visualization, L.L. and Y.G.; supervision, L.L. and Y.G.; project administration, J.W.; funding acquisition, L.L. and Y.G. All authors have read and agreed to the published version of the manuscript.

Funding: This research was funded by the Reginal Collaboration R&D Program of Sichuan Province under Grant (No. 2024YFHZ0209), the Fund Project of the Guangxi Key R&D Programme Projects (No. AB24010346), the National Natural Science Foundation of China (Nos. 22462003, 52161033, and 22262005), the Guangxi Key Laboratory of Automobile Components and Vehicle technology (Nos. 2023GKLACVTZZ02, 2022GKLACVTZZ04, 2023GKLACVTZZ10),), and the Guangxi University Young and Middle-Aged Teachers' Basic Scientific Research Ability Improvement Project (No. 2021KY1715).

Institutional Review Board Statement: Not applicable.

Informed Consent Statement: Not applicable.

Data Availability Statement: The data are contained within the article.

Conflicts of Interest: The authors declare no conflicts of interest.

References

1. Yamada, A. Hidden Negative Issues and Possible Solutions for Advancing the Development of High-Energy-Density in Lithium Batteries: A Review. *Adv. Sci.* **2024**, *11*, 2401739. [CrossRef] [PubMed]
2. Koohi-Fayegh, S.; Rosen, M.A. A review of energy storage types, applications and recent developments. *J. Energy Storage* **2020**, *27*, 101047. [CrossRef]
3. Zaman, W.; Hatzell, K.B. Processing and manufacturing of next generation lithium-based all solid-state batteries. *Curr. Opin. Solid State Mat. Sci.* **2022**, *26*, 101003. [CrossRef]
4. Sun, N.; Liu, Q.; Cao, Y.; Lou, S.; Ge, M.; Xiao, X.; Lee, W.K.; Gao, Y.; Yin, G.; Wang, J.; et al. Anisotropically Electrochemical–Mechanical Evolution in Solid-State Batteries and Interfacial Tailored Strategy. *Angew. Chem. Int. Ed.* **2019**, *58*, 18647–18653. [CrossRef]
5. Ma, J.; Chen, B.; Wang, L.; Cui, G. Progress and prospect on failure mechanisms of solid-state lithium batteries. *J. Power Sources* **2018**, *392*, 94–115. [CrossRef]
6. Hao, F.; Han, F.; Liang, Y.; Wang, C.; Yao, Y. Architectural design and fabrication approaches for solid-state batteries. *MRS Bull.* **2018**, *43*, 775–781. [CrossRef]
7. Stegmaier, S.; Voss, J.; Reuter, K.; Luntz, A.C. Li^+ Defects in a Solid-State Li Ion Battery: Theoretical Insights with a Li_3OCl Electrolyte. *Chem. Mat.* **2017**, *29*, 4330–4340. [CrossRef]
8. Xu, R.; Zhang, X.Q.; Cheng, X.B.; Peng, H.J.; Zhao, C.Z.; Yan, C.; Huang, J.Q. Artificial Soft–Rigid Protective Layer for Dendrite-Free Lithium Metal Anode. *Adv. Funct. Mater.* **2018**, *28*, 1870049. [CrossRef]

9. Ali, S.; Tan, C.; Waqas, M.; Lv, W.; Wei, Z.; Wu, S.; Boateng, B.; Liu, J.; Ahmed, J.; Xiong, J. Highly Efficient PVDF-HFP/Colloidal Alumina Composite Separator for High-Temperature Lithium-Ion Batteries. *Adv. Mater. Interfaces* **2018**, *5*, 1701147. [CrossRef]
10. Farooqui, U.R.; Ahmad, A.L.; Hamid, N.A. Effect of polyaniline (PANI) on Poly(vinylidene fluoride-co-hexaflouro propylene) (PVDF-co-HFP) polymer electrolyte membrane prepared by breath figure method. *Polym. Test.* **2017**, *60*, 124–131. [CrossRef]
11. Wu, Y.; Li, Y.; Wang, Y.; Liu, Q.; Chen, Q.; Chen, M. Advances and prospects of PVDF based polymer electrolytes. *J. Energy Chem.* **2021**, *64*, 62–84. [CrossRef]
12. Zhong, X.; Han, J.; Chen, L.; Liu, W.; Qin, W. Binding mechanisms of PVDF in lithium ion batteries. *Appl. Surf. Sci.* **2021**, *553*, 149564. [CrossRef]
13. Watanabe, M.; Kanba, M.; Matsuda, H.; Tsunemi, K.; Mizoguchi, K.; Tsuchida, E.; Shinohara, I. High lithium ionic conductivity of polymeric solid electrolytes. *Die Makromol. Chem. Rapid Commun.* **1981**, *2*, 741–744. [CrossRef]
14. Zuofeng, C.; Yanxia, J.; Quanchao, Z.; Quanfeng, D.; Ye, W.; Shigang, S. Preparation and characterization of a novel composite microporous polymer electrolyte for Li-ion batteries. *Sci. Bull.* **2005**, *50*, 6.
15. Gonçalves, R.; Miranda, D.; Almeida, A.M.; Silva, M.M.; Meseguer-Dueñas, J.M.; Ribelles, J.L.G.; Lanceros-Méndez, S.; Costa, C.M. Solid polymer electrolytes based on lithium bis(trifluoromethanesulfonyl)imide/poly(vinylidene fluoride -co-hexafluoropropylene) for safer rechargeable lithium-ion batteries. *Sustain. Mater. Technol.* **2019**, *21*, e00104. [CrossRef]
16. Liang, Y.F.; Deng, S.J.; Xia, Y.; Wang, X.L.; Xia, X.H.; Wu, J.B.; Gu, C.D.; Tu, J.P. A superior composite gel polymer electrolyte of $Li_7La_3Zr_2O_{12}$- poly(vinylidene fluoride-hexafluoropropylene) (PVDF-HFP) for rechargeable solid-state lithium ion batteries. *Mater. Res. Bull.* **2018**, *102*, 412–417. [CrossRef]
17. Sohn, J.Y.; Im, J.S.; Shin, J.; Nho, Y.C. PVDF-HFP/PMMA-coated PE separator for lithium ion battery. *J. Solid State Electrochem.* **2012**, *16*, 551–556. [CrossRef]
18. Farina, M.; Duff, B.B.; Tealdi, C.; Pugliese, A.; Blanc, F.; Quartarone, E. Li^+ Dynamics of Liquid Electrolytes Nanoconfined in Metal-Organic Frameworks. *ACS Appl. Mater. Interfaces* **2021**, *13*, 53986–53995. [CrossRef]
19. Wang, L.; Han, Y.; Feng, X.; Zhou, J.; Qi, P.; Wang, B. Metal-organic frameworks for energy storage: Batteries and supercapacitors. *Coord. Chem. Rev.* **2016**, *307*, 361–381. [CrossRef]
20. Wang, S.M.; Mu, X.T.; Liu, H.R.; Zheng, S.T.; Yang, Q.Y. Pore-Structure Control in Metal-Organic Frameworks (MOFs) for Capture of the Greenhouse Gas SF. *Angew. Chem.* **2022**, *61*, e202207066. [CrossRef]
21. Quartarone, E.; Mustarelli, P. Review-Emerging Trends in the Design of Electrolytes for Lithium and Post-Lithium Batteries. *J. Electrochem. Soc.* **2020**, *167*, 50508. [CrossRef]
22. Lv, X.L.; Yuan, S.; Xie, L.H.; Darke, H.F.; Chen, Y.; He, T.; Dong, C.; Wang, B.; Zhang, Y.Z.; Li, J.R. Ligand Rigidification for Enhancing the Stability of Metal-Organic Frameworks. *J. Am. Chem. Soc.* **2019**, *141*, 10283–10293. [CrossRef] [PubMed]
23. Liu, R.; Xue, J.; Xie, L.; Chen, H.; Deng, Z.; Yin, W. A Flexible Yet Robust 3D-Hybrid Gel Solid-State Electrolyte Based on Metal–Organic Frameworks for Rechargeable Lithium Metal Batteries. *Gels* **2024**, *10*, 812. [CrossRef] [PubMed]
24. Wang, G.; He, P.; Fan, L.Z. Asymmetric Polymer Electrolyte Constructed by Metal-Organic Framework for Solid-State, Dendrite-Free Lithium Metal Battery. *Adv. Funct. Mater.* **2020**, *31*, 2007198. [CrossRef]
25. Sun, C.C.; Yusuf, A.; Li, S.W.; Qi, X.L.; Wang, D.Y. Metal Organic Frameworks Enabled Rational Design of Multifunctional PEO-Based Solid Polymer Electrolytes. *Chem. Eng. J.* **2021**, *414*, 128702. [CrossRef]
26. Dong, P.; Zhang, X.H.; Hiscox, W.; Liu, J.; Zamora, J.; Li, X.; Su, M.; Zhang, Q.; Guo, X.; Mccloy, J. Towards High-Performance Metal-Organic-Framework-Based Quasi-Solid-State Electrolytes: Tunable Structures and Electrochemical Properties. *Adv. Mater.* **2023**, *35*, e2211841. [CrossRef]
27. Çakmakçı, E.; Uğur, M.H.; Güngör, A. UV-Cured polypropylene mesh-reinforced composite polymer electrolyte membranes. *E-Polymers* **2015**, *15*, 103–110. [CrossRef]
28. Kim, K.; Chae, W.; Kim, J.; Kim, C.; Earmme, T. Gel Polymer Electrolytes for Lithium-Ion Batteries Enabled by Photo Crosslinked Polymer Network. *Gels* **2023**, *9*, 975. [CrossRef]
29. Wu, X.; Yu, X.; Zhang, Z.; Liu, H.; Ling, S.D.; Liu, X.; Lian, C.; Xu, J. Anisotropic ZnS Nanoclusters/Ordered Macro-Microporous Carbon Superstructure for Fibrous Supercapacitor toward Commercial-Level Energy Density. *Adv. Funct. Mater.* **2023**, *33*, 2300329. [CrossRef]
30. Pan, J.; Zhang, L.; Wang, Z.; Sun, S.P.; Cui, Z.; Tavajohi, N. Poly(vinylidene fluoride-co-hexafluoro propylene) membranes prepared via thermally induced phase separation and application in direct contact membrane distillation. *Front. Chem. Sci. Eng.* **2022**, *16*, 720–730. [CrossRef]
31. Zeng, Q.; Shi, L.; Wang, J.; Zha, X.; Yang, W.; Yang, Y. High-performance MOF-derived polymer electrolytes with modified ionic transport for solid-state lithium metal batteries. *J. Energy Storage* **2025**, *110*, 115314. [CrossRef]
32. Huang, C.; Li, H.; Teng, Z.; Luo, Y.; Chen, W. MOF-modified dendrite-free gel polymer electrolyte for zinc-ion batteries. *RSC Adv.* **2024**, *14*, 15337–15346. [CrossRef] [PubMed]

33. Liu, K.; Chang, X.; Chen, X.; Liu, X.; Jia, M.; Bi, Z.; Guo, X. Hierarchical-structural design of ultrathin composite electrolytes for high-stability solid-state lithium batteries: From "polymer-in-salt" to "polymer-in-ceramic". *Nano Energy* **2025**, *135*, 110644. [CrossRef]
34. Xu, B.; Mei, Y.; Xiao, Z.; Kang, Z.; Wang, R.; Sun, D. Monitoring thermally induced structural deformation and framework decomposition of ZIF-8 through in situ temperature dependent measurements. *Phys. Chem. Chem. Phys.* **2017**, *19*, 27178–27183. [CrossRef]
35. Wang, S.; Zhang, S. Study on the Structure Activity Relationship of ZIF-8 Synthesis and Thermal Stability. *J. Inorg. Organomet. Polym. Mater.* **2017**, *27*, 1317–1322. [CrossRef]
36. Zhang, X.; Su, Q.; Du, G.; Xu, B.; Wang, S.; Chen, Z.; Wang, L.; Huang, W.; Pang, H. Stabilizing Solid-state Lithium Metal Batteries through In Situ Generated Janus-heterarchical LiF-rich SEI in Ionic Liquid Confined 3D MOF/Polymer Membranes. *Angew. Chem. Int. Ed.* **2023**, *62*, e202304947. [CrossRef]
37. Yang, M.; Liu, Y.; Mo, Y. Lithium crystallization at solid interfaces. *Nat. Commun.* **2023**, *14*, 2986. [CrossRef]
38. Yang, M.; Liu, Y.; Nolan, A.M.; Mo, Y. Interfacial Atomistic Mechanisms of Lithium Metal Stripping and Plating in Solid-State Batteries. *Adv. Mater.* **2021**, *33*, e2008081. [CrossRef]
39. Liu, Y.; Zeng, Q.; Chen, P.; Li, Z.; Chen, A.; Guan, J.; Wang, A.; Zhang, L. Modified MOF-Based Composite All-Solid-State Polymer Electrolyte with Improved Comprehensive Performance for Dendrite-Free Li-Ion Batteries. *Macromol. Chem. Phys.* **2021**, *223*, 2100325. [CrossRef]
40. Zugmann, S.; Fleischmann, M.; Amereller, M.; Gschwind, R.M.; Wiemhoefer, H.D.; Gores, H.J. Measurement of transference numbers for lithium ion electrolytes via four different methods, a comparative study. *Electrochim. Acta* **2011**, *56*, 3926–3933. [CrossRef]

gels

MDPI

Article

Preparation and Performance of PVDF-HFP/PAN-Based Gel Polymer Electrolytes

Xiubing Yao [1,†], Lingxiao Lan [1,†], Qiankun Hun [1], Xuanan Lu [1], Jianghua Wei [1,*], Xinghua Liang [1,*], Pengcheng Shen [1], Ying Long [2] and Yifeng Guo [3]

[1] Guangxi Key Laboratory of Automobile Components and Vehicle Technology, Guangxi University of Science & Technology, Liuzhou 545006, China; 18300084899@163.com (X.Y.); llx2685062@163.com (L.L.); 13383371921@163.com (Q.H.); a15007751851@163.com (X.L.); 18852066673@163.com (P.S.)

[2] School of Information Science and Engineering, Liuzhou Institute of Technology, Liuzhou 545616, China; 18178806880@163.com

[3] School of Mechanical Engineering, Chengdu University, Chengdu 610106, China; guobujia2000@163.com

* Correspondence: 15314157652@163.com(J.W.); lxh304@126.com(X.L.)

† These authors contributed equally to this work.

Abstract: Solid-state electrolytes are widely expected to enhance the performance of lithium-ion batteries, providing higher energy density and improved safety. However, challenges still need to be solved in their practical application due to low ionic conductivity and high interfacial resistance at room temperature. In this study, we successfully developed a high-performance gel polymer electrolyte (GPEs) by blending poly(vinylidene fluoride-co-hexafluoropropylene)(PVDF-HFP) and polyacrylonitrile (PAN) through UV curing, cross-linking with ethoxylated trimethylolpropane triacrylate (ETPTA), and incorporating $Li_{6.4}La_3Zr_{1.4}Ta_{0.6}O_{12}$ (LLZTO). At room temperature, the ionic conductivity of the GPEs was 2.8×10^{-4} S/cm, with a lithium-ion transference number of 0.6. Moreover, during lithium plating/stripping tests, the assembled Li/PPEL/Li symmetric cell exhibited stable cycling for up to 600 h at a current density of 0.1 mA/cm^2. Notably, the GPEs enabled the $LiFePO_4$/GPEs/Li battery to achieve excellent performance, delivering high discharge capacities at room temperature (164.3 mAh g^{-1} at 0.1 C and 88.8 mAh g^{-1} at 1 C), with a capacity retention of 89.4% after 200 cycles at 0.5 C. Therefore, solid-state batteries using this electrolyte exhibit excellent performance, including adequate capacity and cycling stability.

Keywords: UV curing process; blending; LLZTO; lithium-ion batteries

1. Introduction

With the rapid development of portable electronic devices and smart grids, the demand for higher energy density and safety has become increasingly critical. Lithium metal anodes, with their lowest reduction potential and highest theoretical specific capacity, have become a key component in advancing lithium-ion batteries [1–3]. Meanwhile, the rapid emergence of flexible and wearable electronics has imposed unprecedented requirements on energy storage systems, particularly in terms of mechanical adaptability and safety under dynamic deformation [4–8]. Solid-state batteries, which use lithium metal as the anode, offer significant advantages in higher energy density and improved safety, making them a growing focus in energy storage systems [9–11]. However, traditional lithium-ion batteries are liquid-based, and their liquid electrolytes present severe toxicity and poor electrochemical stability, which pose significant safety risks and hinder their commercialization [12–14]. Replacing conventional organic electrolytes with gel polymer electrolytes (GPEs) has emerged as an effective strategy to address these safety concerns,

 https://doi.org/10.3390/gels11050317

garnering widespread attention and extensive research [15,16]. The development of gel polymer electrolytes (GPEs) is increasingly focused on sustainability and scalability, with bio-based materials gaining traction [17,18]. Biopolymer-based GPEs show promise in applications like Li-ion batteries by suppressing side reactions and dendrite growth. Yet, these emerging sustainable GPEs generally lag behind established formulations in performance. Balancing eco-friendliness with maintaining high performance remains a key challenge as the field of GPEs advances [19–21].

Polymer solid-state electrolytes (SPEs) offer advantages, such as flexibility and ease of processing, allowing for good contact with both the anode and cathode, which helps reduce interfacial resistance [16]. However, SPEs also have notable drawbacks, including high crystallinity that limits ion conductivity and weak molecular interactions that make them unstable at high voltages [22]. GPEs combine the advantages of liquid and solid electrolytes, provide a promising solution [23,24]. These materials exhibit higher ionic conductivity and excellent mechanical properties, effectively suppressing lithium dendrite growth and enhancing the optimized contact at the electrode interfaces, thus opening new avenues for advancing solid-state electrolytes [25–27]. Poly(vinylidene fluoride-co-hexafluoropropylene)(PVDF-HFP), a polymer electrolyte with a partially amorphous and partially crystalline structure, has a relatively high ionic conductivity and low crystallinity compared to other polymers [28–30]. Yadav et al. successfully incorporated LLZTO ceramic particles into PVDF-HFP/LiTFSI to prepare a composite solid polymer electrolyte with an ionic conductivity of 1.04×10^{-4} S/cm and a lithium-ion transference number of 0.26, with an electrochemical window of 4.7 V. Furthermore, a solid-state $LiFePO_4$/CSPE-20/Li battery delivered a high first discharge capacity of 131 mAh g^{-1}. The as-prepared Li | | 0.2PESF-0.8LLZTO CPEs | | $LiFePO_4$ full cells deliver an initial discharge capacity of 159.8 mAh g^{-1} at 0.1 C and successfully operated for over 80 cycles [31]. With its high antioxidant capability, polyacrylonitrile(PAN) is well-suited for use with high-voltage cathode materials, offering high ionic conductivity, thermal stability, and compatibility with lithium electrodes. Among the inorganic fillers used in polymers, $Li_{6.4}La_3Zr_{1.4}Ta_{0.6}O_{12}$(LLZTO) has emerged as the most thoroughly studied filler, with an ionic conductivity as high as 10^{-3} S cm^{-1} [32,33]. Gao et al. show a large-scale feasible method to prepare a succinonitrile(SN)/PAN coated $Li_{6.4}La_3Zr_{1.4}Ta_{0.6}O_{12}$ (LLZTO) with flexibility and high ionic conductivity by tape-casting. At room temperature, it has an ion conductivity of 4×10^{-4} S/cm, a lithium-ion transference number of 0.65, and an electrochemical window of 4.8 V. In this way, a long lifespan of over 500 cycles and a high discharge capacity (163 mAh g^{-1}) are achieved based on LiFePO4 (LFP) cathodes at 0.2 C [34]. Additionally, LLZTO powder helps reduce crystallinity, thereby improving the migration efficiency of lithium ions in the electrolyte [35,36].

In this study, we modified PVDF-HFP-based gel polymer electrolytes using a blending technique supplemented with inorganic active fillers. A novel porous flexible GPEs was successfully prepared using UV curing. The membrane was constructed with a PVDF-HFP and PAN blend as the matrix, with ethoxylated trimethylolpropane triacrylate (ETPTA) added for cross-linking and LLZTO incorporated as the inorganic active filler. This approach holds great promise for application in next-generation high-performance energy storage systems.

2. Results and Discussion

Figure 1a,b shows the surface SEM images and the magnified view of the PPEL. The surface of the solid-state electrolyte exhibits uniformly distributed micro-sized pores rich in voids. This is attributed to the evaporation of the organic solvent during the drying process, which leads to the formation of evenly distributed channels on the electrolyte surface. These micropores provide additional pathways for lithium-ion transport, enhancing electrochemical performance. Additionally, the smooth surface of the electrolyte membrane facilitates better interface contact between the electrode and the electrolyte [37,38]. It has been reported that porous structures are a promising approach to redistributing lithium-ion flux, enabling more uniform lithium deposition [39]. Figure 1c,d shows the cross-sectional SEM images and magnified views of the PPEL. The cross-section reveals abundant interconnected pores, which are beneficial for enhancing ionic conductivity. EDS analysis of the GPEs was performed to examine the distribution of LLZTO in the electrolyte. Figure 1e–j presents the surface EDS spectra for six elements: Zr, Ta, La, C, N, and F. The distribution of these elements confirms that the inorganic filler LLZTO is uniformly dispersed within the polymer matrix.

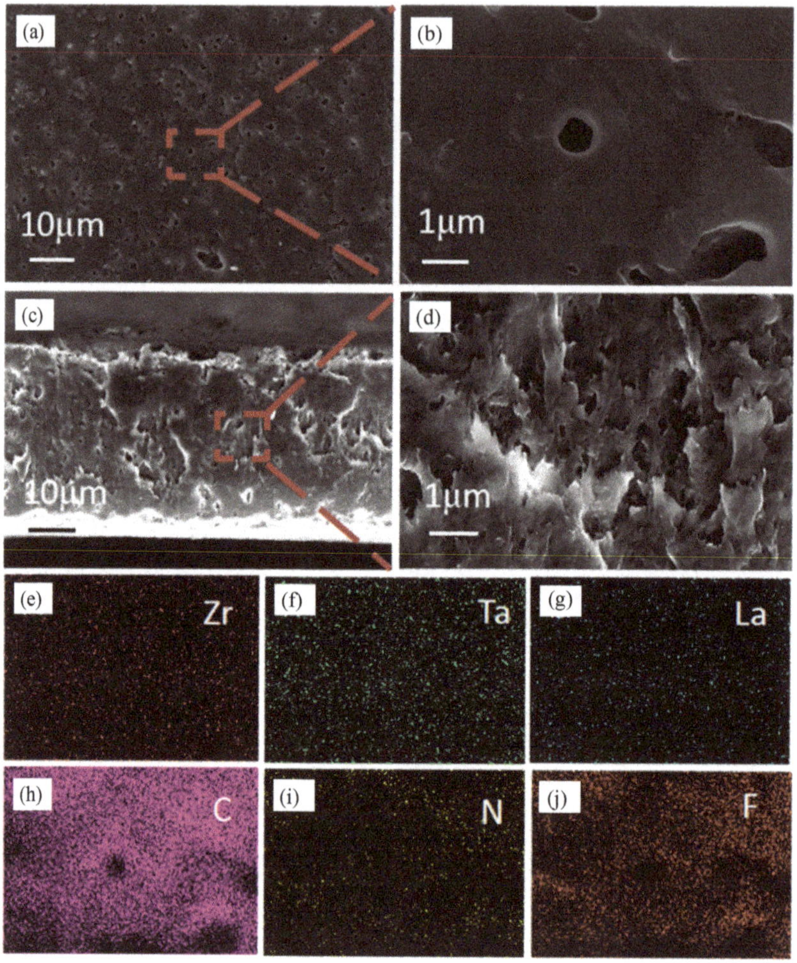

Figure 1. SEM map of solid electrolyte (**a**,**b**) PPEL surface; (**c**,**d**) SEM map of PPEL cross-section; (**e**–**j**) EDS element mapping; analyzed the distribution of Zr, Ta, La, C, N, F elements in the sample electrolyte.

Lithium-ion batteries' mechanical stability largely depends on the strength of the solid-state electrolyte [40]. When the internal stress exceeds the strength of the solid electrolyte, mechanical failure occurs, negatively impacting the battery's electrochemical performance.

As shown in Figure 2a, the tensile strength of PE is 2.21 MPa with an elongation of 39%; the tensile strength of PPE is 3.1 MPa with an elongation of 42.4%; and the tensile strength of PPEL is 4.3 MPa with an elongation of 81.4%. LLZTO exhibits a high modulus, high aspect ratio, and strong interfacial bonding with the matrix, facilitating multi-directional fracture and tensile stress sharing, which is critical for suppressing lithium dendrite growth during cycling [41].

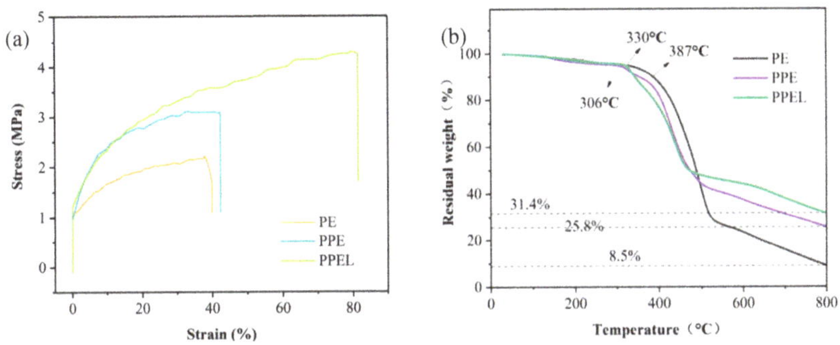

Figure 2. (**a**) Stress–strain curves of PE, PPE, and PPEL membranes; (**b**) thermogravimetric analysis of GPEs of PE, PPE, and PPEL.

We also performed thermogravimetric analysis (TGA) to assess the thermal stability of the solid-state electrolyte membranes, with the results shown in Figure 2b. The weight loss observed at temperatures below 180 °C is attributed to the GPEs' absorption of moisture from the air. After 180 °C, a slight weight loss is due to the residual DMF solvent in the solid electrolyte. PE decomposes at around 390 °C, PPE at around 310 °C, and PPEL at around 330 °C. At a decomposition temperature of 800 °C, the mass loss rates for PE, PPE, and PPEL are 91.5%, 74.2%, and 68.6%, respectively. By comparing the decomposition temperatures, it can be inferred that the lower decomposition temperature of PPE, compared to PE, is due to the lower decomposition temperature of PAN (around 250 °C) when blended with PVDF-HFP. Comparing the weight loss rates shows that PPEL has the least mass loss, indicating that the incorporation of LLZTO improves the thermal stability of the GPEs. With a high decomposition temperature of around 300 °C, the GPEs significantly exceed the typical operating temperature range of lithium batteries in commercial applications, demonstrating that their performance meets the requirements for lithium-ion battery commercialization.

Figure 3a shows the XRD patterns of LLZTO powder, PVDF-HFP, PAN, ETPTA, and composite solid-state electrolytes (PE, PPE, and PPEL). The results indicate that PVDF-HFP exhibits a semi-crystalline structure, with three diffraction peaks at approximately 18.8°, 20.5°, and 39.2°. After UV irradiation, the characteristic peaks of PVDF-HFP almost disappear, and the overall peak around 2θ = 19° becomes broadened. This suggests that the ETPTA monomer has polymerized within the PVDF-HFP, reducing its crystallinity. When PAN is added to PE, a peak similar to the PE peak but with lower intensity remains, indicating that adding PAN further reduces the crystallinity of the solid-state electrolyte. After the introduction of LLZTO, the overall peak intensity of PPEL is noticeably lower than that of PE and PPE, demonstrating that the inorganic filler LLZTO reduces the crystallinity of the blended polymer, leading to a more disordered state of the polymer chains. Compared to LLZTO powder, the characteristic peak positions of LLZTO in the PPEL membrane do not show significant changes, indicating that LLZTO powder is stably integrated into the PVDF-HFP/PAN matrix.

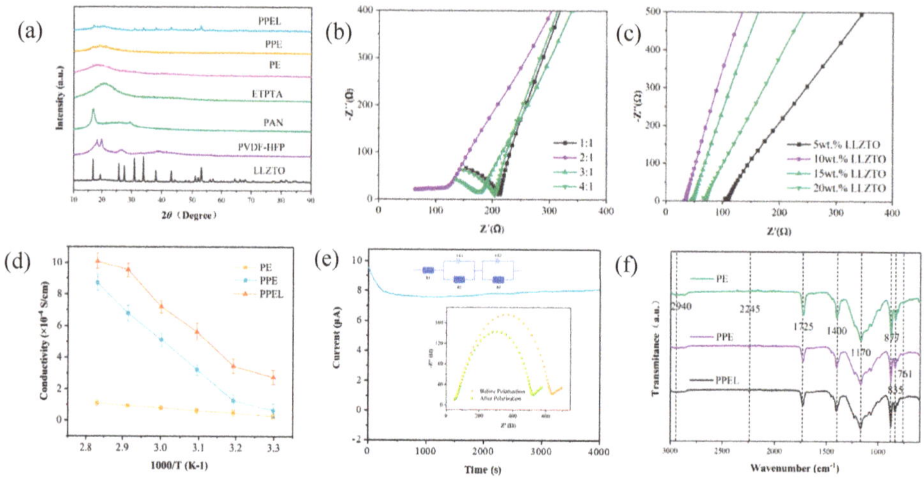

Figure 3. (**a**) XRD patterns of GPEs. EIS impedance diagram of solid electrolyte at room temperature (**b**) PPE; (**c**) PPEL; (**d**) Ionic conductivity values of PE, PPE, and PPEL at different temperatures; (**e**) DC polarization curves of the Li/PPEL/Li batteries at a polarization voltage of 10 mV; (**f**) PE, PPE and PPEL infrared spectrum characterization diagram.

Ion conductivity is a crucial indicator of GPEs. Figure 3b,c shows the Nyquist plots for the PPE and PPEL samples. Figure 3b shows that when the PVDF-HFP: PAN ratio is 2:1, the bulk impedance is minimized, and the ion conductivity reaches 6.1×10^{-5} S/cm. The introduction of LLZTO reduces the polymer crystallinity, expands the amorphous region [42], and creates new pathways for lithium-ion transport, allowing more Li^+ to move freely within the solid-state electrolyte. Figure 3c presents the ion conductivity of PVDF-HFP/PAN-based solid-state electrolytes with varying amounts of LLZTO (5 wt.%, 10 wt.%, 15 wt.%, and 20 wt.%). The highest ion conductivity of 2.8×10^{-4} S/cm is achieved when LLZTO content is 10 wt.%. However, as LLZTO content increases beyond this point, ion conductivity begins to decrease due to the agglomeration of excess LLZTO, which hinders Li^+ transport and reduces the overall conductivity [43]. Figure 3d shows the ion conductivity values of PE, PPE, and PPEL at different temperatures, confirming that the highest ion conductivity is achieved with 10% LLZTO. LLZTO doped with Ta exhibits a cubic phase [44], which can provide more ion channels and result in higher ion conductivity.

The lithium-ion transference number reflects the effective transference of lithium ions in GPEs. A high lithium-ion transference number contributes to anion transference, reducing concentration polarization [45]. Figure 3e shows the DC polarization curves and the electrochemical impedance spectroscopy (EIS) data before and after polarization for the symmetric Li/PPEL/Li cell. The lithium-ion transference number of polymer solid-state electrolytes (SPEs) typically ranges from 0.2 to 0.4 [46]. In contrast, the lithium-ion transference number of PPEL reaches 0.6, significantly higher than traditional SPEs. This improvement is attributed to the abundant F-groups in PVDF-HFP, which interact with lithium ions and allow for faster ion transport, thus enhancing the lithium-ion transference number of the solid-state electrolyte membrane and promoting lithium-ion migration. The significant improvement can also be ascribed to the LLZTO filler, an active lithium-ion conductor with a transference number of 1 [47]. Therefore, the incorporation of LLZTO powder as an inorganic filler not only increases ion conductivity but also has a positive impact on the lithium-ion transference number. In addition, the –C≡N groups in PAN can interact with free Li^+ ions, which also contribute to the enhancement of the lithium-ion transference number [48].

Fourier-transform infrared spectroscopy (FTIR) was used to confirm the quantitative changes in the bands of the blended polymer (Figure 3f). The FTIR spectra reveal the chemical interactions in the three GPEs: PE, PPE, and PPEL. The peaks at 761 cm^{-1} and 877 cm^{-1} correspond to the α-phase of PVDF-HFP, while the peaks at 836 cm^{-1} and 1170 cm^{-1} are associated with the -CF_2 stretching vibration, and the peak at 1400 cm^{-1} corresponds to C-F stretching. These results confirm that PVDF-HFP is a semi-crystalline copolymer. After blending with PAN, the intensity of the characteristic PVDF-HFP peaks weakens, indicating polymer interactions within the PPE. Additionally, two peaks at 2940 cm^{-1} and 2245 cm^{-1} in the PPE spectrum are attributed to the symmetric -CH_2 and -CN stretching vibrations of the PAN polymer, respectively, confirming the successful preparation of the PVDF-HFP/PAN binary composite membrane. Upon introducing LLZTO powder into the PPE, the resulting PPEL electrolyte membrane shows similar spectral features, with no new or disappearing peaks observed. This suggests that the incorporation of nanoparticles into the polymer matrix results in simple physical property modification without inducing significant chemical changes.

The X-ray photoelectron spectroscopy (XPS) spectrum of the cycled PPEL electrolyte is shown in Figure 4a. In the C1s spectrum (Figure 4b), four peaks are observed at 284.78 eV, 286.20 eV, 288.68 eV, and 290.52 eV, corresponding to C-C, C-O/C-N, C=O, and C-F bonds, respectively. The N1s spectrum (Figure 4c) can be fitted into three peaks, with characteristic peaks at 398.80 eV and 399.97 eV assigned to C=N and C-N bonds, respectively. Additionally, the peak at 401.63 eV is attributed to the nitrile group (-C≡N) from PAN. Similarly, in the O1s spectrum (Figure 4d), a strong peak at 532.19 eV and a weaker peak at 533.41 eV are attributed to C=O and C-O bonds, respectively. These oxygen species primarily originate from the LLZTO ceramic fillers incorporated in the polymer matrix [49]. Figure 4e shows the F1s peak at 685.1 eV, which reveals a high content of LiF (56.2%) in the generated solid electrolyte interphase (SEI) layer. A significant amount of LiF plays a crucial role in the passivation of the lithium metal surface and in delaying dendrite growth. The formation of the passivation layer helps protect the lithium metal surface and prevents further reactions with the electrolyte, thereby extending the battery's cycle life. It has been reported that a high LiF content promotes the formation of the SEI layer, and LiF can effectively increase the mechanical strength of the SEI, suppressing the growth of lithium dendrites. Moreover, the LiF layer exhibits high interfacial energy and low surface diffusion barriers for Li-ions, allowing rapid lithium-ion transport to the lithium anode during deposition. As an excellent electronic insulator, LiF inhibits the transmission of electrons across the interface, effectively preventing electrolyte decomposition [50].

The long-term interface stability between the lithium metal anode and the solid-state electrolyte is critical for suppressing the formation and growth of lithium dendrites during lithium plating/stripping cycles [51]. To investigate the compatibility and dynamic stability of the PPEL GPEs with lithium metal, constant current stripping/plating tests were conducted on symmetric cells of Li/PE/Li, Li/PPE/Li, and Li/PPEL/Li, with the results shown in Figure 5a. For the Li/PE/Li symmetric cell, the polarization voltage initially reached 0.58 V. However, it decreased afterward due to the reaction of a small amount of electrolyte added at the beginning, which reacted with the electrodes. The polarization voltage stabilized at 0.26 V between 100 and 250 h (Figure 5b). However, after 250 h, the voltage started to rise, indicating the formation of lithium dendrites and an increase in interface resistance. In the Li/PPE/Li symmetric cell, the polarization voltage remained stable at 0.23 V for up to 350 h (Figure 5c), but with continued cycling, the overpotential rapidly increased. The poor mechanical properties of the GPEs (PE and PPE) led to uneven lithium plating and stripping, preventing stable polarization voltage. In contrast, the Li/PPEL/Li symmetric cell demonstrated a highly stable polarization voltage of 0.16 V over 600 h of cycling at a current density of 0.1 mA cm^{-2} (Figure 5d). This suggests a stable SEI (solid electrolyte interphase) layer formed between the electrode and electrolyte. Additionally, it can be concluded that adding LLZTO improves the interface stability between the solid-state electrolyte and the lithium metal anode. This is likely due to the LLZTO particles acting as a barrier, preventing reactions between the lithium anode and the polymer PVDF-HFP. These electrochemical performance results are highly competitive compared to previously reported GPEs designs.

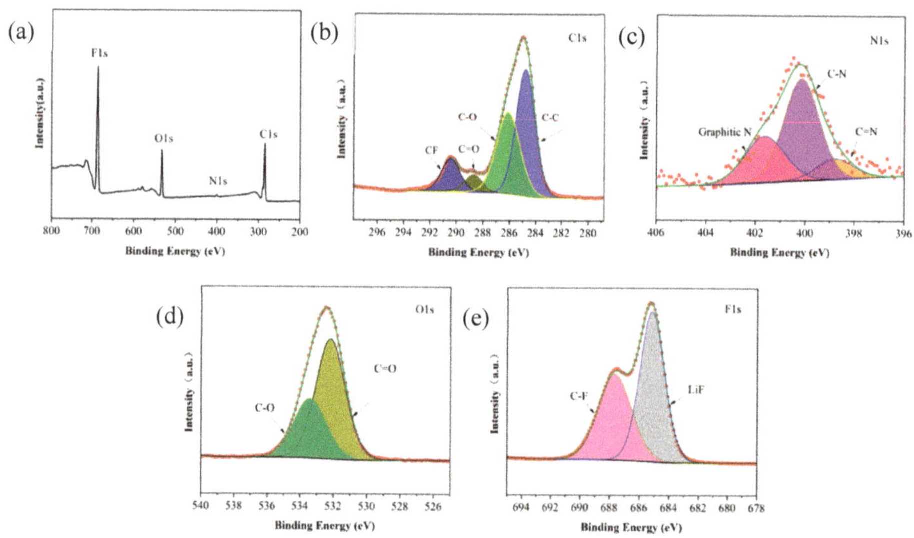

Figure 4. XPS spectra of PPEL GPEs after electrolyte membrane recycling. (**a**) Full spectrum diagram; (**b**) C1s; (**c**) C1s; (**d**) O1s; (**e**) N1s.

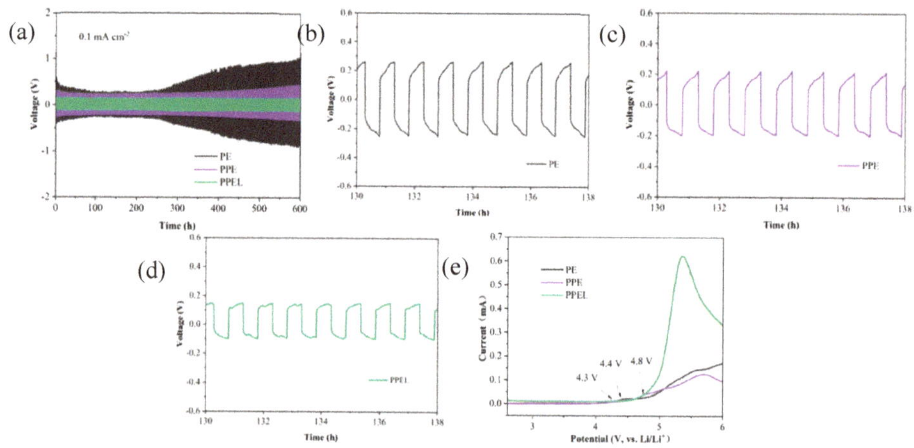

Figure 5. (**a**) Cycling performance of plating / lithium stripping at Li/PE, PE/PE, and PPEL/Li at 0.1 mA/cm^2 current density; polarization voltage at 130–138 h (**b**) PPE; (**c**) PPE; (**d**) PPEL; (**e**) LSV profiles of GPEs membranes.

In pursuing high stability and safety in lithium-ion battery technology, the electrochemical stability of GPEs plays a crucial role. As the voltage increases, electrolyte decomposition occurs. When a certain threshold is reached, intense redox reactions inside the battery are triggered, leading to a rapid increase in current. The results are shown in Figure 5e. For the PE electrolyte, the current only starts to increase gradually at 4.3 V. For PPE, the current begins to rise at 4.4 V, while for PPEL, the current sharply increases after 4.8 V. The sharp rise in current indicates the onset of oxidative decomposition of the electrolyte. This demonstrates that PPEL has a larger electrochemical window and superior electrochemical stability, making it highly promising for high-voltage lithium batteries.

To evaluate the practical feasibility of various GPEs, LFP/GPEs/Li coin batteries were assembled and tested at 25 °C. The rate performance of the batteries with three different electrolytes was studied over a current density range from 0.1 C to 1 C, as shown in Figure 6a. At 0.1 C, the PPEL battery delivered the highest discharge-specific capacity (164.3 mAh g^{-1}). In contrast, the PE and PPE cells showed initial discharge capacities of 120.1 and 135.2 mAh g^{-1}, respectively, with a difference of 44.2 and 29.1 mAh g^{-1} compared to PPEL. The capacity difference between the three solid-state electrolytes widened as the current density increased. The PPEL cell provided discharge capacities of 150.1, 129.4, and 88.8 mAh g^{-1} at 0.2 C, 0.5 C, and 1 C, respectively, while the PE cell showed capacities of 89.5 and 61.2 mAh g^{-1} at 0.2 C and 0.5 C, with no measurable capacity at 1 C. The PPE cell showed discharge capacities of 101.2, 68.5, and 20.3 mAh g^{-1} at the corresponding current densities. Upon returning to 0.1 C, the PPEL cell nearly regained its initial discharge capacity (163.5 mAh g^{-1}). In contrast, the PE and PPE cells maintained 115.9 and 134.4 mAh g^{-1}, representing 96.5% and 99.4% of their initial discharge capacity, respectively. The superior performance of PPEL over PE and PPE is attributed to the combined properties of PAN's high absorptivity and the specific F-groups on the surface of PVDF-HFP, which facilitate lithium-ion transport. The charge–discharge curves of the PPEL cell at various current densities are shown in Figure 6b, where the observed low overpotentials indicate minimal polarization, which is beneficial for a long cycle life. These results demonstrate that the PVDF-HFP/PAN double polymer matrix significantly enhances Li^+ transport, making it highly suitable for high-power-density energy storage systems. Additionally, the room temperature cycling performance of batteries with the three electrolytes was tested at a load of 2 mA cm^{-2} over 200 cycles at a current density of 0.5 C (Figure 6c). The initial discharge capacities were 60.4, 68.4, and 129.8 mAh g^{-1} for the PE, PPE, and PPEL cells, respectively. The PE cell short-circuited after 143 cycles, while the PPE cell retained a discharge capacity of 36.5 mAh g^{-1} after 200 cycles, with a capacity retention of 53.4%. In contrast, the PPEL cell maintained a discharge capacity of 116 mAh g^{-1} after 200 cycles, with a capacity retention of 89.4%. This indicates that the PPEL electrolyte provides superior long-term cycling stability due to the chemical/electrochemical equilibrium at the interface. Figure 6d shows the cyclic voltammetry (CV) curves of the LFP/PPEL/Li battery in the voltage range of 2.8 V–4.0 V (vs. Li/Li^+) at a scan rate of 0.2 mV s^{-1}. Two characteristic peaks were observed at 3.1 V and 3.7 V during the reduction and oxidation scans, respectively. The minimal change in the curves during cycling indicates that no significant polarization occurred, demonstrating that the PPEL-based battery exhibits high reversible charge–discharge behavior and excellent electrochemical stability. To better visualize the overall improvement, Table 1 provides a comparative summary of the key performance metrics for PE, PPE, and PPEL electrolytes.

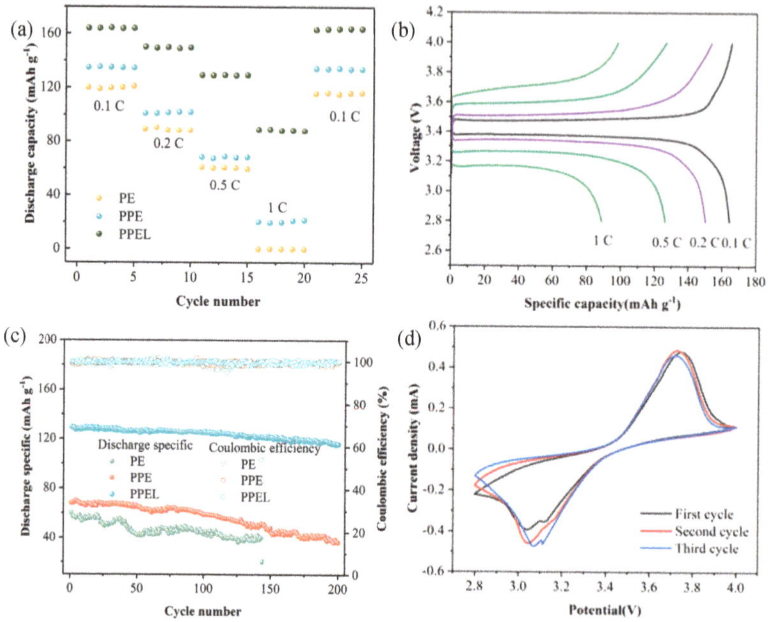

Figure 6. (**a**) The rate capability of PE-based LFP/Li, PPE 3:7-based LFP/Li, and PPEL-based LFP/Li batteries at 25 °C; (**b**) galvanostatic charge-discharge curves of PPEL-based LFP/Li batteries at 0.1, 0.2, 0.5, and 1 C; (**c**) cycling performances of PE-based LFP/Li, PPE-based LFP/Li, and PPEL-based LFP/Li batteries at 25 °C; (**d**) CV curves of lithium-ion batteries with PPEL electrolyte at a scan rate of 0.2 mV s^{-1}.

Table 1. Performance comparison of PE, PPE, and PPEL.

GPEs	Mechanical Stability (MPa)	Ion Conductivity (S/cm)	Electrochemical Window (V)	Discharge Capacities (mAh/g)	Capacity Retention (0.2 C, 200 Cycles)
PE	2.16	9.1×10^{-5}	4.3	120.1	short-circuited after 143 cycles
PPE	3.1	6.1×10^{-5}	4.4	135.2	53.4%
PPEL	4.35	2.8×10^{-4}	4.8	164.3	89.4%

3. Conclusions

In summary, we successfully prepared high-performance GPEs by blending PVDF-HFP and PAN via UV curing, cross-linking with ETPTA, and incorporating LLZTO. This design significantly enhances lithium-ion batteries' safety, functionality, and electrochemical performance. At room temperature, the GPEs exhibits an ionic conductivity of 2.8×10^{-4} S/cm, a lithium-ion transference number of 0.6, an electrochemical window of 4.8 V, and a tensile strength of 4.3 MPa. Furthermore, in the lithium plating/stripping test, the assembled Li/PPEL/Li symmetrical cell demonstrated a stable cycling performance over 600 h at a current density of 0.1 mA cm^{-2}. Notably, the GPEs enabled the $LiFePO_4$/GPEs/Li lithium battery to achieve excellent performance, including high discharge capacities under high current densities (164.3 mAh g^{-1} at 0.1 C and 88.8 mAh g^{-1} at 1 C), and good cycling stability (capacity retention of 89.4% after 200 cycles at 0.5 C, with an average coulombic efficiency of 99.6%). Compared with other PVDF-HFP-based electrolyte, our system exhibits higher ionic conductivity and longer cycling stability, highlighting the advantage of our UV-curing and filler dispersion strategy. This work provides a practical approach for high-performance, safe solid-state lithium-ion batteries. Therefore, GPEs based on this design have the potential to exhibit promising electrochemical performance and contribute to the practical development of safe, high-energy lithium-ion batteries.

4. Materials and Methods

4.1. Preparation of Gel Polymer Electrolytes (GPEs)

Preparation of PE: 2 g of PVDF-HFP (Mn = 600,000, Arkema, Colombes, France) was dissolved in DMF solvent, and $LiClO_4$ (99.99%, Aladdin Chemical Co., Shanghai, China) was added and thoroughly stirred. Then, an appropriate amount of ETPTA and HMPP was introduced, and the mixture was stirred for 5 min. The resulting precursor solution was poured onto a PTFE plate and subjected to UV curing. Finally, the cured membrane was placed in a drying oven for 6 h. The prepared GPE membrane is named PE and has a thickness of approximately 50 μm.

Preparation of PPE: 2 g of PVDF-HFP was dissolved in DMF solvent along with poly(acrylonitrile) (PAN, Mw = 85000, Shanghai, China) in varying ratios to PVDF-HFP (1:1, 2:1, 3:1, 4:1). After thorough mixing, $LiClO_4$ was added. An appropriate amount of Plasticizers trimethylolpropane ethoxylate triacrylate (ETPTA, Mn = 428, Macklin, Shanghai, China) and 2-hydroxy-2-methylpropiophenone (HMPP, Cambridge, Beijing, China) was then introduced, and the mixture was stirred for 5 min. The precursor solution was poured onto a PTFE plate, subjected to UV curing, and dried in an oven for 6 h. The resulting GPEs film is named PEL and is approximately 50 μm thick.

Preparation of PPEL: 2 g of PVDF-HFP and PAN (in a weight ratio 2:1) were dissolved in DMF solvent. After adding $LiClO_4$ and mixing thoroughly, LLZTO was incorporated at different weight ratios (5 wt.%, 10 wt.%, 15 wt.%, 20 wt.) into the solution. An appropriate amount of ETPTA and HMPP was then added, stirring the mixture for 5 min. The precursor solution was poured onto a PTFE plate and subjected to UV curing. Finally, the membrane was dried in an oven for 6 h. The resulting GPE membrane is named PPEL and is approximately 50 μm thick, and a schematic of its preparation is shown in Figure 7.

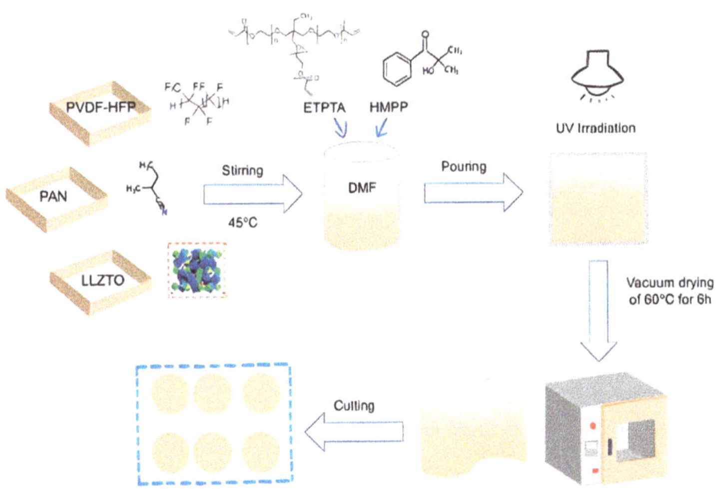

Figure 7. Preparation of PPEL GPEs membrane.

4.2. Material Characterization

The cut Gel Polymer Electrolytes (GPEs) were characterized using X-ray diffraction (XRD, DX-2700, Dandong, China) with Cu-Kα radiation (40 kV × 30 mA). Surface and cross-sectional morphology of the GPEs was examined using scanning electron microscopy (SEM, Phenom Spectra G2, Shanghai, China). Cross-sectional samples were prepared by embrittling the GPEs with liquid nitrogen, followed by gold coating for imaging. The tensile strength of the rectangular GPEs samples was measured using a universal testing machine (WDW-5, Tenson, Jinan, China). Thermogravimetric analysis (TGA) was conducted under a nitrogen atmosphere using a Netzsch F3 Tarsus system (Beyern, Germany), with a temperature range of 30–800 °C and a heating rate of 10 °C/min. Fourier transform infrared spectroscopy (FTIR, Spectrum 100, PerkinElmer, Waltham, MA, USA) was used to identify the functional groups on the surface of the cut GPEs, within the wavenumber range of 400–4000 cm^{-1}. X-ray photoelectron spectroscopy (XPS, Thermo Fisher Scientific K-Alpha+, Waltham, MA, USA) was employed to assess the bonding structure of the GPEs after cycling.

4.3. Electrochemical Properties

Electrochemical impedance spectroscopy (EIS) was performed at various temperatures, with an open-circuit voltage (OCV) ranging from 0.1 Hz to 1 MHz and an amplitude of 10 mV. GPEs were placed between two smooth stainless steel (SS) electrodes. Ionic conductivity (σ) of the electrolytes was determined from the EIS data using Equation (1):

$$\sigma = \frac{L}{RS} \tag{1}$$

In this equation, R represents the bulk resistance measured by EIS, while L and S denote the thickness and effective area of the GPEs, respectively. The activation energy (E_a) of each solid electrolyte is determined using the Arrhenius equation (Equation (2)):

$$\sigma(T) = Aexp\left[-\frac{E_a}{RT}\right] \tag{2}$$

Here, A is the exponential prefactor, E_a represents the activation energy associated with the ion skipping conduction process, and T is the absolute temperature. The electrochemical stability windows of different GPEs were assessed using linear scanning voltammetry (LSV) across a voltage range of 2 V to 6 V, the scan rate is 5 mV/s. Stainless steel (SS) was used as the working electrode and lithium metal as the counter/reference electrode. SS/GPEs/Li assemblies were tested in CR2025 coin cells for LSV measurements. To evaluate the stability of various solid-state electrolytes in contact with the lithium anode, two polished lithium foils were used in a symmetrical cell configuration. To measure the lithium-ion transference number (${t_{Li}}^+$), the same symmetrical cells containing different electrolytes were tested using the chronoamperometry method, with a 10 mV polarization applied for 4000 s. Simultaneously, the AC impedance spectra before and after polarization were recorded at an oscillation voltage of 10 mV, with a frequency range of 0.1 Hz to 1 MHz, and ${t_{Li}}^+$ was calculated using Equation (3):

$$t_{Li^+} = \frac{I_s(\Delta V - R_0 I_0)}{I_0(\Delta V - R_s I_s)} \tag{3}$$

In this equation, I_0 and I_S represent the initial and steady-state currents, respectively, R_0 and R_S are the charge transfer resistances of the electrolyte system before and after polarization, and ΔV is the applied oscillation voltage of 10 mV. LFP/GPEs/Li configurations were assembled in CR2025 coin cells for Cyclic Voltammetry (CV) testing. The CV tests were performed at a scan rate of 0.2 mV/s.

Rate and cycle tests of the batteries were performed using a battery testing system (Neware, Dongguan, China). Commercial lithium iron phosphate (LFP) was used as the positive active material, and a lithium sheet served as the negative electrode. The positive electrode mixture consisted of $LiFePO_4$ (LFP, ≥99.5%, Krud, Shanghai, China), PVDF, and conductive carbon (Super-P, ≥99.5%, Krud, Shanghai, China) in a weight ratio of 8:1:1. This mixture was dissolved in a solvent, and the resulting solution was applied to aluminum foil. The coated foil was then vacuum-dried for 36 h and cut into 14 mm discs for use in the tests. The mass loading was 2.4 mg/cm^2. The electrochemical performance was evaluated within a voltage range of 2.8 to 4.0 V.

Author Contributions: Conceptualization, X.Y. and L.L.; methodology, Q.H. and P.S.; software, X.Y. and J.W.; validation, X.Y., Y.L. and X.L. (Xinghua Liang); formal analysis, X.L. (Xuanan Lu) and P.S.; investigation, Q.H. and X.L. (Xuanan Lu); resources, X.L. (Xinghua Liang); data curation, Q.H.; writing—original draft preparation, X.Y.; writing—review and editing, L.L. and J.W.; visualization, X.L. (Xinghua Liang) and Y.L.; supervision, Q.H. and Y.G.; project administration, Y.L.; funding acquisition, X.L. (Xinghua Liang), Y.L. and Y.G. All authors have read and agreed to the published version of the manuscript.

Funding: This research was funded by the Reginal Collaboration R&D Program of Sichuan Province under Grant (No. 2024YFHZ0209), Fund Project of the Guangxi Key R&D Programme Projects (No. AB24010346), the National Natural Science Foundation of China (No. 22462003, 52161033, 22262005), Guangxi key laboratory of Automobile Components and Vehicle technology (2023GK-LACVTZZ02), Guangxi Key Laboratory of Automobile Components and Vehicle Technology (No. 2022GKLACVTZZ04, 2023GKLACVTZZ10), and Guangxi University Young and Middle-Aged Teachers' Basic Scientific Research Ability Improvement Project (2021KY1715).

Institutional Review Board Statement: Not applicable.

Informed Consent Statement: Not applicable.

Data Availability Statement: Data are contained within the article.

Conflicts of Interest: The authors declare no conflicts of interest.

References

1. Niu, H.; Zhang, N.; Lu, Y.; Zhang, Z.; Li, M.; Liu, J.; Zhang, N.; Song, W.; Zhao, Y.; Miao, Z. Strategies toward the development of high-energy-density lithium batteries. *Microelectron. J.* **2024**, *88*, 111666. [CrossRef]
2. Xu, J.; Cai, X.; Cai, S.; Shao, Y.; Hu, C.; Lu, S.; Ding, S. High-Energy Lithium-Ion Batteries: Recent Progress and a Promising Future in Applications. *Energy Environ. Mater.* **2023**, *6*, e12450. [CrossRef]
3. Liu, R.; Xue, J.; Xie, L.; Chen, H.; Deng, Z.; Yin, W. A Flexible Yet Robust 3D-Hybrid Gel Solid-State Electrolyte Based on Metal–Organic Frameworks for Rechargeable Lithium Metal Batteries. *Gels* **2024**, *10*, 812. [CrossRef]
4. Gao, X.; Guo, C.; Xu, S.; Song, H. Stretchable ionic conductive gels for wearable human-activity detection. *Chem. Eng. J.* **2024**, *489*, 151231. [CrossRef]
5. Lu, J.; Chen, Y.; Lei, Y.; Jaumaux, P.; Tian, H.; Wang, G. Quasi-Solid Gel Electrolytes for Alkali Metal Battery Applications. *Nano-Micro Lett.* **2025**, *17*, 194. [CrossRef] [PubMed]
6. Lu, X.; Mo, Z.; Liu, Z.; Hu, Y.; Du, C.; Liang, L.; Liu, Z.; Chen, G. Robust, Efficient, and Recoverable Thermocells with Zwitterion-Boosted Hydrogel Electrolytes for Energy-Autonomous and Wearable Sensing. *Angew. Chem. Int. Ed.* **2024**, *63*, e202405357. [CrossRef] [PubMed]
7. Sha, D.; Ding, D.; Tang, S.; Ma, Z.; Liu, C.; Yuan, Y. Solvent-Triggered, Ultra-Adhesive, Conductive, and Biocompatible Transition Gels for Wearable Devices. *Small* **2024**, *20*, 2310731. [CrossRef]
8. Tordi, P.; Tamayo, A.; Jeong, Y.; Bonini, M.; Samorì, P. Multiresponsive Ionic Conductive Alginate/Gelatin Organohydrogels with Tunable Functions. *Adv. Funct. Mater.* **2024**, *34*, 2410663. [CrossRef]
9. Kainat, S.; Anwer, J.; Hamid, A.; Gull, N.; Khan, S.M. Electrolytes in Lithium-Ion Batteries: Advancements in the Era of Twenties (2020's). *Mater. Chem. Phys.* **2024**, *313*, 13. [CrossRef]
10. Ji, W.; Luo, B.; Wang, Q.; Yu, G.; Zhang, Z.; Tian, Y.; Zhao, Z.; Zhao, R.; Wang, S.; Wang, X. Interface engineering enabling thin lithium metal electrodes down to 0.78 μm for garnet-type solid-state batteries. *Nat. Commun.* **2024**, *15*, 9920. [CrossRef]

11. Hu, J.K.; Gao, Y.C.; Yang, S.J.; Wang, X.L.; Chen, X.; Liao, Y.L.; Li, S.; Liu, J.; Yuan, H.; Huang, J.Q. High Energy Density Solid-State Lithium Metal Batteries Enabled by In Situ Polymerized Integrated Ultrathin Solid Electrolyte/Cathode. *Adv. Funct. Mater.* **2024**, *34*, 2311633. [CrossRef]
12. Gao, M.; Li, H.; Xu, L.; Xue, Q.; Wang, X.; Bai, Y.; Wu, C. Lithium metal batteries for high energy density: Fundamental electrochemistry and challenges. *J. Energy Chem.* **2021**, *59*, 666–687. [CrossRef]
13. Deng, K.; Zeng, Q.; Wang, D.; Liu, Z.; Meng, Y. Nonflammable organic electrolytes for high-safety lithium-ion batteries. *Energy Storage Mater.* **2020**, *32*, 425–447. [CrossRef]
14. Castillo, J.; Robles-Fernandez, A.; Cid, R.; González-Marcos, J.A.; Armand, M.; Carriazo, D.; Zhang, H.; Santiago, A. Dehydrofluorination Process of Poly(vinylidene difluoride) PVdF-Based Gel Polymer Electrolytes and Its Effect on Lithium-Sulfur Batteries. *Gels* **2023**, *9*, 336. [CrossRef] [PubMed]
15. Huang, J.; Li, C.; Jiang, D.; Gao, J.; Cheng, L.; Li, G.; Luo, H.; Xu, Z.L.; Shin, D.M.; Wang, Y. Solid-State Electrolytes for Lithium Metal Batteries: State-of-the-Art and Perspectives. *Adv. Funct. Mater.* **2025**, *35*, 2411171. [CrossRef]
16. Zhao, Q.; Stalin, S.; Zhao, C.Z.; Archer, L.A. Designing solid-state electrolytes for safe, energy-dense batteries. *Nat. Rev. Mater.* **2020**, *5*, 229–252. [CrossRef]
17. Ding, J.; Yang, Y.; Poisson, J.; He, Y.; Zhang, H.; Zhang, Y.; Bao, Y.; Chen, S.; Chen, Y.M.; Zhang, K. Recent Advances in Biopolymer-Based Hydrogel Electrolytes for Flexible Supercapacitors. *ACS Energy Lett.* **2024**, *9*, 1803–1825. [CrossRef]
18. Hao, Q.; Ma, X.; Gao, Y.; Chen, F.; Chen, X.; Qi, Y.; Li, N. Commercial carbonate based gel polymer electrolytes enable safe and stable high-voltage Li-metal batteries. *Energy Storage Mater.* **2024**, *70*, 103509. [CrossRef]
19. Jia, R.; Wei, C.; Ma, B.; Li, L.; Yang, C.; Wang, B.; Tan, L.; Feng, J. Biopolymer-Based Gel Electrolytes for Advanced Zinc Ion Batteries: Progress and Perspectives. *Adv. Funct. Mater.* **2025**, *35*, 2417498. [CrossRef]
20. Tordi, P.; Ridi, F.; Samorì, P.; Bonini, M. Cation-Alginate Complexes and Their Hydrogels: A Powerful Toolkit for the Development of Next-Generation Sustainable Functional Materials. *Adv. Funct. Mater.* **2025**, *35*, 2416390. [CrossRef]
21. Yang, W.; Yang, W.; Zeng, J.; Chen, Y.; Huang, Y.; Liu, J.; Gan, J.; Li, T.; Zhang, H.; Zhong, L.; et al. Biopolymer-based gel electrolytes for electrochemical energy Storage: Advances and prospects. *Prog. Mater. Sci.* **2024**, *144*, 101264. [CrossRef]
22. St-Onge, V.; Cui, M.; Rochon, S.; Daigle, J.-C.; Claverie, J.P. Reducing crystallinity in solid polymer electrolytes for lithium-metal batteries via statistical copolymerization. *Commun. Mater.* **2021**, *2*, 83. [CrossRef]
23. Islam, M.; Ahmed, M.S.; Faizan, M.; Ali, B.; Bhuyan, M.M.; Bari, G.A.K.M.R.; Nam, K.W. Review on the Polymeric and Chelate Gel Precursor for Li-Ion Battery Cathode Material Synthesis. *Gels* **2024**, *10*, 586. [CrossRef] [PubMed]
24. Ahmed, M.S.; Islam, M.; Raut, B.; Yun, S.; Kim, H.Y.; Nam, K.-W. A comprehensive review of functional gel polymer electrolytes and applications in lithium-ion battery. *Gels* **2024**, *10*, 563. [CrossRef]
25. Li, S.; Zhang, S.Q.; Shen, L.; Liu, Q.; Ma, J.B.; Lv, W.; He, Y.B.; Yang, Q.H. Progress and Perspective of Ceramic/Polymer Composite Solid Electrolytes for Lithium Batteries. *Adv. Sci.* **2020**, *7*, 1903088. [CrossRef]
26. Li, S.; Li, L.; Yang, H.; Zhao, Y.; Shan, Y. A review of composite polymer electrolytes for solid-state lithium-sulfur batteries: Synthesis methods, optimal design, and critical challenges. *Chem. Eng. J.* **2024**, *484*, 149433. [CrossRef]
27. Ma, X.; Liu, M.; Wu, Q.; Guan, X.; Wang, F.; Liu, H.; Xu, J. Composite Electrolytes Prepared by Improving the Interfacial Compatibility of Organic–Inorganic Electrolytes for Dendrite-Free, Long-Life All-Solid Lithium Metal Batteries. *ACS Appl. Mater. Interfaces* **2022**, *14*, 53828–53839. [CrossRef]
28. Halder, B.; Mohamed, M.G.; Kuo, S.-W.; Elumalai, P. Review on composite polymer electrolyte using PVDF-HFP for solid-state lithium-ion battery. *Mater. Today Chem.* **2024**, *36*, 101926. [CrossRef]
29. Jie, J.; Liu, Y.; Cong, L.; Zhang, B.; Lu, W.; Zhang, X.; Liu, J.; Xie, H.; Sun, L. High-performance PVDF-HFP based gel polymer electrolyte with a safe solvent in Li metal polymer battery. *Joural Energy Chem.* **2020**, *29*, 9. [CrossRef]
30. Liu, L.; Wang, X.; Yang, C.; Han, P.; Zhang, L.; Gao, L.; Wu, Z.; Liu, B.; Liu, R. PVdF-HFP-Based Gel Polymer Electrolyte with Semi-Interpenetrating Networks For Dendrite-Free Lithium Metal Battery. *Acta Metall. Sin.* **2021**, *34*, 417–424. [CrossRef]
31. Yadav, P.; Hosen, M.S.; Dammala, P.K.; Ivanchenko, P.; Van Mierlo, J.; Berecibar, M. Development of composite solid polymer electrolyte for solid-state lithium battery: Incorporating LLZTO in PVDF-HFP/LiTFSI. *Solid State Ion.* **2023**, *399*, 116308. [CrossRef]
32. Lee, S.; Park, H.; Kim, J.Y.; Kim, J.; Choi, M.J.; Han, S.; Kim, S.; Kim, W.; Jang, H.W.; Park, J. Unveiling crystal orientation-dependent interface property in composite cathodes for solid-state batteries by in situ microscopic probe. *Nat. Commun.* **2024**, *15*, 7947. [CrossRef]
33. Lei, J.; Lu, H.; Chen, J.; Yang, J.; Nuli, Y.; Wang, J. Crosslinked polyacrylonitrile precursor for S@pPAN composite cathode materials for rechargeable lithium batteries. *J. Energy Chem.* **2022**, *65*, 186–193. [CrossRef]
34. Gao, Y.; Sun, Z.; Cui, C.; Wang, H.; Cao, W.; Hou, Z.; Zhu, D.; Yang, Y.; Zhang, T. An Ultrathin, Flexible Solid Electrolyte with High Ionic Conductivity Enhanced by a Mutual Promotion Mechanism. *ACS Appl. Mater. Interfaces* **2022**, *14*, 45373–45381. [CrossRef]
35. Wu, Y.; Zhu, T.; Lv, Y.; Fang, J.; Dong, S.; Yao, S. "Polymer in ceramic" type LLZTO/PEO/PVDF composite electrolyte with high lithium migration number for solid-state lithium batteries. *Ionics* **2024**, *30*, 787–798. [CrossRef]

36. Shan, C.; Wang, R.; Zhang, X.; Li, L.; Sun, W.; Gao, M.; Pan, H.; Liu, Y. Remarkably improved ionic conductivity and lithium metal compatibility of LLZTO-Li4 (BH4) 3I composite solid-state electrolyte. *J. Alloys Compd.* **2024**, *980*, 173554. [CrossRef]
37. Busche, M.R.; Weiss, M.; Leichtweiss, T.; Fiedler, C.; Drossel, T.; Geiss, M.; Kronenberger, A.; Weber, D.A.; Janek, J. The Formation of the Solid/Liquid Electrolyte Interphase (SLEI) on NASICON-Type Glass Ceramics and LiPON. *Adv. Mater. Interfaces* **2020**, *7*, 2000380. [CrossRef]
38. Seol, J.-c.; Balasubramaniam, R.; Aravindan, V.; Thangavel, R.; Lee, Y.-S. Ameliorating the electrode/electrolyte interface compatibility in Li-ion solid-state batteries with plasticizer. *J. Alloys Compd.* **2022**, *927*, 167077. [CrossRef]
39. Jia, H.; Li, X.; Song, J.; Zhang, X.; Zhang, J.G. Hierarchical porous silicon structures with extraordinary mechanical strength as high-performance lithium-ion battery anodes. *Nat. Commun.* **2020**, *11*, 1474. [CrossRef]
40. Hun, Q.; Lan, L.; Lu, X.; Hu, Q.; Liang, X.; Guo, Y.; Wang, Y. Bilayer Heterostructure Electrolytes Were Prepared by a UV-Curing Process for High Temperature Lithium-Ion Batteries. *Polymers* **2024**, *16*, 2972. [CrossRef]
41. Guan, D.; Huang, Y.; He, M.; Hu, G.; Du, K. Multilayer PEO/LLZTO composite electrolyte enables high-performance solid-state Li-ion batteries. *Ionics* **2021**, *27*, 4127–4134. [CrossRef]
42. Zhuang, H.; Ma, W.; Xie, J.; Liu, X.; Li, B.; Jiang, Y.; Huang, S.; Chen, Z.; Zhao, B. Solvent-free synthesis of PEO/garnet composite electrolyte for high-safety all-solid-state lithium batteries. *J. Alloys Compd.* **2021**, *860*, 157915. [CrossRef]
43. Yang, X.; Liu, J.; Pei, N.; Chen, Z.; Li, R.; Fu, L.; Zhang, P.; Zhao, J. The critical role of fillers in composite polymer electrolytes for lithium battery. *Nano-Micro Lett.* **2023**, *15*, 74. [CrossRef]
44. Yoon, S.A.; Oh, N.R.; Yoo, A.R.; Lee, H.G.; Lee, H.C. Preparation and Characterization of Ta-substituted Li7La3Zr2-xO12 Garnet Solid Electrolyte by Sol-Gel Processing. *J. Korean Ceram. Soc.* **2017**, *54*, 278–284. [CrossRef]
45. Tong, B.; Song, Z.; Wu, H.; Wang, X.; Feng, W.; Zhou, Z.; Zhang, H. Ion transport and structural design of lithium-ion conductive solid polymer electrolytes:a perspective. *Mater. Future* **2022**, *1*, 042103. [CrossRef]
46. Zhou, P.; Zhang, X.; Xiang, Y.; Liu, K. Strategies to enhance Li^(+) transference number in liquid electrolytes for better lithium batteries. *Nano Res.* **2023**, *16*, 8055–8071. [CrossRef]
47. Lee, Y.G.; Hong, S.; Pan, B.; Wu, X.; Dickey, E.C.; Whitacre, J.F. Improving Bulk and Interfacial Lithium Transport in Garnet-Type Solid Electrolytes through Microstructure Optimization for High-Performance All-Solid-State Batteries. *ACS Appl. Mater. Interfaces* **2024**, *16*, 60340–60347. [CrossRef]
48. Xu, K.; Xu, C.; Jiang, Y.; Cai, J.; Ni, J.; Lai, C. Sandwich structured PVDF-HFP-based composite solid electrolytes for solid-state lithium metal batteries. *Ionics* **2022**, *28*, 3243–3253. [CrossRef]
49. Duan, T.; Cheng, H.; Liu, Y.; Sun, Q.; Nie, W.; Lu, X.; Dong, P.; Song, M.-K. A multifunctional Janus layer for LLZTO/PEO composite electrolyte with enhanced interfacial stability in solid-state lithium metal batteries. *Energy Storage Mater.* **2024**, *65*, 103091. [CrossRef]
50. Zhang, Y.; Qiao, R.; Nie, Q.; Zhao, P.; Li, Y.; Hong, Y.; Chen, S.; Li, C.; Sun, B.; Fan, H. Synergetic regulation of SEI mechanics and crystallographic orientation for stable lithium metal pouch cells. *Nat. Commun.* **2024**, *15*, 4454. [CrossRef]
51. Hood, Z.D.; Chen, X.; Sacci, R.L.; Liu, X.; Veith, G.M.; Mo, Y.; Niu, J.; Dudney, N.J.; Chi, M. Elucidating Interfacial Stability between Lithium Metal Anode and Li Phosphorus Oxynitride via In Situ Electron Microscopy. *Nano Lett.* **2021**, *21*, 151–157. [CrossRef] [PubMed]

gels

MDPI

Article

Preparation and Failure Behavior of Gel Electrolytes for Multilayer Structure Lithium Metal Solid-State Batteries

Chu Chen [1,†], Wendong Qin [1,†], Qiankun Hun [1], Yujiang Wang [1,*], Xinghua Liang [1,*], Renji Tan [1], Junming Li [1] and Yifeng Guo [2]

1 Guangxi Key Laboratory of Automobile Components and Vehicle Technology, Guangxi University of Science & Technology, Liuzhou 545006, China; 13434119278@163.com (C.C.); qinwendong@163.com (W.Q.); 13383371921@163.com (Q.H.); 18068533778@163.com (R.T.); jmli@gxust.edu.cn (J.L.)
2 School of Mechanical Engineering, Chengdu University, Chengdu 610106, China; guobujia2000@163.com
* Correspondence: 13152528815@163.com (Y.W.); lxh304@126.com (X.L.)
† These authors contributed equally to this work.

Abstract

High safety gel polymer electrolyte (GPE) is used in lithium metal solid state batteries, which has the advantages of high energy density, wide temperature range, high safety, and is considered as a subversive new generation battery technology. However, solid-state lithium batteries with multiple layers and large capacity currently have poor cycle life and a large gap between the actual output cycle capacity retention rate and the theoretical level. In this paper, polyvinylidene fluoride-hexafluoropropylene (PVDF-HFP)/polyacrylonitrile (PAN)—lithium perchlorate ($LiClO_4$)—lithium lanthanum zirconium tantalate (LLZTO) gel polymer electrolytes was prepared by UV curing process using a UV curing machine at a speed of 0.01 m/min for 10 s, with the temperature controlled at 30 °C and wavelength 365 nm. In order to study the performance and failure mechanism of multilayer solid state batteries, single and three layers of solid state batteries with ceramic/polymer composite gel electrolyte were assembled. The results show that the rate and cycle performance of single-layer solid state battery with gel electrolyte are better than those of three-layer solid state battery. As the number of cycles increases, the interface impedance of both single-layer and three-layer electrolyte membrane solid-state batteries shows an increasing trend. Specifically, the three-layer battery impedance increased from 17 Ω to 42 Ω after 100 cycles, while the single-layer battery showed a smaller increase, from 2.2 Ω to 4.8 Ω, indicating better interfacial stability. After 100 cycles, the interface impedance of multi-layer solid-state batteries increases by 9.61 times that of single-layer batteries. After 100 cycles, the corresponding capacity retention rates were 48.9% and 15.6%, respectively. This work provides a new strategy for large capacity solid state batteries with gel electrolyte design.

Keywords: gel polymer electrolyte; multilayer structure; lithium metal; performance and failure behavior; solid-state batteries

1. Introduction

With the rapid development of electronic technology, the energy density of traditional batteries can no longer meet the modern demand for high-energy batteries. Currently, the development of batteries faces two main challenges: how to improve the energy density of batteries and how to enhance battery safety. Compared with traditional liquid lithium-ion batteries, the most remarkable feature of solid-state batteries is the use of solid electrolytes to replace flammable liquid electrolytes. In terms of safety, when liquid batteries encounter overcharging, overheating, or external impacts, the flammable liquid electrolytes are extremely prone to causing fires or even explosions. Solid-state batteries, however, fundamentally avoid this risk, significantly improving operational safety. In terms of energy density, solid-state batteries perform even more impressively. Under the same volume or weight, they can store more electrical energy. Additionally, solid-state batteries have a longer cycle life, capable of withstanding more charge-discharge cycles, which effectively reduces battery performance degradation and lowers the cost and frequency of battery replacement. Precisely due to these outstanding advantages, solid-state batteries have become the focus of research and development for numerous scientific teams and enterprises [1–6]. Currently, polymer electrolytes [7–13], ceramics materials [14–20] and ceramic/polymer composite electrolytes [21–26] have been intensively studied. Although rubbery plastic electrolytes like that are easy to process, they are more than 60 microns in thickness, resulting in more useless space in the battery. Besides, used on the market. Ceramic materials have good electrical conductivity, but are brittle and hard, and they are prone to cracking when slightly bent. As for all-solid-state lithium batteries, although their safety has been improved, they still cannot beat the current liquid batteries in terms of fast charging speed and energy storage capabilities. There is a very interesting contradiction here. It seems that one of them must be sacrificed for safety and high performance. How to find the balance point has become the focus of research. However, the larger thickness of the gel polymer electrolyte (>60 μm) increases the volume proportion of inactive components in the battery, and its electrical conductivity is much worse than the liquid electrolytes. As for all-solid-state lithium batteries (ASSLB), although their safety has been improved, they still cannot beat the current liquid batteries in terms of fast charging speed and energy storage capabilities [27–30].

Gel polymer electrolyte membranes (GPEs) are similar to existing commercial separators in terms of thickness and cycle life, which lays the foundation for the rapid promotion of long-life solid-state batteries [31]. This type of material also has the characteristics of light weight and low processing equipment cost, especially its relatively simple preparation process and an electrochemical window that can be adjusted. These characteristics make it an ideal option for large-scale production of solid-state batteries [32,33]. Specifically, polyvinylidene fluoride (PVDF) materials have a promoting effect on lithium-ion transmission because they contain strong polar C-F bonds and high dielectric constants. The PVDF-HFP copolymer formed by the introduction of hexafluoropropylene can not only effectively dissolve the lithium salt, but also exhibit excellent chemical stability and film-forming characteristics. Polyacrylonitrile (PAN) exhibits unique advantages when comparing different polymer materials. This material has a low molecular orbital energy level without occupies a low molecular orbital energy level, has a wide electrochemical window and high voltage stability, and is especially suitable for lithium metal battery systems that match high voltage positive electrodes. What is more worthy of attention is that the C≡N group in the PAN molecular chain can form a coordination effect with Li^+. This characteristic not only promotes the dissociation of lithium salts, but also effectively inhibits the growth of lithium dendrites in the composite gel electrolyte. It can be seen that the coordinated use of PVDF-HFP/PAN provides a new direction for the development of lithium metal battery electrolytes that combine high energy density and safety. Zhang et al. [34] dispersed lithium lanthanum zirconium tantalate (LLZTO) ceramic powder in a polymer matrix and successfully prepared PVDF/LLZTO composite electrolyte. In the $LiCoO_2$/Li battery composed of this composite electrolyte, the initial specific discharge capacity reached 150 mAh·g^{-1}. After 120 cycles at 0.4 C, the capacity retention rate of the battery can reach 98%. Gu et al. [35] PVDF-HFP/LLZTO composite GPEs with reported a simple synthesis route and readily available raw materials. The composite electrolyte exhibited a wide electrochemical stability window (about 5 V), with a specific capacity of about 150 mAh·g^{-1} at 0.1 C, and a coulombic efficiency of about 99% after 50 cycles. Professor Liang Feng's team from Kunming University of Science and Technology has proposed a new strategy for the high-efficiency ultraviolet (UV) curing preparation of composite solid-state electrolytes with high ionic conductivity. This strategy enables the rapid preparation of composite solid-state electrolytes within 45 s. Through solvation structure regulation and in-situ curing, it simultaneously constructs a stable interfacial phase and a tight interface, effectively inhibiting battery dendrite growth and interfacial reactions, reducing battery interface impedance, and achieving long cycle life of the battery [36]. The UV curing technology provides a new idea for the commercialization of solid-state electrolyte membranes. However, it has relatively insufficient experience in industrial applications, requiring further verification of long-term stability and integration with other strategies (such as three-dimensional framework current collectors) to enhance practicality.

Multilayer pouch cells have gained widespread adoption in consumer electronics, electric vehicles, and energy storage systems, primarily owing to their high energy density, design flexibility, and relatively low weight [37]. However, they are inherently challenged by several critical issues. The utilization of aluminum laminate film as the casing material, while lightweight, results in significantly inferior mechanical strength compared to metallic (steel or aluminum) hard-case batteries. Furthermore, gas generation during cycling or under elevated temperatures leads to cell swelling ("pouch bulge") [38,39]. The flexible pouch structure provides inadequate constraint against this expansion, and persistent swelling can progressively stretch and weaken the aluminum laminate film, thereby increasing the risk of rupture and potentially compromising the integrity of the enclosing module structure. Thermal management presents another significant challenge within the multilayer stacked configuration. Heat generated internally must dissipate through the stacked cell layers to the surface and then to the external environment, encountering relatively high thermal resistance. This impedes efficient heat dissipation from the core region, promoting the formation of localized hotspots [40,41]. The heat sealing process of the aluminum laminate film is critical for ensuring battery safety. The seal edges are susceptible to failure due to various factors, including suboptimal sealing parameters (e.g., improper temperature, pressure, or duration control), material defects, long-term electrolyte corrosion, and mechanical stresses (e.g., vibration, swelling-induced stress). Such seal integrity breaches pose substantial safety and reliability risks. While multilayer pouch cells offer distinct advantages in energy density and design versatility, the aforementioned challenges—particularly concerning safety, longevity, cost reduction, and manufacturability at scale—remain the focus of intensive research and engineering efforts aimed at enabling their broader application.

In this paper, PVDF-HFP/PAN-$LiClO_4$-LLZTO composite gel electrolyte was prepared using a UV curing process, single-layer soft-pack batteries and three-layer soft-pack batteries were assembled, and the cycle life was compared and tested. The different reasons for the capacity attenuation of three-layer soft-pack batteries and single-layer soft-pack batteries were analyzed, and the capacity attenuation of multi-layer soft-pack batteries and single-layer soft-pack batteries was investigated. The electrochemical behavior of soft-pack batteries is studied to explore ways to improve the cycle life of multilayer soft-pack batteries in order to promote their industrialization.

2. Results and Discussion

Figure 1a,b are the experimental pictures of the ignition of GPEs film for 1 s and 5 s respectively. It can be seen from the experiment that the film can not be ignited, indicating that the GPEs film is nonflammable. Figure 1c,d are the comparison diagrams of GPEs film before and after bending respectively. It can be seen from the diagram that the film has basically no change before and after bending, indicating that the film has good flexibility.

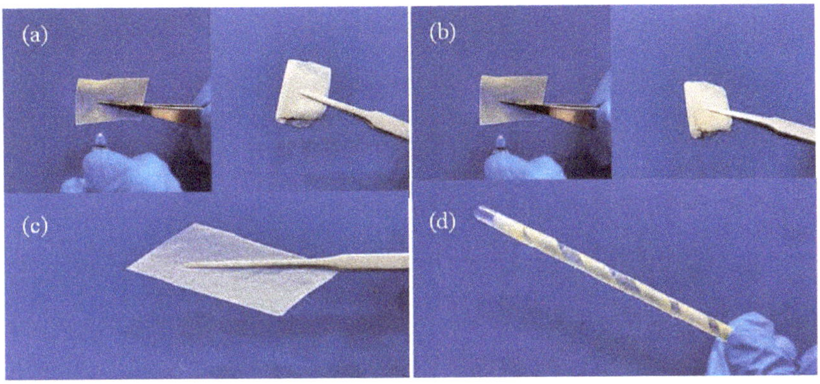

Figure 1. (**a**) GPEs ignited for 1 s; (**b**) GPEs ignited for 5 s; (**c**) GPEs tile; (**d**) After GPEs bending.

Observing the phase analysis data in Figure 2 can reveal more important information. In the X-ray diffraction (XRD) pattern of Figure 2a, the PVDF-HFP matrix in the original state shows a typical crystal diffraction peak at 20.7° in 2θ, and the characteristic peak of the PAN component is located near 17° in 2θ. It is worth noting that in the three-layer soft-pack battery samples after 100 charge and discharge cycles, these characteristic peaks significantly weakened or even disappeared. This change in phase structure suggests that the ion transport channel inside the electrolyte may have been damaged and the stability of the gel system has dropped significantly. The Raman spectrum in Figure 2b further confirms this conclusion. The characteristic vibration peaks that should appear in the three-layer structure sample completely disappear in the single-layer sample, which is mutually verified with the XRD detection results. In addition, the thermogravimetric analysis curve of Figure 2c shows the decomposition behavior of the material at high temperatures, and its thermal stability characteristics also provide an important reference for subsequent research. The above experimental results generally show that although GPEs materials have excellent flame retardancy and mechanical properties in the initial state, their internal structure will undergo irreversible changes after long-term use. The stability of this material during recycling, especially the disintegration of the crystal structure and changes in chemical composition, may be key factors that restrict their practical application. Follow up research needs to conduct a more in depth discussion on these failure mechanisms.

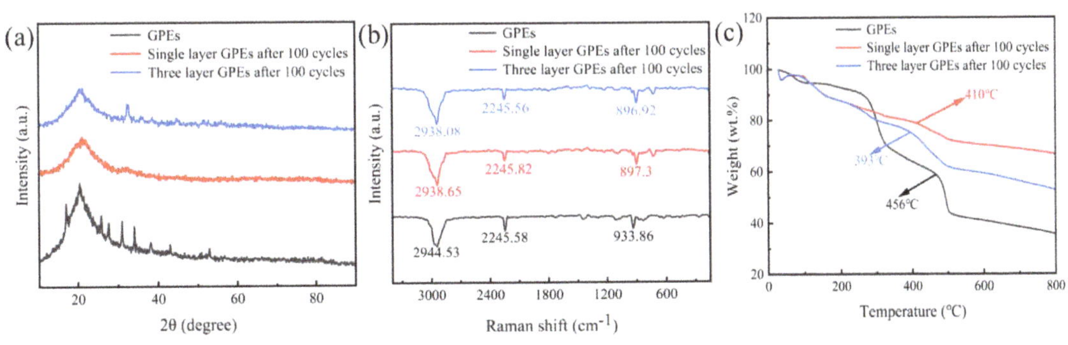

Figure 2. (**a**) XRD, (**b**) Raman, and (**c**) TG of GPEs.

Thermogravimetric analysis (TGA) results are shown in Figure 2c that in the temperature range of 30~800 °C, the weight loss temperature of GPEs before charge and discharge cycle is about 456 °C, while the weight loss temperature of single-layer and multi-layer GPEs after 100 charge and discharge cycles is about 400 °C. When the temperature reaches 800 °C, the weight loss of single-layer and multi-layer GPEs before charge and discharge cycle is 33.37% and 47.28%, respectively, indicating that the weight loss rate decreases after 100 charge and discharge cycles. The slope of the TGA curve becomes less steep after 100 cycles, which implies a slower decomposition or evaporation rate of the components under heating. This reduced weight loss rate suggests that the structure of the GPEs or the interface between GPEs and electrode has become more stable, likely due to the formation of a stable passivation layer or improved interfacial contact during cycling. After 100 charge and discharge cycles, the TGA curve shows a more gradual weight loss, indicating a reduced thermal decomposition rate, which may be attributed to the enhanced stability of the GPEs structure or the formation of a stable interfacial layer. Combined with the analysis of XRD and Raman spectroscopy results, it is shown that GPEs has decomposed at this time.

Figure 3 shows the scanning electron microscopy (SEM) observation results of polymer electrolyte membranes under different states. In the initial samples before the experiment (Figure 3a–d), it can be observed that there are a large number of solid particles and uneven morphological characteristics on the surface of the sample. This situation may be related to the local aggregation of ceramic fillers (LLZTO) during the slurry drying process, and the crystallization precipitation formed during the curing stage can easily lead to an increase in surface roughness. It is worth noting that this structural feature will have an important impact on the mechanical properties of the electrolyte membrane. After 100 charge and discharge cycles, the electrolyte membrane (Figure 3e–h) in the single-layer battery showed significant structural deterioration. The surface image shows a large number of cracks and holes, especially the peeling phenomenon in the edge area of the film is more prominent. Judging from the cross-sectional image, the originally dense layered structure begins to appear loosely layered, which indicates that the electrolyte material may expand or shrink during long-term charging and discharging. This change in physical structure will directly weaken the material's ability to inhibit lithium dendrites, thereby affecting the safety of the battery. Interestingly, in the test samples of multi-layer soft-pack batteries (Figure 3i–l), although the surface morphology looks relatively complete, abnormal grain growth still exists at the microscopic scale. The cross-sectional image shows obvious structural looseness, and the inter-layer bonding intensity in some areas is significantly reduced. This situation may be due to the uneven distribution of stresses generated in the multi-layer lamination process and the interface degradation caused by stress release during the cycle. In addition, the surface texture of the PTFE mold will affect the micromorphology of the electrolyte membrane through transfer, and the surface fluctuations formed by this transfer effect may change the contact state of the electrolyte with the electrode material. Through auxiliary analysis of element distribution map (Figure 4a–f), it can be found that the dispersion of LLZTO ceramic particles in the electrolyte membrane is relatively uniform. The design of this heterostructure has a dual mechanism: the smooth surface facing the positive electrode is rich in polymer components, which is conducive to reducing the interface impedance; while the rough surface facing the negative electrode is rich in ceramic fillers, which improves the contact performance with metal lithium through physical anchoring. However, experimental data also show that as the number of cycles increases, the synergy between the two interfaces gradually weakens, which may be one of the important reasons for the overall performance of the battery to decay. Figure 4a–f shows the dispersion of each element inside the GPEs material. Through the element distribution map, the existence and distribution characteristics of carbon, chlorine, fluorine, oxygen and tantalum elements in the material system can be visually observed, which provides basic spatial information support for subsequent analysis.

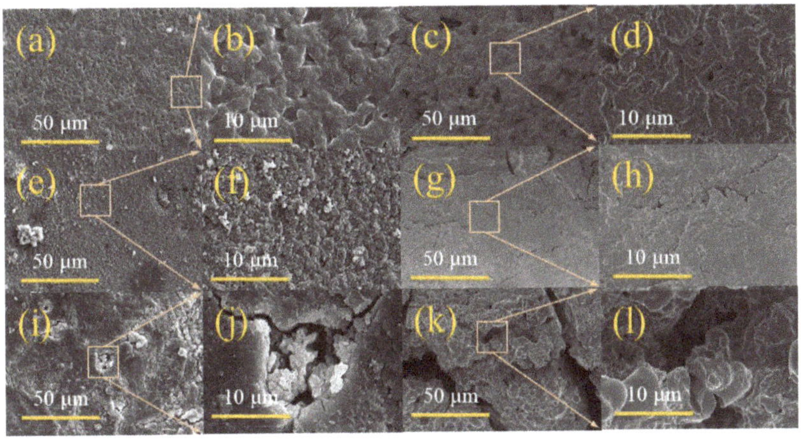

Figure 3. SEM images of GPEs before 100 cycles: surface (**a**,**b**); cross-section (**c**,**d**). SEM images of GPEs after 100 cycles in single-layer battery: surface (**e**,**f**); cross-section (**g**,**h**). SEM images of GPEs after 100 cycles in three-layer battery: surface (**i**,**j**); cross-section (**k**,**l**).

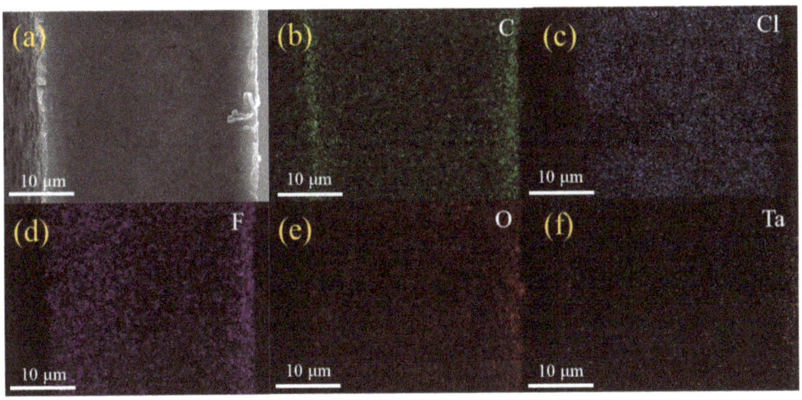

Figure 4. (**a**–**f**) Elemental distribution of C, Cl, F, O and Ta in GPEs.

The Figure 5 shows the C1s, F1s, N1s, O1s spectra of X-ray photoelectron spectroscopy (XPS) of GPEs membrane battery before charging and discharging cycle, after 100 charging and discharging cycles of single-layer GPEs membrane battery, and after 100 charging and discharging cycles of three-layer GPEs membrane battery. From the C1s energy spectrum (Figure 5a), it can be found that there are characteristic peaks in four locations: 284.78 eV, 286.20 eV, 288.68 eV and 290.52 eV. These characteristic peaks correspond to different types of carbon chemical bond structures, which belong to C-C single bond, C-O/C-N mixed bond, C=O double bond and C-F bond type, respectively. The O1s energy spectrum (Figure 5b) shows the main peak at 532.19 eV, and combined with the secondary peak at 533.41 eV, it fully proves that there are a large number of oxygen-containing groups in the material system, which is mainly related to the LLZTO ceramic filler added to the polymer matrix. Of particular concern are the two obvious peaks in the N1s energy spectrum (Figure 5c). After analysis, the former corresponds to the C=N double bond structure, while the latter comes from the C-N single bond system. In the F1s energy spectrum (Figure 5d), a strong signal at 686.8 eV indicates that the solid electrolyte interface (SEI) layer is rich in LiF components. By comparing the samples after different cycles (Figure 5f,g), it can be observed that with the complexity of the battery structure, the LiF content in the SEI layer showed a downward trend, while in contrast, the LiF content in the single-layer structure decreased more significantly. This change may affect the passivation effect of the lithium metal surface, because experimental data show that the presence of LiF has a positive effect on slowing down the formation of lithium dendrites. It inhibits the continuous reaction between lithium metal and electrolyte by forming a stable protective layer, which is of great significance to improving the cycle life of the battery.

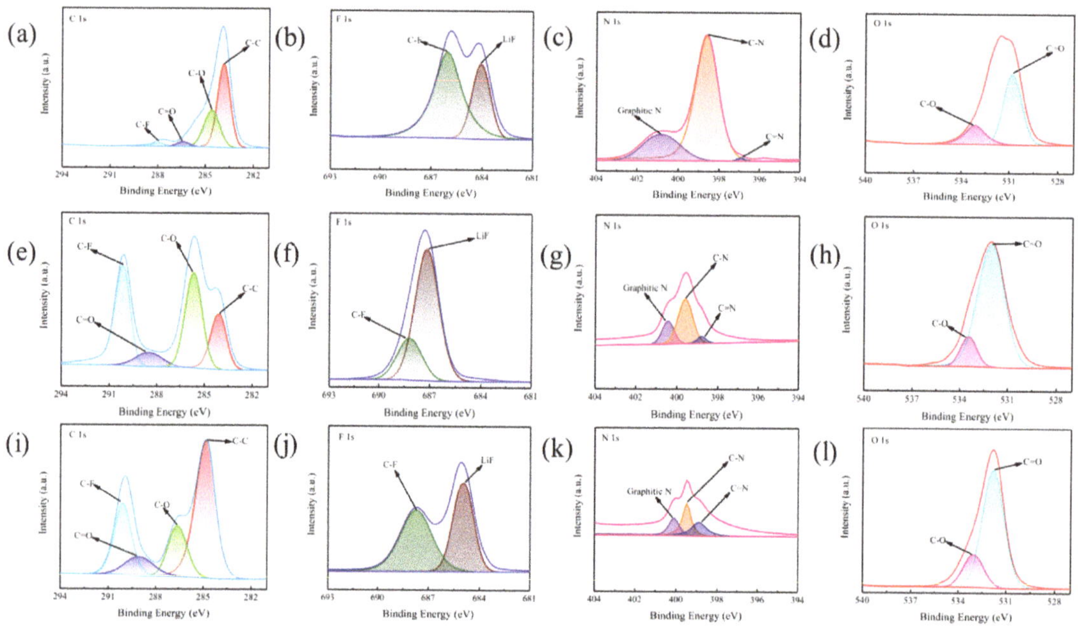

Figure 5. XPS spectra of GPEs under different states. (**a–d**), Before charge discharge cycle C1s, F1s, N1s and O1s; (**e–h**), After single-layer battery charging and discharging cycle C1s, F1s, N1s and O1s; (**i–l**), After three-layer battery charging and discharging cycle C1s, F1s, N1s and O1s.

Further observations found that after multiple charge and discharge cycles, not only did the LiF content change, but the peak intensity of other chemical bonds also had a certain deviation. For example, in the region corresponding to the C=O bond (Figure 5e), the cycled samples show a wider peak-shaped feature, which may imply that oxidative reconstruction of the surface layer of the material is performed. The peak difference in the N1s energy spectrum (Figure 5i,k) suggests that the chemical environment of nitrogen elements has undergone slight changes during the cycle, and this structural evolution may affect the stability of the ion transport channel. These phenomena together show that the electrode/electrolyte interface will undergo dynamic adjustment during repeated charging and discharge, and this interface reconstruction behavior is an important factor restricting the degradation of battery performance.

Electrochemical Impedance Spectroscopy (EIS) was conducted in situ for the full cell by performing EIS every 10 cycles during galvanostatic cycling. The measurement was done at the end of discharge after an hour of resting at the open circuit voltage. Deconvolution of the EIS data by DRT (distribution of relaxation times) transformation in the relaxation time space, as shown in Figure 6.

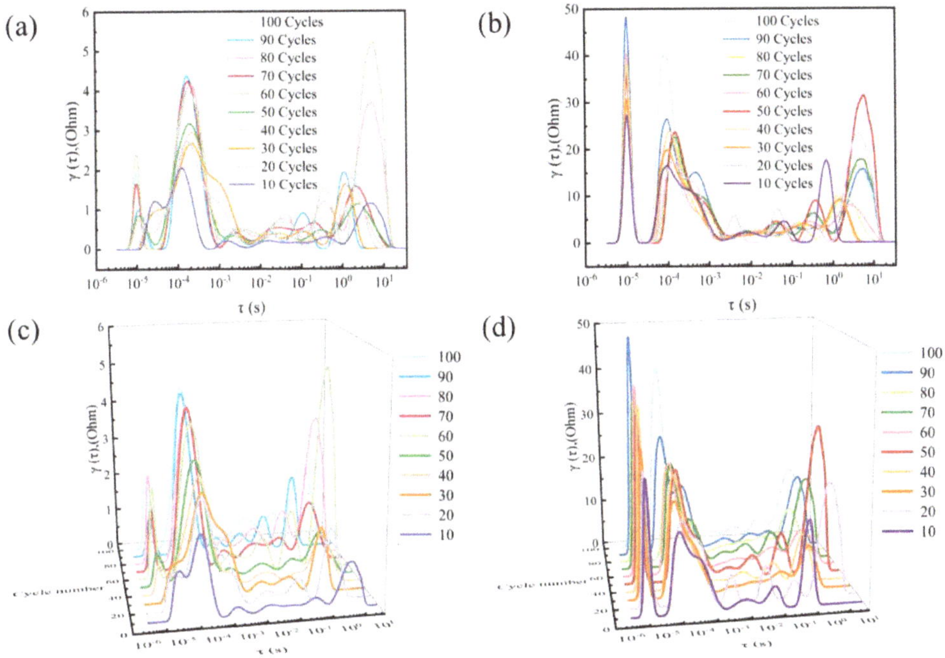

Figure 6. DRT analysis in EIS ((**a**,**c**) single-layer battery; (**b**,**d**) three-layer battery). The corresponding 3D evolution of the (**a**,**b**) is shown in the (**c**,**d**).

Quantitative assessment of the interfacial impedance evolution in batteries was performed via in-situ EIS. EIS enables real-time tracking of the evolution of ionic transport resistance at the interface under varying charge-discharge cycle numbers, providing critical evidence for understanding the rate and mechanisms of interfacial degradation. The three-layer pouch cell exhibited a significant increase in interfacial impedance after extended cycling, as shown in Figure 6b, where the ionic transport pathways became gradually obstructed with progressive cycling, leading to rapid capacity fade. In contrast, the single-layer pouch cell demonstrated a markedly slower growth in interfacial impedance after multiple cycles (Figure 6a). The arc-shaped evolution of the EIS spectra indicates more stable interfacial contact and sustained unimpeded ion transport.

Figure 7 system shows the multi-dimensional performance characterization of gel polymer electrolytes (GPEs). The impedance spectrum test results of Figure 7a show that the electrolyte membrane incorporated with LLZTO inorganic filler exhibits better conduction characteristics. From the perspective of microstructure, this may be because the addition of rigid ceramic particles destroys the regular arrangement of the polymer matrix, causing their crystallinity to significantly reduce. This disordered crystal structure can open up more migration paths for lithium ions, just like opening up multiple shortcuts for passage in dense forests. The impedance measured by EIS is 4.773 Ω and the ionic conductivity calculated by Formula (1) is 3.86×10^{-4} S·cm^{-1}. This numerical comparison is more intuitive in the lateral comparison of Table 1.

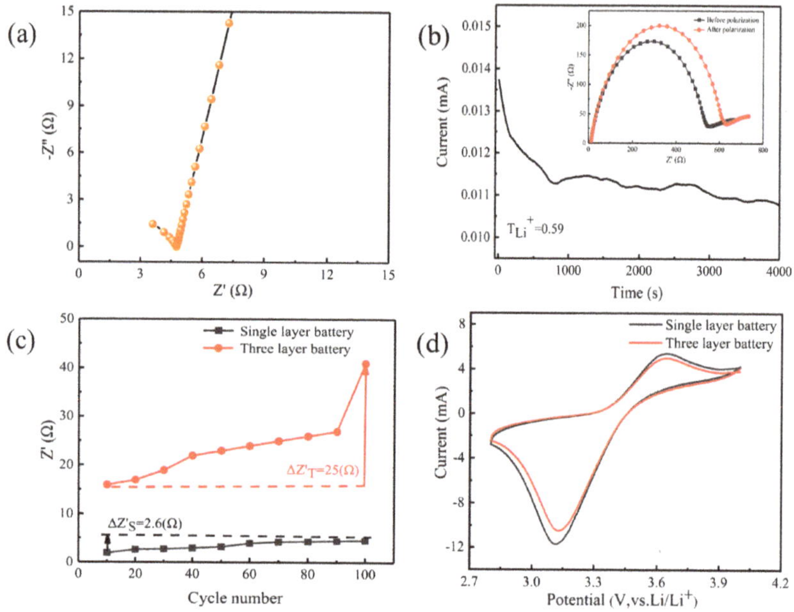

Figure 7. (**a**) Impedance diagram, (**b**) DC polarization and AC impedance curves of GPEs, (**c**) Impedance and charge discharge cycles, (**d**) CV curves.

Table 1. Contrastive Evaluation of Emerging and Established Gel Polymer Electrolytes.

GPEs	Ionic Conductivity ($\times 10^{-4}$ S/cm)	Ref.
PAP/PEP	4.27 (RT)	[42]
CEPs	0.81(25 °C)	[43]
CSE (15 wt.% LLZTO)	0.55 (30 °C)	[44]
PPPL-10	4 (25 °C)	[45]
PEO/PVDF/NaNO3	0.93 (25 °C)	[46]
PEO-PVDF-LITFSI	2.46 (25 °C)	[47]
SPEs (PLiZ)	1.19 (25 °C)	[48]
PPLG (5 wt.% graphene)	3.39 (25 °C)	[49]
SSCEs	2.91 (25 °C)	[50]
GPEs	3.86 (RT)	This work

Figure 7b displays the composite solid polarization curve and initial/steady-state impedance profile of the electrolyte membrane. Parameters: $I_0 = 0.01492$ mA, $I_{SS} = 0.01076$ mA, $R_0 = 552.123\ \Omega$, $R_{SS} = 633.598\ \Omega$. The lithium-ion transference number calculated using Equation (2) is 0.59. Furthermore, studies indicate that incorporating LLZTO ceramic fillers induces relaxation of polymer chains. Interactions between the inorganic fillers and polymer chains promote segmental motion and accelerate dynamic processes [51]. These interactions not only enhance ionic conductivity but also improve mechanical properties and interfacial stability of the electrolyte membrane. Battery impedance increases progressively during charge/discharge cycles. Comparative analysis reveals that the three-layer cell exhibits significantly greater impedance growth (Figure 7c).

Figure 7d presents cyclic voltammetry (CV) results for single-layer and three-layer GPEs within 2.8–4 V at a scan rate of 0.2 mV s^{-1}. For the single-layer pouch cell, a distinct reduction peak at 3.15 V and a strong oxidation peak at 3.75 V correspond to lithium-ion insertion into and extraction from the LFP cathode, respectively. The substantially larger integrated redox peak area indicates higher specific capacity in the single-layer cell.

Conversely, the three-layer pouch cell exhibits weak redox peaks, demonstrating inferior electrochemical performance.

The above differences are further verified by the comparison of charge and discharge curves in Figure 8. In the 2.5–4 V operating voltage range, batteries of both structures exhibit typical voltage platform characteristics during the initial charging and discharging process, which is consistent with the reversible deintercalation mechanism of lithium ions in lithium iron phosphate materials. However, under different current density conditions, the behavior of the two shows significant differences: when a small magnification of 0.1 C is used, the discharge capacity of the three-layer structure reaches 126.2 mAh g^{-1}, which is higher than 106.3 mAh g^{-1} in the single-layer structure; however, after increasing the current to 0.5 C, the discharge capacity of the three-layer structure drops sharply to 71.1 mAh g^{-1}, which is significantly lower than 83.4 mAh g^{-1} in the single-layer structure. This performance inversion phenomenon caused by changes in current density may be related to the differences in its internal ion transmission path.

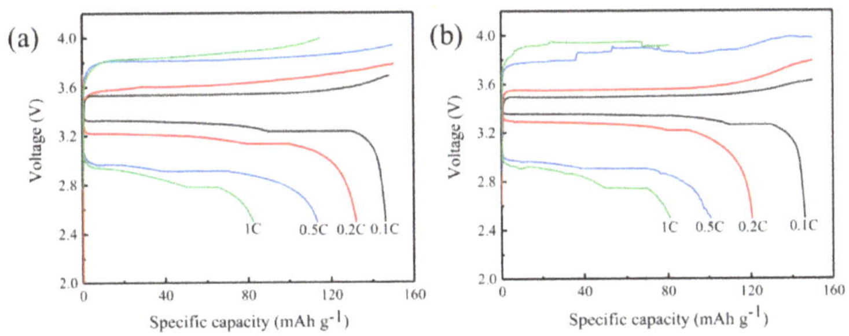

Figure 8. Charge-discharge curves. (**a**) single-layer soft-pack battery; (**b**) three-layer soft-pack battery.

The rate performance test results of Figure 9 show more fully this characteristic change trend. In the process of gradually increasing the discharge rate (0.1 C→0.2 C→0.5 C→1 C), although the single-layer soft-pack battery also exhibits capacity attenuation, its capacity retention rate at each stage is relatively stable. Especially when the 0.1 C ratio is adjusted back again, its capacity is basically restored to the initial value, showing good reversibility. In contrast, the capacity of the three-layer structure shows a cliff-like decline with the increase in magnification. Even under the conditions of high current of 1 C, its capacity is only comparable to that of a single-layer structure. This obvious performance decay may be related to the increase in charge transport impedance caused by its stacked structure, and the contact loss between multi-layer interfaces may be an important factor restricting its high current working ability.

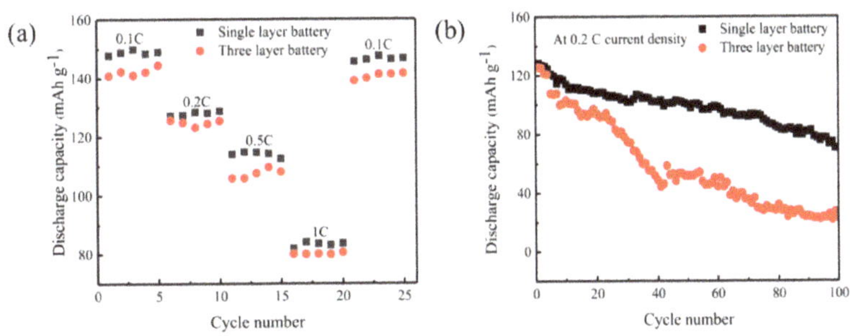

Figure 9. (**a**) Rate capability, (**b**) cycling performances.

It should be noted that there is a good mutual proof relationship between these electrochemical test results. The peak area difference in the voltammetric curve corresponds to the charge and discharge capacity data, and the performance deterioration trend of the three-layer structure in the magnification test also forms a logical chain with its smaller response peak intensity in the voltammetric test. These experimental phenomena together show that although multilayer structures may exhibit capacity advantages under certain specific conditions, their overall electrochemical stability and kinetic performance are still significantly inferior to those of single-layer structure design.

Figure 9b shows the cycle stability test results of assembled solid-state batteries under different structural conditions. The experiment selected a 0.2 C ratio for constant current charging and discharge, and compared the 100 cycles performance of single-layer and three-layer soft-pack batteries in the voltage range of 2.8 to 4.0 volts. Observing the data, it can be found that the initial discharge capacity of the two types of batteries is not much different. The starting capacity of the single-layer structure is 144.5 mAh g^{-1}, and the three-layer structure is slightly higher than 151.1 mAh g^{-1}. However, as the number of cycles increases, the capacity attenuation trend has significantly differentiated: by the 50th week, the single-layer battery still maintained a capacity of 100.3 mAh g^{-1} (about 69.4% of the initial value), while the three-layer structure has dropped sharply to 52.1 mAh g^{-1} (only 34.5% left); after continuous charging and discharging to 100 weeks, the single-layer battery capacity remained at 70.7 mAh g^{-1} (48.9% retention rate), but only 23.6 mAh g^{-1} (15.6% retention rate) remained at the three-layer battery. This shows that the monolayer configuration still has good electrochemical stability in the medium term, and the rapid decline of the three-layer structure indicates that there may be interface failure problems inside it.

Figure 10 shows the LED lamp test photos of single-layer GPEs membrane soft pack battery and three-layer GPEs membrane soft pack battery under the conditions of cutting, bending at different angles, needling, etc. It can be seen from the figure that the LED can still light up after 90°, 180° bending, cutting and needling of both single-layer and three-layer GPEs film soft pack batteries, which indicates that the soft pack battery assembled based on GPEs has high safety.

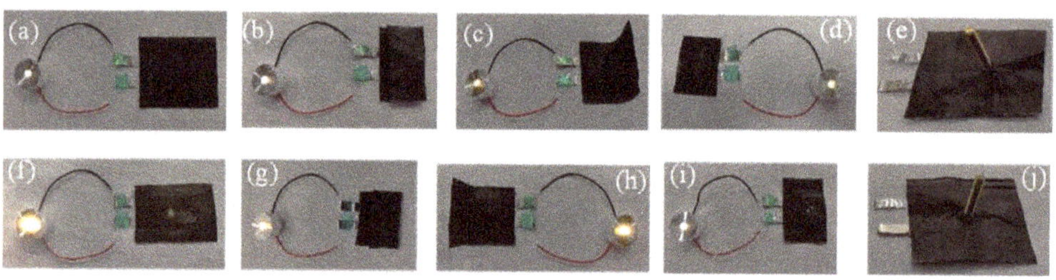

Figure 10. (**a**) Successfully working with the LED light before cutting or bending or needle puncture of the single-layer GPEs membrane soft pack battery, (**b**) After 180° bending, (**c**) After 90° bending, (**d**) After cutting half, (**e**) after needle puncture; (**f**) Successfully working with the LED light before cutting or bending or needle puncture of three-layer GPEs membrane soft pack battery, (**g**) After 180° bending, (**h**) After 90° bending, (**i**) After cutting half, (**j**) after needle puncture.

3. Conclusions

In this paper, PVDF-HFP/PAN-$LiClO_4$-LLZTO composite gel electrolyte was prepared through UV curing technology, and the performance differences between single-layer and three-layer soft-pack batteries were systematically compared. In the low-rate discharge test of 0.1 C, the three-layer structure battery exhibited a discharge specific capacity of

126.2 mAh g^{-1}, which was significantly better than the 106.3 mAh g^{-1} of the single-layer structure. When the high magnification of 0.5 C was reached, the situation reversed, and the capacity of the three-layer structure dropped sharply to 71.1 mAh g^{-1}, which was about 15% lower than the 83.4 mAh g^{-1} of the single-layer structure. This phenomenon shows that in application scenarios where rapid discharge is required, the design of a single-layer structure may have more advantages.

It is worth noting that under the medium magnification 0.2 C, the initial capacity of the two structures is basically at the same level. The single-layer battery is measured to be 144.5 mAh g^{-1}, and the third-layer battery is 151.1 mAh g^{-1}. However, as the number of cycles increases, the performance differentiation between the two begins to appear. After 50 charge and discharge cycles, the single-layer battery can still maintain a capacity of 100.3 mAh g^{-1}, while the capacity of the three-layer battery has dropped significantly to 52.1 mAh g^{-1}. In terms of capacity retention, the single-layer structure has a stability of 69.4%, but the three-layer structure has a residual capacity of only 34.5%. This rapid attenuation may be due to interface contact problems caused by multi-layer stacking.

In view of the existing experimental results, subsequent research needs to focus on solving the interfacial impedance problem in multi-layer structures, try polymer-ceramic composite materials with different ratios, and perhaps find a combination solution with better conductivity. At the same time, the battery's high and low temperature adaptability should be systematically evaluated, after all, the actual use environment is much more complicated than that in the laboratory. It should also be noted that the current test is mainly carried out at the laboratory level. To advance to the industrial application stage, more comprehensive performance verification is also required in the complete battery system. In particular, in-depth exploration is needed in terms of electrode material matching and packaging process optimization.

4. Materials and Methods

4.1. Preparation of PVDF-HFP/PAN-LiClO$_4$-LLZTO Gel Polymer Electrolyte

The preparation of GPEs adopts a UV curing process, and the specific operation process is shown in Figure 10. First, 12 g of N, N-dimethylformamide (DMF, Aladdin, Shanghai, China, 99.8%) was weighed into a beaker, and a magnetic stirrer was activated at 45 °C and 300 rpm. Subsequently, 1.4 g of PVDF-HFP (Mn = 600,000, Zhanyang, Dongguan, China) and 0.6 g of PAN (Zhanyang, Dongguan, China) were dissolved/dispersed in the DMF, and the stirring speed was increased to 500 rpm. After thorough mixing, 0.6 g of $LiClO_4$ (Aladdin, Shanghai, China, 99%) was added while maintaining the temperature at 45 °C, followed by 5 min of stirring. Next, 0.2 g of LLZTO (Kcrystal, Shenzhen, China, 99%) was added and stirred for 5 h. After this period, 0.8 g of ETPTA and 0.5 g of 425 were introduced and stirred for 5 min. Finally, 3 drops of HMPP were added and stirred for an additional 5 min.

After cooling for 5 min, the mixed slurry was transferred to a polytetrafluoroethylene (PTFE) mold for static casting. Following coating onto a PTFE plate, the gel electrolyte membrane was cured using a UV curing machine at a speed of 0.01 m/min for 10 s under controlled conditions (30 °C, wavelength 365 nm). After vacuum drying at 60 °C for 40 min, the resulting PVDF-HFP/PAN-$LiClO_4$-LLZTO GPEs could be easily peeled off. Finally, the electrolyte membrane was cut into rectangular pieces measuring 48 mm long and 28 mm wide, and stored in an argon-filled glove box (<0.01 ppm H_2O and O_2). The thickness of the PVDF-HFP/PAN-$LiClO_4$-LLZTO membrane prepared by solution casting was measured as 37 μm using a micrometer screw gauge. The preparation flowchart of GPEs is shown in Figure 11.

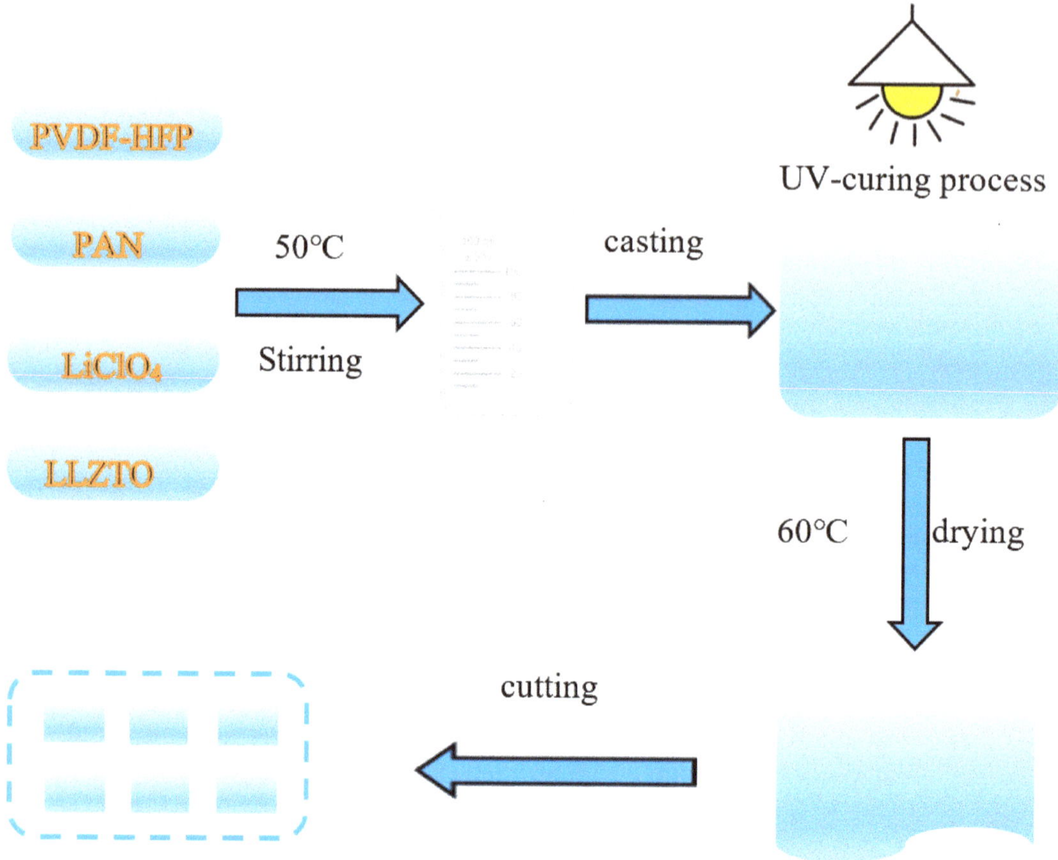

Figure 11. Schematic pattern of the preparation process of GPEs.

4.2. Material Characterization

Phase composition analysis employed XRD (DX-2700, Manufacturer: Dandong Haoyuan Instrument Co., Dandong, China; Cu-Kα radiation). Surface morphology and chemistry were characterized via SEM (Phenom Pharos G2, Manufacturer: Shanghai FEI Ltd. Shanghai, China) integrated with XPS spectroscopy. Raman spectra acquisition utilized an ATR8000 confocal microscope (Manufacturer: Fujian FocusTek, Fuzhou, China). TGA measurements proceeded on a Netzsch F3 Tarsus system (Bavaria, Germany) under argon purge, implementing a dynamic temperature program (30→800 °C, β = 10 °C/min).

4.3. Electrode Preparation

N-methylpyrrolidone (3 drops) was introduced into a borosilicate beaker. Polyvinylidene fluoride (0.2 g) was dissolved under magnetic stirring (300 rpm, RT). Separately, a blended powder of $LiFePO_4$ (1.6 g) and carbon black (0.2 g) was ground in an agate mortar (40 min). The powder mixture was dispersed into the PVDF/NMP solution and stirred (4 h, RT). The resultant slurry was coated via doctor blade (50 μm gap) onto current collectors, followed by vacuum drying (60 °C, 36 h). Dried electrodes were cut to 44 mm × 24 mm with 12 mm × 8 mm tabs and stored under argon atmosphere.

4.4. Battery Assembly

The soft-pack battery is assembled by stacking. First, the negative electrode copper-lithium composite strip is cut into a 44 mm long, 24 mm wide, 12 mm long, and 8 mm wide electrode, which is matched with the cut positive electrode. The first layer is the negative electrode, followed by the cut gel electrolyte membrane of 48 mm long and 28 mm wide, and then the positive electrode of the same size as the negative electrode. The multi-layer soft-pack battery is stacked in this way. After the stacking is completed, tape is attached to the four sides of the soft-pack battery, and then 0.8 MPa and 1.0 MPa are applied on the sheet press (Manufacturer: ZJ-D33-YP15 Beijing Jingke Zhichuang Technology Development Co., Ltd. Beijing, China) respectively for single-layer and three-layer soft-pack batteries, so that the interface between the soft-pack battery electrode sheet and the gel electrolyte membrane is better contacted. After the sheeting is completed, the pole ear is welded on the ultrasonic spot welding machine (Manufacturer: Shenzhen Kejing Equipment Co., Ltd., Shenzhen, China, MSK-800). The size of the pole ear is 0.1 * 8 * 46 mm. The positive pole ear is an aluminum pole ear, and the negative pole ear is a nickel pole ear, which is welded on the positive current collector aluminum foil and the negative current collector copper foil respectively. Then cut the aluminum plastic film with a length of 65 mm and a width of 55 mm, set the parameters on the vacuum pre-sealing machine (Manufacturer: Xinwei Equipment Co., Ltd., Shenzhen, China,), the temperature is 150 °C, the time is 3 s, and one side of the aluminum plastic film is encapsulated. Then put the battery batteries with the pole ear welded into the aluminum plastic film and encapsulate it on the top. Then inject $LiPF_6$ electrolyte into the soft pack battery, and finally set the vacuum degree to −93 kPa in the vacuum pre-sealing machine to encapsulate the last side. The design capacity of the single-layer soft pack battery is 20 mAh, and the design capacity of the three-layer soft pack battery is 60 mAh. The assembled soft-pack battery was aged at 55 °C for 8 h, then left at room temperature for 16 h, then charged at 0.05 C for 2 h, then charged at 0.1 C for 3 h, and finally aged at 55 °C for 24 h. After aging, the soft-pack battery was vented and repackaged for testing. Figure 12a,b show the schematic diagrams of the assembled single-layer and three-layer pouch batteries, respectively.

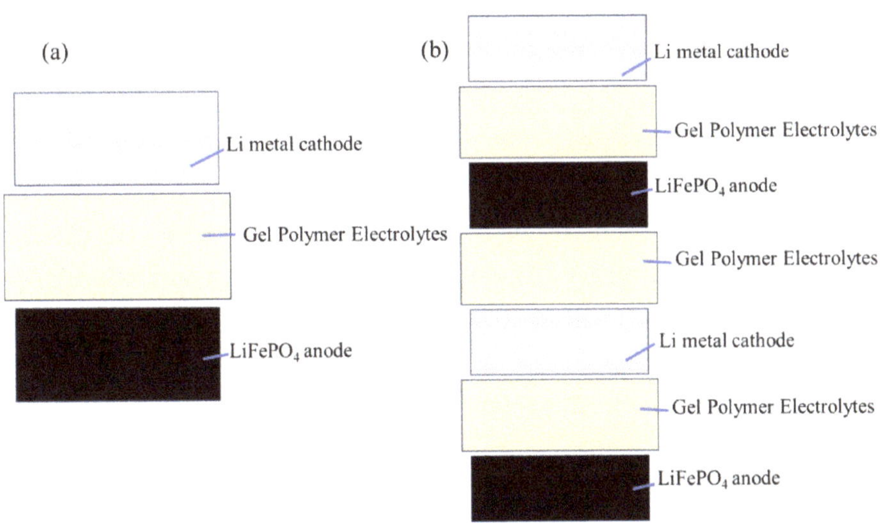

Figure 12. The schematic diagrams of the assembled. (**a**) single-layer batter, (**b**) three-layer battery.

4.5. Electrochemical Properties Test

EIS was performed on an electrochemical workstation (DH 7000, Donghua, Nanjing, China) using a symmetric button battery assembled with a stainless steel SS/GPEs/SS structure. The frequency range of the EIS test was 10^6 Hz to 0.1 Hz with an amplitude of 10 mV.

The electrochemical stability window of the GPE was evaluated by linear sweep voltammetry (LSV) using a semi-symmetric SS/GPEs/Li cell within a voltage range of 2.5 V to 6.0 V at a scan rate of 5 mV/s. The lithium-ion transference number ($T_{Li}{}^{+}$) of the GPE was quantified via Bruce-Vincent method using symmetric Li | | GPEs | | Li cells, combining chronoamperometry (DC polarization at 10 mV) with EIS (100 kHz–0.1 Hz) using the same electrochemical workstation.

The calculation formula of ionic conductivity σ is as follows: [34]

$$\sigma = \frac{L}{RS} \tag{1}$$

where:

σ is the ionic conductivity (S/cm).

L denotes the thickness of the membrane (cm).

R represents the local minimum resistance (Ω) in the full impedance spectrum (typically the bulk resistance R_b from EIS Nyquist plot).

S indicates the contact area between the membrane and stainless steel (SS) electrodes (cm^2).

The Lithium ion transference number is an important parameter for evaluating the ion migration rate [52].

The calculation formula of lithium ion transfer numbers as follows: [53,54]

$$T_{Li^+} = \frac{I_{SS}(\Delta V - I_0 R_0)}{I_0(\Delta V - I_{SS} R_{SS})} \tag{2}$$

where:

T_{Li^+} denotes the lithium-ion transference number in the electrolyte.

I_0 and I_{SS} represent the current values before and after DC polarization stabilization (mA).

R_0 and R_{SS} are the resistances before and after DC polarization (Ω).

ΔV is the voltage of 10 mV applied across the battery.

Author Contributions: Conceptualization, C.C. and X.L.; methodology, Q.H. and R.T.; software, W.Q.; validation, X.L., C.C. and Q.H.; formal analysis, C.C.; investigation, Y.W.; resources, X.L.; data curation, C.C.; writing—original draft preparation, C.C.; writing—review and editing, X.L.; visualization, W.Q. and Y.G.; supervision, W.Q. and Y.G.; project administration, Y.W. and J.L.; funding acquisition, J.L. and Y.G. All authors have read and agreed to the published version of the manuscript.

Funding: This research was funded by the Reginal Collaboration R&D Program of Sichuan Province under Grant (No. 2024YFHZ0209), Fund Project of the Guangxi Key R&D Programme Projects (No. AB24010346), National Natural Science Foundation of China (No. 22462003, 22362005, 22262005), Guangxi key laboratory of Automobile Components and Vehicle technology (2023GKLACVTZZ02), Guangxi Key Laboratory of Automobile Components and Vehicle Technology (No. 2022GKLACVTZZ04, 2023GKLACVTZZ10), Innovation Project of Guangxi Graduate education (YCSW2025576), Ph.D. Programs Foundation of Guangxi University of Science and Technology (Grant No. 23Z04).

Institutional Review Board Statement: Not applicable.

Informed Consent Statement: Not applicable.

Data Availability Statement: The data are contained within the article.

Acknowledgments: We are thankful to the following authors for their contributions, as well as for the pur-chase of materials and conducting the relative experiments: Chu Chen, Wendong Qin, Qiankun Hun, Yujiang Wang, Xinghua Liang, Renji Tan, Junming Li and Yifeng Guo.

Conflicts of Interest: The authors declare no conflicts of interest.

References

1. Yuan, B.; Wen, K.; Chen, D.; Liu, Y.; Dong, Y.; Feng, C.; Han, Y.; Han, J.; Zhang, Y.; Xia, C.; et al. Composite separators for robust high rate lithium ion batteries. *Adv. Funct. Mater.* **2021**, *31*, 2101420. [CrossRef]
2. Zhou, S.; Zhong, S.; Dong, Y.; Liu, Z.; Dong, L.; Yuan, B.; Xie, H.; Liu, Y.; Qiao, L.; Han, J.; et al. Composition and structure design of poly (vinylidene fluoride)-based solid polymer electrolytes for lithium batteries. *Adv. Funct. Mater.* **2023**, *33*, 2214432. [CrossRef]
3. Li, Z.; Dong, Y.; Feng, J.; Xu, T.; Ren, H.; Gao, C.; Li, Y.; Cheng, M.; Wu, W.; Wu, M. Controllably enriched oxygen vacancies through polymer assistance in titanium pyrophosphate as a super anode for Na/K-ion batteries. *ACS Nano* **2019**, *13*, 9227–9236. [CrossRef] [PubMed]
4. Goodenough, J.B. How we made the Li-ion rechargeable battery. *Nat. Electron.* **2018**, *1*, 204. [CrossRef]
5. Zhong, S.; Yuan, B.; Guang, Z.; Chen, D.; Li, Q.; Dong, L.; Ji, Y.; Dong, Y.; Han, J.; He, W. Recent progress in thin separators for upgraded lithium ion batteries. *Energy Storage Mater.* **2021**, *41*, 805–841. [CrossRef]
6. Lv, Z.; Kang, Y.; Tang, R.; Yang, J.; Chen, G.; Hu, Y.; Lin, P.; Yang, H.; Wu, Q.; Zhang, M.; et al. Stabilizing layered superlattice $MoSe_2$ anodes by the rational solvation structure design for low-temperature aqueous zinc-ion batteries. *Electron* **2023**, *1*, e5. [CrossRef]
7. Wu, Y.; Li, Y.; Wang, Y.; Liu, Q.; Chen, Q.; Chen, M. Advances and prospects of PVDF based polymer electrolytes. *J. Energy Chem.* **2022**, *64*, 62–84. [CrossRef]
8. Chattopadhyay, J.; Pathak, T.S.; Santos, D.M. Applications of polymer electrolytes in lithium-ion batteries: A review. *Polymers* **2023**, *15*, 3907. [CrossRef]
9. Liu, X.; Jia, H.; Li, H. Flame-retarding quasi-solid polymer electrolytes for high-safety lithium metal batteries. *Energy Storage Mater.* **2024**, *67*, 103263. [CrossRef]
10. Zhou, X.; Zhou, Y.; Yu, L.; Qi, L.; Oh, K.; Hu, P.; Lee, S.; Chen, C. Gel polymer electrolytes for rechargeable batteries toward wide-temperature applications. *Chem. Soc. Rev.* **2024**, *53*, 5291–5337. [CrossRef]
11. Zou, S.; Yang, Y.; Wang, J.; Zhou, X.; Wan, X.; Zhu, M.; Liu, J. In-situ polymerization of solid-state polymer electrolytes for lithium metal batteries: A review. *Energy Environ. Sci.* **2024**, *17*, 4426–4460. [CrossRef]
12. Liu, Q.; Yang, G.; Li, X.; Zhang, S.; Chen, R.; Wang, X.; Gao, Y.; Wang, Z.; Chen, L. Polymer electrolytes based on interactions between [solvent-Li^+] complex and solvent-modified polymer. *Energy Storage Mater.* **2022**, *51*, 443–452. [CrossRef]
13. An, Y.; Han, X.; Liu, Y.; Azhar, A.; Na, J.; Nanjundan, A.K.; Wang, S.; Yu, J.; Yamauchi, Y. Progress in solid polymer electrolytes for lithium-ion batteries and beyond. *Small* **2022**, *18*, 2103617. [CrossRef]
14. Sand, S.C.; Rupp, J.L.; Yildiz, B. A critical review on Li-ion transport, chemistry and structure of ceramic–polymer composite electrolytes for solid state batteries. *Chem. Soc. Rev.* **2024**, *54*, 178–200. [CrossRef] [PubMed]
15. Jia, M.; Zhao, N.; Huo, H.; Guo, X. Comprehensive investigation into garnet electrolytes toward application-oriented solid lithium batteries. *Electrochem. Energy Rev.* **2020**, *3*, 656–689. [CrossRef]
16. Yan, B.; Li, X.; Xiao, W.; Hu, J.; Zhang, L.; Yang, X. Design, synthesis, and application of metal sulfides for Li–S batteries: Progress and prospects. *J. Mater. Chem. A* **2020**, *8*, 17848–17882. [CrossRef]
17. Xu, R.; Han, F.; Ji, X.; Fan, X.; Tu, J.; Wang, C. Interface engineering of sulfide electrolytes for all-solid-state lithium batteries. *Nano Energy* **2018**, *53*, 958–966. [CrossRef]
18. Bi, Z.; Zhao, N.; Ma, L.; Fu, Z.; Xu, F.; Wang, C.; Guo, X. Interface engineering on cathode side for solid garnet batteries. *Chem. Eng. J.* **2020**, *387*, 124089. [CrossRef]
19. Huang, X.; Lu, Y.; Song, Z.; Rui, K.; Wang, Q.; Xiu, T.; Badding, M.E.; Wen, Z. Manipulating Li_2O atmosphere for sintering dense $Li_7La_3Zr_2O_{12}$ solid electrolyte. *Energy Storage Mater.* **2019**, *22*, 207–217. [CrossRef]
20. Lu, W.; Xue, M.; Zhang, C. Modified Li7La3Zr2O12 (LLZO) and LLZO-polymer composites for solid-state lithium batteries. *Energy Storage Mater.* **2021**, *39*, 108–129. [CrossRef]
21. Dirican, M.; Yan, C.; Zhu, P.; Zhang, X. Composite solid electrolytes for all-solid-state lithium batteries. *Mater. Sci. Eng. R Rep.* **2019**, *136*, 27–46. [CrossRef]
22. Xu, L.; Li, G.; Guan, J.; Wang, L.; Chen, J.; Zheng, J. Garnet-doped composite polymer electrolyte with high ionic conductivity for dendrite-free lithium batteries. *J. Energy Storage* **2019**, *24*, 100767. [CrossRef]
23. Keller, M.; Varzi, A.; Passerini, S. Hybrid electrolytes for lithium metal batteries. *J. Power Sources* **2018**, *392*, 206–225. [CrossRef]

24. Li, Y.; Xu, B.; Xu, H.; Duan, H.; Lü, X.; Xin, S.; Zhou, W.; Xue, L.; Fu, G.; Manthiram, A.; et al. Hybrid polymer/garnet electrolyte with a small interfacial resistance for lithium-ion batteries. *Angew. Chem. Int. Ed.* **2017**, *56*, 753–756. [CrossRef] [PubMed]
25. Li, L.; Deng, Y.; Duan, H.; Qian, Y.; Chen, G. LiF and $LiNO_3$ as synergistic additives for PEO-PVDF/LLZTO-based composite electrolyte towards high-voltage lithium batteries with dual-interfaces stability. *J. Energy Chem.* **2022**, *65*, 319–328. [CrossRef]
26. Li, S.; Zhang, S.Q.; Shen, L.; Liu, Q.; Ma, J.B.; Lv, W.; He, Y.B.; Yang, Q.H. Progress and perspective of ceramic/polymer composite solid electrolytes for lithium batteries. *Adv. Sci.* **2020**, *7*, 1903088. [CrossRef]
27. Zheng, Y.; Yao, Y.; Ou, J.; Li, M.; Luo, D.; Dou, H.; Li, Z.; Amine, K.; Yu, A.; Chen, Z. A review of composite solid-state electrolytes for lithium batteries: Fundamentals, key materials and advanced structures. *Chem. Soc. Rev.* **2020**, *49*, 8790–8839. [CrossRef]
28. Yang, S.J.; Jiang, F.N.; Hu, J.K.; Yuan, H.; Cheng, X.B.; Kaskel, S.; Zhang, Q.; Huang, J.Q. Life cycle safety issues of lithium metal batteries: A perspective. *Electron* **2023**, *1*, e8. [CrossRef]
29. Zhong, L.; Li, Z.; Wang, S.; Huang, S. Ultra-Thin Solid Electrolyte in Lithium-Ion Batteries. *Sustain. Polym. Energy* **2023**, *1*, 10004. [CrossRef]
30. Wang, Q.; Wang, S.; Lu, T.; Guan, L.; Hou, L.; Du, H.; Wei, H.; Liu, X.; Wei, Y.; Zhou, H. Ultrathin Solid Polymer Electrolyte Design for High-Performance Li Metal Batteries: A Perspective of Synthetic Chemistry. *Adv. Sci.* **2023**, *10*, 2205233. [CrossRef]
31. Chen, D.; Liu, Y.; Feng, C.; He, Y.; Zhou, S.; Yuan, B.; Dong, Y.; Xie, H.; Zeng, G.; Han, J.; et al. Unified throughout-pore microstructure enables ultrahigh separator porosity for robust high-flux lithium batteries. *Electron* **2023**, *1*, e1. [CrossRef]
32. Shi, P.; Ma, J.; Liu, M.; Guo, S.; Huang, Y.; Wang, S.; Zhang, L.; Chen, L.; Yang, K.; Liu, X.; et al. A dielectric electrolyte composite with high lithium-ion conductivity for high-voltage solid-state lithium metal batteries. *Nat. Nanotechnol.* **2023**, *18*, 602–610. [CrossRef] [PubMed]
33. Han, S.; Wen, P.; Wang, H.; Zhou, Y.; Gu, Y.; Zhang, L.; Shao-Horn, Y.; Lin, X.; Chen, M. Sequencing polymers to enable solid-state lithium batteries. *Nat. Mater.* **2023**, *22*, 1515–1522. [CrossRef] [PubMed]
34. Zhang, X.; Liu, T.; Zhang, S.; Huang, X.; Xu, B.; Lin, Y.; Xu, B.; Li, L.; Nan, C.; Shen, Y. Synergistic coupling between $Li_{6.75}La_3Zr_{1.75}Ta_{0.25}O_{12}$ and poly (vinylidene fluoride) induces high ionic conductivity, mechanical strength, and thermal stability of solid composite electrolytes. *J. Am. Chem. Soc.* **2017**, *139*, 13779–13785. [CrossRef] [PubMed]
35. Gu, Y.; She, S.; Hong, Z.; Huang, Y.; Wu, Y. Enabling lithium metal battery with flexible polymer/garnet type solid oxide composite electrolyte. *Solid State Ion.* **2021**, *368*, 115710. [CrossRef]
36. Li, F.; Ren, K.; Hou, M.; Lin, M.; Yang, X.; Zhou, Y.; Xiong, S.; Liang, F. Ultrafast UV Curing Enabling A Stable Interphase and Interface for Solid-State Sodium–Metal Batteries. *ACS Energy Lett.* **2024**, *10*, 195–204. [CrossRef]
37. Galos, J.; Pattarakunnan, K.; Best, A.S.; Kyratzis, I.L.; Wang, C.H.; Mouritz, A.P. Energy storage structural composites with integrated lithium-ion batteries: A review. *Adv. Mater. Technol.* **2021**, *6*, 2001059. [CrossRef]
38. Shateri, N.; Auger, D.J.; Fotouhi, A.; Brighton, J.; Du, W.; Owen, R.E.; Brett, D.J.; Shearing, P.R. Investigation of the effect of temperature on lithium-sulfur cell cycle life performance using system identification and x-ray tomography. *Batter. Supercaps* **2022**, *5*, e202200035. [CrossRef]
39. Stock, S.; Diller, F.; Böhm, J.; Hille, L.; Hagemeister, J.; Sommer, A.; Daub, R. Operando analysis of the gassing and swelling behavior of lithium-ion pouch cells during formation. *J. Electrochem. Soc.* **2023**, *170*, 60539. [CrossRef]
40. Li, Y.; Zhao, S.; Zhang, K.; Lu, G.; Li, Y. Extremely high heat flux dissipation and hotspots removal with nature-inspired single-phase microchannel heat sink designs. *Appl. Therm. Eng.* **2023**, *234*, 121282. [CrossRef]
41. Wang, Q.; Tao, J.; Cui, Z.; Zhang, T.; Chen, G. Passive enhanced heat transfer, hotspot management and temperature uniformity enhancement of electronic devices by micro heat sinks: A review. *Int. J. Heat Fluid Flow* **2024**, *107*, 109368. [CrossRef]
42. Hun, Q.; Lan, L.; Lu, X.; Hu, Q.; Liang, X.; Guo, Y.; Wang, Y. Bilayer Heterostructure Electrolytes Were Prepared by a UV-Curing Process for High Temperature Lithium-Ion Batteries. *Polymers* **2024**, *16*, 2972. [CrossRef] [PubMed]
43. Liang, X.; Zhang, Y.; Ning, Y.; Huang, D.; Lan, L.; Li, S. Quasi-solid-state lithium-sulfur batteries assembled by composite polymer electrolyte and nitrogen doped porous carbon fiber composite cathode. *Nanomaterials* **2022**, *12*, 2614. [CrossRef] [PubMed]
44. Liang, X.; Ning, Y.; Lan, L.; Yang, G.; Li, M.; Tang, S.; Huang, J. Electrochemical performance of a PVDF-HFP-$LiClO_4$-$Li_{6.4}La_{3.0}Zr_{1.4}Ta_{0.6}O_{12}$ composite solid electrolyte at different temperatures. *Nanomaterials* **2022**, *12*, 3390. [CrossRef]
45. Wang, Y.; Chen, Z.; Wu, Y.; Li, Y.; Yue, Z.; Chen, M. PVDF-HFP/PAN/PDA@ LLZTO composite solid electrolyte enabling reinforced safety and outstanding low-temperature performance for quasi-solid-state lithium metal batteries. *ACS Appl. Mater. Interfaces* **2023**, *15*, 21526–21536. [CrossRef]
46. Mallaiah, Y.; Jeedi, V.R.; Swarnalatha, R.; Raju, A.; Reddy, S.N.; Chary, A.S. Impact of polymer blending on ionic conduction mechanism and dielectric properties of sodium based PEO-PVdF solid polymer electrolyte systems. *J. Phys. Chem. Solids* **2021**, *155*, 110096. [CrossRef]
47. Wu, Y.; Zhu, T.; Lv, Y.; Fang, J.; Dong, S.; Yao, S. "Polymer in ceramic" type LLZTO/PEO/PVDF composite electrolyte with high lithium migration number for solid-state lithium batteries. *Ionics* **2024**, *30*, 787–798. [CrossRef]

48. Nguyen, M.C.; Nguyen, H.L.; Duong, T.P.M.; Kim, S.H.; Kim, J.Y.; Bae, J.H.; Kim, H.K.; Lim, S.N.; Ahn, W. Highly Safe, Ultra-Thin MOF-Based Solid Polymer Electrolytes for Superior All-Solid-State Lithium-Metal Battery Performance. *Adv. Funct. Mater.* **2024**, *34*, 2406987. [CrossRef]
49. Liang, X.; Huang, D.; Lan, L.; Yang, G.; Huang, J. Enhancement of the Electrochemical Performances of Composite Solid-State Electrolytes by Doping with Graphene. *Nanomaterials* **2022**, *12*, 3216. [CrossRef]
50. Zou, K.; Cai, Z.; Ke, X.; Wang, K.; Tan, X.; Luo, D.; Huang, F.; Wang, C.; Cheng, J.; Xiao, R. Electrochemical properties of LATP ceramic electrolyte doped with $LiBiO_3$ sintering additive and its derived sandwich structure composite solid electrolyte. *Ionics* **2023**, *29*, 2665–2678. [CrossRef]
51. Xu, C.; Xiang, W.; Wu, Z.; Xu, Y.; Li, Y.; Wang, Y.; Xiao, Y.; Guo, X.; Zhong, B. Highly stabilized Ni-rich cathode material with Mo induced epitaxially grown nanostructured hybrid surface for high-performance lithium-ion batteries. *ACS Appl. Mater. Interfaces* **2019**, *11*, 16629–16638. [CrossRef]
52. Shen, X.; Tian, Z.; Fan, R.; Shao, L.; Zhang, D.; Cao, G.; Kou, L.; Bai, Y. Research progress on silicon/carbon composite anode materials for lithium-ion battery. *J. Energy Chem.* **2018**, *27*, 1067–1090. [CrossRef]
53. Lin, L.; Chen, C. Accurate characterization of transference numbers in electrolyte systems. *J. Power Sources* **2024**, *603*, 234236. [CrossRef]
54. Meng, J.; You, Y.; Lin, M.; Wu, J.; Song, Z. Multi-scenarios transferable learning framework with few-shot for early lithium-ion battery lifespan trajectory prediction. *Energy* **2024**, *286*, 129682. [CrossRef]

gels

MDPI

Review

Gel Electrolytes in the Development of Textile-Based Power Sources

Ana Isabel Ribeiro *, Cátia Alves, Marta Fernandes, José Abreu, Fábio Pedroso de Lima, Jorge Padrão and Andrea Zille

Centre for Textile Science and Technology (2C2T), Department of Textile Engineering, University of Minho, Campus of Azurém, 4800-058 Guimarães, Portugal; catia.alves@2c2t.uminho.pt (C.A.); marta.fernandes@det.uminho.pt (M.F.); id11556@uminho.pt (J.A.); fabioplima.conceicao@gmail.com (F.P.d.L.); padraoj@2c2t.uminho.pt (J.P.); azille@2c2t.uminho.pt (A.Z.)
* Correspondence: afr@2c2t.uminho.pt

Abstract: The interest in flexible and wearable electronics is increasing in both scientific research and in multiple industry sectors, such as medicine and healthcare, sports, and fashion. Thus, compatible power sources are needed to develop secondary batteries, fuel cells, supercapacitors, sensors, and dye-sensitized solar cells. Traditional liquid electrolytes pose challenges in the development of textile-based electronics due to their potential for leakage, flammability, and limited flexibility. On the other hand, gel electrolytes offer solutions to these issues, making them suitable choices for these applications. There are several advantages to using gel electrolytes in textile-based electronics, namely higher safety, leak resistance, mechanical flexibility, improved interface compatibility, higher energy density, customizable properties, scalability, and easy integration into manufacturing processes. However, it is also essential to consider some challenges associated with these gels, such as lower conductivity and long-term stability. This review highlights the application of gel electrolytes to textile materials in various forms (e.g., fibers, yarns, woven, knit, and non-woven), along with the strategies for their integration and their resulting properties. While challenges remain in optimizing key parameters, the integration of gel electrolytes into textiles holds immense potential to enhance conductivity, flexibility, and energy storage, paving the way for advanced electronic textiles.

Keywords: gel electrolytes; wearable electronics; e-textiles; energy storage; sensors

1. Introduction

The rapid growth of wearable and portable smart technologies has significantly increased the demand for efficient and miniaturized energy storage systems [1–6]. In this field, fiber-based wearable electronics represent a promising intersection between textiles and technology, offering a wide range of opportunities for creating intelligent and interactive materials that can be integrated into daily life and technical products. These materials have attracted attention from scientists and industries in several areas due to their high flexibility, stretchability, and conformal contact with the skin [7,8]. Produced using well-established and cost-effective textile production methods, they offer the potential to create innovative and functional wearable devices with diverse applications in flexible sensors, personalized healthcare, entertainment, and thermal management [9]. Traditional energy storage devices, such as batteries, can hold significant energy but struggle to deliver it rapidly due to their high internal resistance. In contrast, capacitors can charge and discharge quickly but have limited energy storage capacity. To bridge this gap, the development of high-energy-density capacitors, known as supercapacitors, has emerged as a key research focus in electrochemical energy storage and conversion [10]. Supercapacitors stand out as one of the most promising energy storage solutions, relying on non-Faradaic and/or Faradaic processes occurring at the electrode–electrolyte interface. Faradaic processes involve electron transfer and redox reactions at the electrode surface, while non-Faradaic processes involve no electron transfer, only physical phenomena like double-layer charging or adsorption [11]. Compared to conventional batteries, supercapacitors offer notable advantages, including an extended lifespan of over 10,000 charge/discharge cycles, rapid charging within seconds, and superior power density. However, their practical use remains constrained by their relatively low energy density compared to batteries. Despite this limitation, ongoing research actively explores innovative strategies to enhance supercapacitor performance [1,12]. Supercapacitors can be classified into three types based on their energy storage mechanisms: electrical double-layer capacitors (EDLC), pseudocapacitors, and hybrid supercapacitors [12]. EDLCs function similarly to lithium-ion (Li-ion) batteries, storing energy through static charge accumulation caused by the adsorption of anions and cations onto the electrode surface. Typically, they are made from carbon-based materials like graphene or carbon nanotubes (CNTs), as their porous structures provide a high surface area, enabling greater electron activity and increased capacitance. In contrast, pseudocapacitors utilize Faradaic and electrochemical reactions between the electrolyte and the active electrode material for energy storage and are often composed of conductive polymers or metal oxides. Hybrid supercapacitors integrate EDLC and pseudocapacitor properties, making them particularly suitable for textile-based energy storage applications. Thus, selecting the appropriate materials for textile-based energy storage is crucial, as the final device must exhibit high conductivity, surface area, efficient electron transport, and strong power output while maintaining flexibility, stretchability, mechanical durability, and washability [13,14]. Electrolytes are essential components of electrochemical energy storage devices, where their physical and chemical characteristics significantly influence performance metrics such as capacity, power density, charge/discharge rate, cycle stability, and overall safety [15]. A widely accepted approach to enhancing the safety and flexibility of wearable energy storage systems is replacing liquid electrolytes with solid or quasi-solid-state electrolytes. Those electrolytes offer several advantages over other types, including high ionic conductivity, straightforward packaging, and the absence of liquid leakage. In the context of supercapacitors, polymer-based solid electrolytes are widely utilized and can be categorized into three main types: solid polymer electrolytes, gel polymer electrolytes, and polyelectrolytes. Gel polymer electrolytes, often referred to as quasi-solid-state electrolytes, contain a liquid phase [16,17]. These electrolytes consist of a polymer matrix, such as poly(ethyl oxide) (PEO), poly(ethylene glycol) (PEG), or poly(vinyl alcohol) (PVA), combined with an aqueous electrolyte like potassium hydroxide (KOH), sulfuric acid (H_2SO_4), or potassium sulfate (K_2SO_4), or a conducting salt dissolved in a solvent. Gel electrolytes commonly have the highest ionic conductivity compared to the three types of solid-state electrolytes due to a liquid phase [15,18]. However, inadequate wetting at the interface between the polymer gel electrolyte and the electrode may lead to poor electrochemical performance [19]. Selecting a host polymer compatible with the redox couples is key; this involves factors such as its pH sensitivity, low solubility, thermal stability, and mechanical properties [20]. Various techniques and material combinations can be employed to incorporate gel electrolytes into textiles, with the selection depending on the specific application requirements and the desired properties of the final product. This review examines recent advancements in textile-based energy storage systems that apply gel electrolytes, focusing on how their design and integration impact system performance. The discussion is structured according to different textile structures, including fibers, yarns, woven fabrics, knitted fabrics, and non-woven materials. Additionally, this review highlights the crucial role of gel electrolytes in preserving the durability of conductive properties, ensuring long-term electrochemical stability, mechanical flexibility, and overall device reliability.

2. Gel Electrolytes Applied in Textile Structures

2.1. Fibers

In the field of wearable energy storage, fiber-form devices have received tremendous interest due to being highly flexible, lightweight, and easily integrated or directly weaved/knitted into textiles [21,22]. In this section, we will describe scientific studies that use fibers as a base for constructing energy storage textiles, focusing on their design, fabrication methods, and performance characteristics (Table 1). Flexible fiber-shaped supercapacitor devices are usually assembled in three different configurations, namely, parallel, twisting, and coaxial, and coated with an electrolyte (Figure 1) [23,24]. Various carbon-based materials have been studied for the development of these devices, including graphene and CNT, and their composites with transition metal oxides or conducting polymers, such as polyaniline (PANI) and poly(3,4-ethylenedioxythiophene) poly(styrenesulfonate) (PEDOT:PSS). These materials endow the fabricated supercapacitors with outstanding electrochemical attributes, including high specific capacitance and electrical conductivity, porosity, and reversible electrochemical doping/de-doping capability [25].

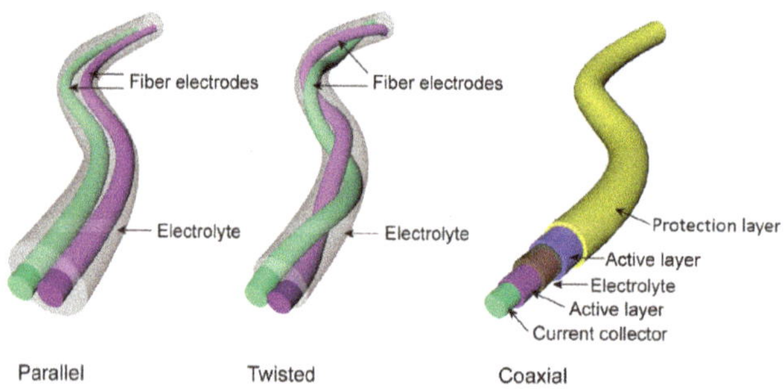

Figure 1. Scheme showing design of three common fiber supercapacitors using gel electrolytes [24].

Among the various electrode materials, graphene has emerged as a promising candidate for the creation of flexible fibrous supercapacitors [26,27]. Gao et al. [28] reported the creation of a wearable fiber-shaped supercapacitor using reduced graphene oxide (rGO) obtained via a wet-spinning method. The graphene oxide (GO) dispersions were injected into a coagulation bath containing calcium chloride ($CaCl_2$) and dried. This was followed by chemical reduction by immersion into a hydriodic acid solution. Then, to prepare the fiber-shaped supercapacitor, two symmetric rGO fibers were coated with PVA/H_2SO_4 gel electrolyte and twisted to form a double-helix-shaped fiber. The produced rGO fibers showed a highly oriented porous structure, facilitating ion transfer and improving ionic site accessibility. The device displayed a volume ratio capacitance of 5.0 mF/cm^3 and observed an increase in the specific capacitance tested at 40 mA/cm^3, which could be attributed to the wetting of graphene with the electrolyte during the charge/discharge cycle. Moreover, the assembled supercapacitors successfully filtered an alternating current (AC) input signal of 1.5 V (60 Hz) to a stable direct current (DC) output signal of 1.04 V. In another work from Park et al. [29], hybrid fibers of porous CNT-rGO were also fabricated by wet spinning. CNTs were used as spacers between the rGO sheets, enhancing the porosity of rGO and the ion accessibility. The hybrid fibers were subsequently activated hydrothermally using hydrogen peroxide (H_2O_2) and ammonium hydroxide (NH_4OH), further enhancing the porosity. Two parallel individual fibers were coated with a thin layer of the PVA/KOH gel electrolyte to form the solid-state fiber-based supercapacitors, where the volumetric capacitance reached 6.1×10^4 mF/cm^3 and the volumetric energy density reached 4.83 mWh/cm^3. The supercapacitors also exhibited excellent cycling stability of 94% after 10,000 cycles and stable capacitive performance under deformation conditions. Zhang et al. [30] demonstrated the application of zinc-ion (Zn-ion) fiber supercapacitors in wearable electronics by hydrothermally assembling rGO and single-walled CNT fiber as a cathode, coating graphite fibers with a uniform Zn layer as an anode, and producing a new neutral zinc sulfate ($ZnSO_4$)-filled polyacrylic acid (PAA) hydrogel as a quasi-solid-state electrolyte. The single-walled CNTs were added to prevent the restacking of rGO nanosheets, so the ion-accessible surface was enlarged, while the hydrogel electrolyte offered high ionic conductivity and stretchability. The zinc (Zn) fiber supercapacitors delivered a high energy density of 4.9×10^{-2} Wh/cm^3 at a power density of 1.8×10^{-1} W/cm^3, a volumetric capacitance of 1.0×10^5 mF/cm^3 at a current density of 400 mA/cm^3, and an outstanding stable cycling stability of 98.5% capacity retention after 10,000 cycles, showing promising applications in wearable electronics. In the work carried out by Li et al. [31], the porosity of rGO was increased by doping with 1T-molybdenum disulfide (MoS_2). MoS_2 prevents the tight restacking of rGO into compact lamellar structures. The fibers were prepared by wet spinning, followed by reduction via immersion in aqueous hydroiodic acid (HI) solution. They were then treated with PVA/phosphoric acid (H_3PO_4) gel electrolytes to produce all-solid-state fiber-shaped supercapacitors. The volumetric capacitance reached 2.2×10^5 mF/cm^3 (at 50 mA). Three supercapacitors were connected in parallel in a fabric, reaching a high capacitance of 2.4×10^2 mF/cm^2 (at 500 mA) and an energy density of 32 mWh/cm^3. However, the specific capacity was reduced by 28% after bending 500 times at different angles. Chen and coworkers [32] fabricated a new type of 1D carbon nanomaterial, carbon nanobranches, covered with abundant carbon dots by the pyrolysis of starch. The conductive fibers composed of carbon nanobranches and thermoplastic polyurethane (PU) were produced using a customized microfluidic spinning approach. For the assembly of a microsupercapacitor, two aligned conductive fibers were twisted together and coated with the PVA/H_3PO_4 gel electrolyte. The produced fiber-shaped supercapacitor displayed a large specific capacitance of 2.0×10^2 mF/cm^2, a high energy density of 4.5×10^{-6} Wh/cm^2 (at 8.0×10^{-5} W/cm^2), and impressive electrochemical stability with a capacitance retention of 98% after 10,000 cycles. Furthermore, after integration into a woven fabric, it successfully powered 19 light-emitting diode (LEDs). The development of supercapacitors based on composites of carbon materials and PANI has also been extensively growing. PANI has appreciable electrochemical properties such as high pseudocapacitance (around 1200 F/g), cost-effectiveness, and environmental stability [33,34]. Storage efficiency depends on the degree of H^+ doping and the oxidation state of the macromolecular chain, and so grafting PANI onto carbonaceous materials enables ion transfer during the redox reaction, which increases the capacitance [35]. For example, Jiang et al [36] designed supercapacitor fibers based on PANI, GO, and CNTs for wearable electrodes. First, a composite film was developed by preparing a CNT film by chemical vapor deposition, followed by immersion in a GO/PANI mixture solution. The composite film exhibited a specific capacitance of 729.3 F/g (5.1×10^2 mF/cm^2) at 1 A/g. Then, for the preparation of the fiber-like supercapacitor, the composite film was over-twisted into two helical fibers and coated with a PVA/H_3PO_4 gel electrolyte as the separator. The two helical fibers were intertwined and coated with the PVA/H_3PO_4 gel electrolyte again, creating stretchable fiber-shaped supercapacitors. This device presented a specific capacitance of 89.5 F/g (1.8×10^2 mF/cm^2) and a relatively good capacitance retention of approximately

80% over 5000 cycles (at 5 A/g). In another study, Adusei et al. [21] developed PANI–graphene–CNT hybrid fiber electrodes by preparing CNT/rGO via the chemical vapor deposition of rGO in CNT fibers, followed by coating with PANI nanorods via oxidation polymerization. The hybrid fibers were used as the supercapacitor electrode using the gel electrolyte polyvinylidene fluoride (PVDF)- 1-ethyl-3-methylimidazolium tetrafluoroborate ($EMIMBF_4$). The fast-redox reactions between the PANI and the electrolyte gave rise to the superior electrochemical performance of these fibers. The supercapacitors achieved a gravimetric energy density of 1.3×10 Wh/kg and a power density of 1.4×10^3 W/kg (at a current density of 1 A/g). In addition, the supercapacitors demonstrated excellent cyclic stability by maintaining a high capacitance retention of 98.4% after 2000 cycles. PEDOT is another conducting polymer that has drawn interest in the development of flexible supercapacitors owing to its high conductivity (up to 103 S/cm), substantial theoretical specific capacitance (210 F/g), and high chemical and physical stability [37]. For example, He et al. [38] synthesized PEDOT:PSS thin-walled hollow fibers by coaxial extrusion via wet spinning to obtain a flexible supercapacitor with a high surface area. The thin fibers displayed a diameter of approximately 125 μm and a wall structure of 8 μm, which allowed for an increase in the specific capacitance of ~ 67%. When used as electrode materials for gel electrolyte fiber-shaped supercapacitors, the fibers displayed a capacitance of 1.1×10^2 mF/cm^2 (at 0.3 mA/cm^2) with a high energy density of 9.0×10^{-6} Wh/cm^2 (at 1.1×10^{-4} W/cm^2). In addition, impressive long-term cycle stability was observed, maintaining 81% of the initial capacitance after 10,000 cycles. The excellent electrochemical performance was attributed to the fiber's structure, the electrochemical characteristics of the PEDOT:PSS, and the stability of the gel electrolyte. In another work from Gibertini et al. [39], an easy method was developed for the preparation of robust and flexible composite fiber-shaped capacitors by wet spinning nanofibrillated Kevlar®, followed by immersion in a gel state with PEDOT:PSS aqueous dispersion. The symmetric device, coupled with PVA/H_3PO_4 electrolyte gel, displayed a capacitance of 1.1 mF, a volumetric energy density of 7.1×10^{-2} mWh/cm^3, and good capacitance retention of 80.5% after 10,000 cycles. When integrated into an e-textile circuit, it was able to power a blue LED for several seconds. Efforts have been made to produce long and continuous fibers electrodes. Hong et al. [40] designed a multichannel spinning method to continuously produce supercapacitor fibers, kilometers in length, at high production rates of up to 118 m/h. A multichannel spinneret, with two internal parallel nozzles for electrode inks (PEDOT:PSS/CNT) and a larger external nozzle carrying gel electrolyte (PVA/chitosan), simultaneously extruded two fiber electrodes into the coagulation bath (sodium hydroxide (NaOH) containing lithium perchlorate ($LiClO_4$)), where the gel electrolyte was quickly solidified. The supercapacitor fibers displayed good electrochemical stability after being bent for 1×10^5 cycles. In particular, they had a volumetric capacitance of 1.6×10^4 mF/cm^3 (at 0.11 A/cm^3) and displayed cycle stability, with 96% capacity remaining after 5000 cycles. Moreover, it could be woven into a flexible and wearable power scarf, reaching a capacitance of 1.34×10^6 F and an energy of 116 Wh. In the work by Lai and coworkers [37], a new strategy was developed to synthesize wire-shaped solid-state supercapacitors with excellent electrochemical and mechanical performances in a facile dip-coating process. Polyacrylonitrile (PAN) nanofibers were electrospun on a titanium (Ti) metal wire using glycerol to form a sacrificing aerogel template with a huge void volume, followed by dip-coating with PEDOT:PSS and the dissolution of the template. PSS was then removed using H_2SO_4, forming a highly porous PEDOT layer. Supercapacitors based on the Ti/PEDOT wire electrodes and PVA/H_2SO_4 gel electrolyte reached an energy density of 5.5 Wh/kg (at 3.5×10^2 W/kg), exhibiting a specific capacitance of 68 F/g (at 0.97 A/g) and a power density of 9.1×10^3 W/kg (at 4.6 Wh/kg). The cycling stability of the fabricated supercapacitor retained 81% of the capacitance after 10,000 cycles (at 25 A/g). Moreover, the use of supercapacitors as a power source in wearables was demonstrated, as they could successfully power a purple LED by knitting it into a fabric and connecting it in series. Transition metal oxides such as vanadium pentoxide (V_2O_5) were incorporated into conductive polymeric fibers to achieve high capacitance and good cyclic stability, protect the structure from damage during charge/discharge cycles, and increase the molecular order and stability. Achieving high performance using wet spinning, Xu et al. [41] adopted a direct mixing process to fabricate supercapacitors successfully based on PEDOT:PSS/V_2O_5 hybrid fibers. The fiber-shaped supercapacitor in the PVA/H_2SO_4 gel electrolyte achieved an energy density of 1.4×10^{-6} Wh/cm^2 (at 2.0×10^{-5} W/cm^2), a specific capacitance of 6.0×10 mF/cm^2, and excellent cycle stability with a retention of 94.02% after 4000 cycles (at 0.1 mA/cm^2). MXenes are 2D transition metal carbonitride layered materials that have attracted considerable interest for the development of energy storage devices because of their advantageous features, such as large specific surface area, hydrophilicity, high conductivity (>6000 S/cm), adjustable functional terminal groups, and fast electron transport capabilities [42]. Qin et al. [43] used $Mo_{1.33}C$ *i*-MXene nanosheets and PEDOT:PSS as the base materials to prepare composite high-performance fibers as electrodes that can be applied to next-generation wearable devices. PEDOT:PSS was used as a conductive binder to promote spinnability and improve flexibility by increasing the interlayer spacing between $Mo_{1.33}C$ *i*-MXene layers. This new fiber was assembled into a solid-state fiber-based asymmetric supercapacitor with reduced rGO fibers and a thin layer of PVA/H_2SO_4 gel electrolyte, reaching a capacitance of 105 F/g and an energy density of 3.7×10^{-2} Wh/g (at 1.6 V). The supercapacitor also demonstrated excellent cycling stability, with a capacitive retention of 94% after 10,000 cycles. In another work, Lu et al. [44] developed ternary manganese dioxide (MnO_2)/MXene-$Ti_3C_2T_x$/rGO composite fiber supercapacitors by wet spinning MXene-$Ti_3C_2T_x$/GO, reducing it in HI/acetic acid (CH_3COOH) solution, and immersing it in potassium permanganate ($KMnO_4$) solution. Two of these fiber electrodes were twisted together and assembled with a PVA/LiCl gel electrolyte in an all-solid-state symmetric fiber supercapacitor. The new fiber supercapacitor displayed a volumetric capacitance of 2.4×10^4 mF/cm^3, a volumetric energy density of 2.1×10^{-3} mWh/cm^3 (at 8.2×10^{-23} W/cm^3), and a good cycle stability of 92% after 10,000 cycles. Wei et al. [45] prepared regenerated cellulose (RC)-based conductive microfibers in a simple, continuous, and scalable wet-spinning process. A mixture of MXene/MnO_2-wood fiber cellulose (WFC) was span in an H_2SO_4 coagulation bath; we obtained a MXene/MnO_2-RC fiber cathode with a highly aligned and stable three-dimensional (3D) nanonetwork structure formed by the self-assembly of RC during molecular chain rearrangement. Due to the strong interaction between cellulose and electroactive nanomaterials and their highly aligned structure, the conductive fibers showed good tensile strength and improved electrolyte permeation. The assembled quasi-solid-state fiber electrodes, with MXene/MnO_2-RC as the cathode, Zn wire as the anode, and zinc sulfate ($ZnSO_4$)/gelatin gel as the electrolyte, showed excellent electrochemical performance with a high volumetric capacitance of 1.1×10^2 mF/cm^3 (at 0.57 mA/cm^3), an energy density of 2.2×10^{-2} mWh/cm^3 (at 0.57 A/cm^3), excellent cycling stability with 90.5% capacity retention, and close to 100% Coulombic efficiency, after 5000 cycles. Recently, Ovhal et al. [46] produced one-meter-long 3D fiber-shaped supercapacitors of MXene-PEDOT:PSS active electrodes using the 3D direct-ink-writing method. MXene/PEDOT:PSS/Ag/MXene/PEDOT:PSS composite fibers with a sandwich structure were fabricated by extruding the MXene/PEDOT:PSS ink in an ion-permeable dialysis membrane separator; then, an Ag current collector was printed on top of the electrode, followed by an additional layer of MXene/PEDOT:PSS. The process was repeated for five cycles. The PVA/H_3PO_4 gel electrolyte was then printed; we finished the process by encapsulating the whole fiber with NOA 63 resin (Figure 2). The Ag current collector was used to facilitate faster charge transport, eliminating the thickness and length problems of electrode-dependent capacitance in fiber-shaped devices. The fiber-shaped 3D supercapacitors showed prominent electrical conductivity of 1.6×10^3 S/cm, a high areal capacitance of 1.1×10^3 mF/cm^2, a gravimetric capacitance of 185.9 F/g, and a high areal energy density of 9.4×10^{-5} Wh/cm^2 (at 1.1×10^{-3} W/cm^2). Moreover, the fibers demonstrated an excellent stability of 92% after 25,000 cycles. After being incorporated into a woven fabric, they were able to successfully power an LED indicator.

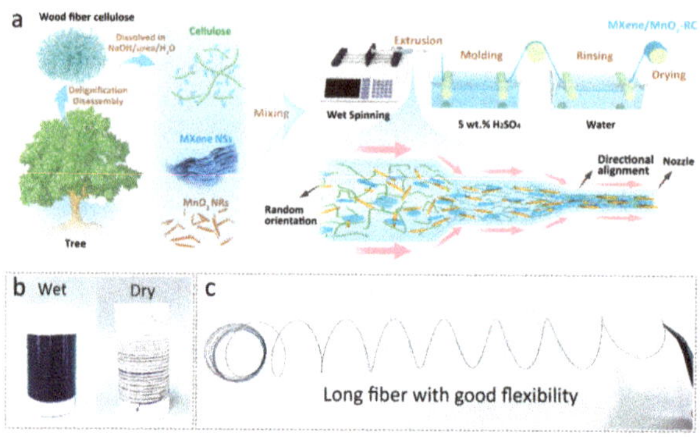

Figure 2. (**a**) Scheme depicting fabrication process of MXene/MnO_2-RC composite fibers. (**b**) Image showing fibers in both wet and dry conditions. (**c**) Long fiber demonstrating excellent flexibility [45].

Smart textiles can behave like batteries by incorporating flexible fiber-shaped rechargeable materials on a single-fiber architecture with high energy density and long-term stability. To ensure that a thin fiber battery works upon stretching, bending, and cleaning, Xiao et al. designed a high-performance rechargeable solid-state Zn/MnO_2 fiber battery with a GO-embedded PVA hydrogel electrolyte. GO supports ion transport, improving the ionic conductivity of the hydrogel electrolyte. This rechargeable solid-state fiber battery showed stable cyclic performance, exceeding the standard performance with up to 500 h of 98.0% capacity after more than 1000 charging/recharging cycles. This Zn/MnO_2 fiber battery could be used as a stable energy power unit for human health monitoring by seamlessly integrating it into a multifunctional e-textile [47]. Zhang et al. [48] developed high-voltage coaxial–fiber aqueous rechargeable Zn-ion batteries (CARZIBs) with outstanding flexibility and high-capacity retention. This novel coaxial structure consisted of Zn nanosheet array on CNT fibers as the core electrode, carboxymethyl cellulose (CMC)-$ZnSO_4$ as the gel electrolyte, and Zn hexacyanoferrate (ZnHCF) on a CNT sheet as the outer electrode. CMC, used as a cathode binder, provided good mechanical properties. The assembled battery achieved a large capacity of 100.2 mAh/cm^3 (at 0.1 A/cm^3) and a high energy density of 195.4 mWh/cm^3 (at 2.0×10^{-1} W/cm^3), a capacity retention of 91.8% after 200 cycles, and a Coulombic efficiency of 96.8%. Additionally, the wearability indicated by the high-capacity retention of 93.2% (after bending 3000 times) was demonstrated by incorporating serial and parallel CARZIB devices into a fabric to illuminate a 3.3 V blue LED. Pan et al. [49] successfully designed advanced cathodes of CNT-stitched pyrovanadate (ZVO) 2D nanosheets, grown on oxidized CNT fibers, for wearable quasi-solid-state ZIBs. CNTs improved electronic conductivity and mechanical robustness across the 2D cathode nanosheets. Moreover, the large open frameworks of ZVO offer superior reversible Zn^{2+} ion deintercalation/intercalation. The hybrid cathode was assembled with Zn nanosheets, electrodeposited onto CNT fibers as the anode and CMC/HSO_4 as the electrolyte gel. The fiber-shaped high-performance ZIB includes a high-rate capability of 69.7% capacity retention at 100 times enhancement in current density (discharge in 25 s), a high stack volumetric energy density of 7.2×10^{-2} mWh/cm^3, and excellent flexibility with cycle stability of 88.9% retention after 2000 cycles (at 1.0 A/cm^3). The fiber-shaped ZIB knitted into a cotton (CO) string glove exhibited outstanding flexibility after bending to different states. Li et al. [50] made a great advance in the further development of Ag-based cathodes for flexible batteries by eliminating issues associated with the structural pulverization and the migration of Ag^+ ion into the electrolyte solution. The cathode of the quasi-solid-state fiber-shaped zinc silver oxide (Zn-Ag_2O) battery was constructed by depositing Ag_2O nanoparticles onto a nitride-doped CNT hybrid fiber and coating a conducting polymer with a protective layer of PEDOT:PSS as the cathode to eliminate the issues mentioned. The anode was obtained by growing Zn nanosheets directly on the hybrid fiber and PVA/KOH flexible hydrogel was employed as an electrolyte. The as-prepared flexible fiber-shaped Zn–Ag_2O battery could be integrated with a wind power generator to be charged solely by wind energy, exhibiting a high capacity of 1.05 mAh/cm^2, a high energy density of 1.6×10^{-3} mWh/cm^2, a remarkable power density of 1.4×10^{-3} W/cm^2, and good electrochemical durability with 79.5% retention after 200 cycles. In another work from Li et al. [51], ZIBs with a double-helix structure were reported and used for highly deformable wearable textiles. Ag fiber was covered in graphene nanosheets to protect it from corrosion and coated with PANI to enhance the electrical conductivity (cathode). Another Ag fiber was coated with Zn nanoflake by electrodeposition (anode). Then, both electrodes and the PVA/$ZnSO_4$ gel electrolyte were encapsulated with a PU film, which was parallel-twisted around a spandex fiber. The resulting fiber-shaped Zn-ion battery displayed an enhanced specific capacity of 32.56 mAh/cm^3, a desirable energy storage performance with an energy density of 3.6×10^{-2} mWh/cm^3 (at 3.0×10^{-3} W/cm^3), excellent cycling, and deformation stability with a capacity of 76.5% after 1000 cycles, and capacities of 99, 93.6, and 91.5% during knotting, bending, and twisting, respectively. Due to the high stretchability of the spandex fiber and the unique helix structure, the battery exhibited a remarkable strain up to 900%, maintaining 71% of its original capacity. Another stretchable fiber-shaped battery was reported by Xiong et al. [52], where aqueous Al-ion batteries were assembled with a manganese hexacyanoferrate (MnHCF) cathode, a GO-modified MoO_3 anode, and a PVA/aluminiumtrifluormethansulfonat ($Al(CF_3SO_3)_3$) hydrogel electrolyte. A CNT fiber was coated with the active materials within the hydrogel electrolyte. Then, both fibers were assembled in parallel, winded on an elastic silicone rubber substrate to form a helical structure, and coated with another layer of the gel electrolyte. PVA was employed due to the good mechanical properties and high-water absorbency, which resulted in a PVA-based hydrogel with good transparency, high modulus (0.55 MPa) and strain (461%), and ionic conductivity (2.2×10^{-2} S/cm), indicating the potential of the hydrogel for improving the electrochemical performance of aluminum (Al) ion batteries. The assembled fiber-based battery achieved a high specific capacity of 42 mAh/cm^3, a high energy density of 3.1×10^{-2} Wh/cm^3 (at 0.5 A/cm^3), and good stability by retaining 91.6% capacity after 100 cycles (at 1 A/cm^3). Finally, when integrated into wearable textiles, the developed fiber-shaped batteries could power one red LED, showing their potential for wearable electronics. Recently, Zhao et al. [53] developed a new method for preparing flexible fiber Li-ion batteries using direct ink writing-based 3D printing. Lithium iron phosphate (LFP) and lithium titanate (LTO) were used as cathode and anode active materials, respectively. First, GO was etched using H_2O_2 and reduced with vitamin C, followed by freeze-drying to obtain rGO with holes (HrGO). This porous structure provides additional channels, increasing the Li^+ ions' transmission rate. An adhesive based on PVDF was prepared by dissolving PVA in *N*-Methyl-2-pyrrolidone (NMP), and then PVDF was added. For the preparation of the printing ink, LFP or LTO and CNTs were dissolved in the PVA/PVDF solution with a specific mass ratio of LFP/LTO: HrGO:CNTs:PVA/PVDF of 12:3:1:5. The flexible fiber electrodes were produced by extruding the printing inks into a sodium metaborate ($NaBO_2$) aqueous solution. BO_2^- ions undergo solvent exchange with NMP and transfer to the interior of the electrode, precipitating the PVDF on the surface of the fibers and cross-linking the PVA, which forms a three-dimensional network structure with borate ester bonds, improving mechanical properties and stability. Moreover, the dynamic reversibility of the borate bond imparts a self-healing property to the material. The fibers then passed through a PVDF/HFP/aluminum oxide (Al_2O_3) solution, forming a separator layer on the surface of the fiber electrode, which prevents the passage of electrons while allowing ions to pass through. The obtained electrodes reached a strain of ~30%, and the assembled fiber Li-ion battery exhibited a volumetric energy density of 1.6×10^{-1} Wh/cm^3. The functional fibers and developed materials were applied in wearable AC line filters [28], wearable and portable miniaturized energy storage devices [29,30,36,44,48–52], and textile capacitors [21,32,39]. The examples presented here, and summarized in Table 1, show the trend towards the use of gel electrolytes in the development of fiber-based electrodes. This fact is explained by PVA's ability to form stable films and its low economic cost [54]. However, due to its low electrical conductivity at ambient temperature, it is usually used in complexes with acids such as H_2SO_4 and H_3PO_4, which provide free ions in gel electrolytes, acting as charge carriers [55]. In the case of PVDF, when compared to other polymers, it presents higher mechanical properties and excellent chemical and thermal stability, being suitable for use as the skeleton of gel electrolytes. PVDF, when mixed with HFP or ionic liquids, has a reduced crystalline phase and increased amorphous region, increasing ion permeation capacity [56].

Analyzing the overall results (Table 1), gel electrolytes based on PVA, PVDF, CMC, or gelatin have been widely utilized as polymeric matrices for the development of fiber-based energy storage materials, with PVA emerging as the most frequently used. PVA is highly favored due to its excellent mechanical flexibility, film-forming ability, high ionic conductivity, water solubility, and environmental friendliness, making it ideal for wearable and flexible energy devices. The PVA matrix can be effectively combined with various acids and salts such as H_2SO_4, H_3PO_4, LiCl, $ZnSO_4$, and KOH, as well as natural polymers like chitosan, to enhance ionic conductivity and electrochemical stability. Additionally, the incorporation of GO further improves performance by increasing ion transport and mechanical strength. These gel electrolytes are mostly applied in conductive fibers such as CNT, GO, and PEDOT:PSS, or in other synthetic fibers like PAN, Kevlar®, and spandex.

2.2. Yarns

The preparation of yarn-based power sources using gel electrolytes has become a prominent research topic in recent years in the development of batteries and supercapacitors (Table 2). Similar to what can be observed for fibers, the use of yarns to assemble energy storage textiles is also linked to their versatility, applicability, and their ease of integration into the device. Here, we describe the latest scientific studies where different yarns were used in combination with gel electrolytes to develop energy storage apparatuses, highlighting the assembly procedure, components, and electrochemical performance. Yarns from different textile substrates can be used, from polyester (PET) to cotton (CO) or carbon fibers, which can then be assembled into yarn supercapacitors (YSCs). The most widely used method to assemble energy storage devices using yarns involves the construction of two electrodes, an anode and a cathode, by dip-coating, electrodeposition, electroless deposition, or in situ polymerization. The yarns are then used to assemble the devices, combining them with different gel electrolytes, resulting in high-energy and high-power-density devices with superior electrochemical performances.

Table 1. Comparison of fiber-based energy storage devices using gel electrolytes, highlighting preparation methods and electrochemical properties.

Gel Electrolyte	Fiber Composition	Electrodes	Preparation Method	Conductivity Tests and Results	Ref.
PVDF/$EMIMBF_4$	CNT	PANI–graphene–CNT	Dry-spinning CNT fibers; CNT/rGO via chemical vapor deposition; PANI nanorod coating; gel electrolyte coating	Energy density of 1.3×10 Wh/kg; power density of 1.4×10^3 W/kg; 98.4% capacitance retention after 2000 cycles	[21]
H_2SO_4	GO	rGO	GO wet spinning; gel electrolyte coating and two-fiber twisting	Excellent performance and fast reponse rate (500 V/s); cycling stability; stable performance at different angles	[28]
KOH	CNT-GO	CNT-rGO	CNT-GO wet spinning; hydrothermal activation; gel electrolyte coating	Volumetric capacitance of 6.1×10^4 mF/cm^3; volumetric energy density of 4.8×10^{-3} Wh/cm^3; 94% stability after 10,000 cycles; stable performance at different angles	[29]
$ZnSO_4$-filled PAA	CNT-GO	CNT/rGO and Zn/graphite	Hydrothermally assemble CNT/rGO; coating of Zn fiber with graphite by electrodeposition; gel electrolyte coating	Ionic conductivity of 21.7 mS/cm; stretchability up to 2500; energy density of 48.5 mWh/cm^3; volumetric capacitance of 1.0×10^5 mF/cm^3; 98.5% stability after 10,000 cycles	[30]
H_3PO_4	GO	rGO-MoS_2	Wet spinning of MoS_2-GO, followed by chemical reduction; gel electrolyte coating	Volumetric capacitance of 2.2×10^5 mF/cm^3; three fibers connected—capacitance of 2.4×10^2 mF/cm^2; energy density of 3.2×10^{-2} Wh/cm^3; 72% stability after bending 500 times at different angles	[31]

Table 1. *Cont.*

Gel Electrolyte	Fiber Composition	Electrodes	Preparation Method	Conductivity Tests and Results	Ref.
H_3PO_4	Carbon nanobranches	Carbon nanobranches/PU	Carbon nanobranches covered with carbon dots by pyrolysis of starch; microfluidic spinning for carbon nanobranches/PU fibers; twisting of fibers and gel electrolyte coating	Specific capacitance of 2.0×10^2 mF/cm^2; energy density of 4.5 μWh/cm^2; 98% stability after 10,000 cycles; powered 19 LEDs	[32]
H_3PO_4	CNT/GO/PANI	CNT/GO/PANI	CNT film by chemical vapor deposition; immersion in a GO/PANI mixture solution; over-twisting of strips of the film into two helical fibers; gel electrolyte coating; intertwining fibers and gel electrolyte coating	CNT/GO/PANI film with a specific capacitance of 5.1×10^2 mF/cm^2; specific capacitance of 1.8×10^2 mF/cm^2; 80% stability after 5000 cycles; stability after bending for 500 cycle at 180°	[36]
H_2SO_4	PAN	Ti/PEDOT	Electrospun PAN nanofibers on a Ti wire with glycerol; sip-coating with PEDOT:PSS; PSS etching with H_2SO_4; gel electrolyte coating, twisting, and coating again with gel electrolyte	Energy density of 5.5 Wh/kg; specific capacitance of 68 F/g; power density of 9.1×10^3 W/kg; 81% stability after 10,000 cycles; stability under different deformations; powered one LED	[37]
LiCl	PEDOT:PSS	PEDOT:PSS	Coaxial PEDOT:PSS wet spinning; gel electrolyte coating	Electrical conductivity of 1514 S/cm; specific areal capacitance of 1.2×10^2 mF/cm^2; energy density of 9×10^{-6} Wh/cm^2; 81% stability after 10,000 cycles; Coulombic efficiency of ~100%; stable specific capacitance after bending 3000 times at 180°	[38]
H_3PO_4	Kevlar® fibers	PEDOT:PSS/Kevlar®	Wet spinning of nanofibrillated Kevlar®; immersion in PEDOT:PSS gel	Capacitance of 1.1 mF; volumetric energy density of 7.1×10^{-2} Wh/cm^3; 80.5% stability after 10,000 cycles; stability under different deformations; Coulombic efficiency of 99.1%; powered one LED	[39]

Table 1. *Cont.*

Gel Electrolyte	Fiber Composition	Electrodes	Preparation Method	Conductivity Tests and Results	Ref.
Chitosan	PEDOT:PSS/CNT	PEDOT:PSS/CNT	Wet spinning of PEDOT:PSS/CNT with internal nozzles for electrode inks and external nozzle carrying the gel electrolyte	Volumetric capacitance of 1.6×10^4 mF/cm^3; 96% stability after 5000 cycles; electrochemical stability after 1×10^5 bending cycles	[40]
H_2SO_4	PEDOT:PSS	PEDOT:PSS/V_2O_5	PEDOT:PSS/V_2O_5 wet spinning; gel electrolyte coating	Energy density 1.4×10^{-6} Wh/cm^2; specific capacitance of 6.0×10 mF/cm^2; 94.02% stability after 4000 cycles	[41]
H_2SO_4	PEDOT:PSS	PEDOT:PSS $Mo_{1.33}$Ci-MXene and rGO fibers	PEDOT:PSS $Mo_{1.33}$Ci-MXene and GO wet spinning; gel electrolyte coating	Capacitance of 105 F/g; energy density of 3.7×10^{-2} Wh/g; 94% stability after 10,000 cycles	[43]
PVA–LiCl	GO	$MnO_2/Ti_3C_2T_x$/rGO	$Ti_3C_2T_x$/GO wet spinning; immersion in HI/CH_3COOH; immersion in $KMnO_4$ solution; twisting of fibers and gel electrolyte coating	Capacitance of 2.4×10^4 mF/cm^3; energy density of 2.1×10^{-3} Wh/cm^3; 92% stability after 10,000 cycles; 100% stability after bending 1000 times at 90°	[44]
$ZnSO_4$/gelatin		MXene/MnO_2-RC (cathode) and Zn wire (anode)	MXene/MnO_2-RC fiber cathode wet spinning; gel eletrolyte coating	Volumetric capacitance of 1.1×10^2 mF/cm^3; energy density of 2.2×10^{-2} Wh/cm^3; 90.5% stability and 100% Coulombic efficiency after 5000 cycles	[45]
PVA/H_3PO_4		MXene/PEDOT:PSS/ Ag/MXene/PEDOT:PSS	MXene/PEDOT:PSS ink by 3D direct-ink writing; printing of Ag on top of the electrode; printing the gel electrolyte; Coating in NOA 63 resin	Electrical conductivity of 1.6×10^3 S/cm; areal capacitance of 1.1×10^3 mF/cm^2; gravimetric capacitance of 185.9 F/g; areal energy density of 9.4×10^{-5} Wh/cm^2; 92% stability after 25,000 cycles; powered one LED	[46]

Table 1. *Cont.*

Gel Electrolyte	Fiber Composition	Electrodes	Preparation Method	Conductivity Tests and Results	Ref.
PVA/GO/$ZnSO_4$/$MnSO_4$	Carbon	Carbon wire coated with nano-MnO_2 (cathode) and Zn wire (anode)	GO gel crosslinking with nano-MnO_2; gel eletrolyte coating; silicone layer coating	Gel electrolyte with ion conductivity of 2.1×10^{-2} S/cm; ~230% stretchability and self-healing; energy density of 91 Wh/L; 98.0% stability after 1000 cycles	[47]
CMC/$ZnSO_4$	CNT fiber	CNT fiber coated with ZnHCF (cathode); CNT fiber coated with Zn nanosheet arrays (anode)	Roll electrodeposition of Zn arrays on CNT fibers; gel electrolyte coating comprising CNT sheets, ZnHCF, and CNT sheets	Capacity of 100.2 mAh/cm^3; energy density of 2.0×10^{-1} Wh/cm^3; 91.8% stability after 200 cycles; 96.8% Coulombic efficiency; 93.2% stability after bending 3000 cycles at 90°; powered one LED	[48]
CMC-SO_4	CNT fiber	CNT fiber coated with stitched ZVO nanosheets (cathode) and CNT fiber coated with Zn (anode)	CNT-stitched ZVO nanosheets hydrothermally; electrodeposition of Zn on CNT fiber; gel electrolyte coating; twisting of electrodes	69.7% stability after 100 cycles; volumetric energy density of 7.2×10^{-2} Wh/cm^3; 88.9% stability retention after 2000 cycles; 100% stability after bending angles from 0 to 180°	[49]
KOH	CNT fiber	Nitride-doped CNT fiber coated with Ag_2O/PEDOT:PSS (cathode) and CNT fiber coated with Zn (anode)	Deposition of Ag_2O on a nitride-doped CNT fiber and coating with PEDOT:PSS; electrodeposition of Zn on CNT fiber; gel electrolyte coating; twisting of two electrodes	Capacity of 1.05 mAh/cm^2; energy density 1.6×10^{-3} Wh/cm^2; power density of 1.4×10^{-3} W/cm^2; 79.5% eletrochemical stabily retention after 200 cycles	[50]
$ZnSO_4$	Ag fiber Spandex fiber	Ag fiber coated with graphene and PANI (cathode) and Ag fiber coated with Zn (anode)	Coating of Ag fibers with graphene and PANI, and Zn nanoflakes by electrodeposition; gel electrolyte coating; encapsulation with PU, and twisting around a spandex fiber	Specific capacity of 32.56 mAh/cm^3; energy density of 3.6×10^{-2} Wh/cm^3; stability of 76.5% after 1000 cycles and 99.0, 93.6 and 91.5% during knotting, bending, and twisting; strain up to 900%	[51]

Table 1. *Cont.*

Gel Electrolyte	Fiber Composition	Electrodes	Preparation Method	Conductivity Tests and Results	Ref.
$Al(CF_3SO_3)_3$	CNT fiber	CNT fiber coated with MnHCF (cathode) and CNT fiber coated with GO/MoO_3 (anode)	Coating of CNT fibers with MnHCF and GO/MoO_3; gel electrolyte coating; assembled on a silicone rubber substrate and gel electrolyte coating	$Al(CF_3SO_3)_3$ hydrogel with ionic conductivity of 2.2×10^{-2} S/cm and strain of 461%; specific capacity of 42 mAh/cm^3; energy density of 3.1×10^{-2} Wh/cm^3; stability of 91.6% after 100 cycles	[52]
$PVDF/HFP/Al_2O_3$		LFP (cathode) and LTO (anode)	Etching of GO with H_2O_2 and reduction with vitamin C; PVA/PVDF adhesive in NMP; printing ink by dissolving LFP or LTO and CNTs in the PVA/PVDF solution; extrusion into a $NaBO_2$ solution; immersion in electrolyte	Specific capacity/capacity retention rate of 153.7 mAh/g/92% and 156.5 mAh/g/86.32% after 100 cycles, LFP/HrGO and LTO/HrGO fiber electrodes; FLIB@HrGO with a discharge capacity/capacity retention rate of 142.2 mAh/g/90.4%, after 100 cycles; specific capacity of 62.42 mAh/cm^3; energy density of 1.6×10^{-1} Wh/cm^3	[53]

Table 2. Comparison of fiber-based energy storage devices using gel electrolytes, highlighting preparation methods and electrochemical properties.

Gel Electrolyte	Fiber Composition	Electrodes	Preparation Method	Conductivity Tests and Results	Ref.
PVA/KCl	PET	Au/Ni-MOF@ carbon yarn	PET dip-coating in dopamine; PET/PDA electroless plating; dip-coating in $Ni\text{-}Ni_3HHTP_2$ solution	Length capacitance of 1.1×10^2 mF/cm; energy density of 3.9×10^{-6} Wh/cm; power density of 2.5×10^{-4} W/cm	[57]
$PVA/ZnSO_4$	Stainless-steel CO blend	$G/Zn\text{-}MnO_2$	MnO_2 (cathode) or Zn (anode) electrodeposition $PVA/ZnSO_4$ coating	Specific capacity of 43.53 mAh/g; energy density of 5.2×10^{-2} Wh/g	[58]

Table 2. *Cont.*

Gel Electrolyte	Yarn Composition	Electrodes	Preparation Method	Conductivity Tests and Results	Ref
PVA/H_2SO_4	Cellulose	Ti_3C_2 MXene	Mxene dip-coating, followed by gel electrolyte coating	Conductivity up to 440.3 ± 0.9 S/cm; specific capacitance of 7.6 × 10^2 mF/cm at 2 mV/s; 2000-cycle stability at ~14% compression strain	[59]
PVA/H_2SO_4	CO/Stainless steel	PEDOT:PSS	PEDOT in situ polymerization; SC assembled and woven	Areal specific capacitance reached a max. 1.6 × 10^2 mF/cm^2; Areal energy density 1.0 × 10^{-5} Wh/cm^2; 81.9% capacitance retention after 10,000 cycles	[60]
PVA/H_2SO_4	Stainless steel/CO	PPy	PPY electrochemical deposition; gel electrolyte dip-coating	High areal capacitance (up to 3.4 × 10^2 mF/cm^2); high cycle stability (apx. 93% capacitance retention))	[61]
PVA/H_3PO_4	CO	PAN/PEDOT:PSS	Ni dip-coating; PAN electroless deposition; PEDOT:PSS dip-coating	Volumetric capacitance 2.7 × 10^4 mF/cm^3; energy density 9.6 × 10^{-3} Wh/cm^3; power density 8.3 × 10^{-1} W/cm^3	[62]
PVA/H_3PO_4	CO/Carbon fiber	Carbon fiber yarn	CO + carbon fiber twisting; yarns fixed into ABS mold; material coated with cellulose-based ionic hydrogel and gel electrolyte	Specific energy density 1.7 × 10^{-8} Wh/cm^2; specific power density 5.3 × 10^{-4} W/cm^2; 92% capacitance retention	[63]
PVA/H_3PO_4	MWCNT/ TiO_2 fibers	MWCNT/ TiO_2	TiO_2 nanofibers (NFs) produced by electrospinning; TiO_2 NF electrophoretic deposition in MWCNT; D gel electrolyte dip-coating; yarn twisting of 2 electrodes + gel electrolyte dip-coating	Good capacitance (3.7 × 10 mF/cm at 0.1 mA/cm); capacitance retention; good energy (1.2 × 10^{-5} Wh/cm) and power (3.7 × 10^{-4} W/cm) densities	[64]
PVA/H_3PO_4/KI	Carbon fibers	Carbon fibers	gel electrolyte dip-coating; yarn twisting of 2 electrodes; second dip-coating in gel electrolyte	Volumetric capacitance 363.9 F/L; energy density 5.1 × 10^{-2} Wh/L; specific capacitance (13.9 mF) using a 7.5 mM KI solution	[65]

Table 2. *Cont.*

Gel Electrolyte	Yarn Composition	Electrodes	Preparation Method	Conductivity Tests and Results	Ref
PVA/H_3PO_4	CO and Nylon®	$Ti_3C_2T_x$ MXene	MXene yarn coating of CO and Nylon®; yarn knitting in fabrics; PET yarn knitted in between; gel electrolyte coating	Capacitance up to 7.1×10^2 mF/cm² and 5.2×10^2 mF/cm², in 1 M H_3PO_4 and H_3PO_4 gel electrolyte; delivery energy 2.5×10^{-5}; power densities 4.7×10^{-7} Wh/cm²; cycling stability	[66]
PVA/H_3PO_4	Carbon nanotubes (CNT)	PEDOT:PSS/CNT-*NMP*	CNT forest and sheet by chemical vapor deposition; NMP drop casting in CNT forest; yarn electrode production via biscrolling process; gel electrolyte coating	Capacitance 112.76 F/g; maximum power and energy density 9.8×10^2 W/kg (1.2 Wh/kg) and 3.8 Wh/kg (1.5×10^2 W/kg); cycling stable performance after 5000 cycles	[67]
PVA/KOH	PET	PPy@NiCo-double hydroxide@stainless-steel electrodes	NiCo-layer double hydroxide via hydrothermal process; PPU electrochemical deposition; gel electrolyte coating	Good specific capacitance 1196 F/g; energy density 6.6 mWh/cm³; power density 1.6×10^{-1} mW/cm³; good cycle performance (85.7% capacity retained)	[68]
PVA/KOH	Au/CO	NiCo2S4@Ni−Co LDH	Hydrothermal synthesis of $NiCO_2S_4$ nanotubes on Au/CO yarns; Ni−Co LDH nanosheets electrodeposition	Capacitance 5.7×10^3 mF/cm; good areal energy density 3.5×10^{-6} Wh/cm²; stable cycling performance small capacitance retention decrease (9%)	[69]
PVA/KOH	PET	rGO/CNT and NiCo-BOH	Ni electroless deposition on PET yarns + Cu electrodeposition; co-self-assembled rGO/CNT hydrothermal reaction on Ni/Cu-plated PET yarns; gel electrolyte dip-coating	Energy density apx. 7.8×10^{-5} Wh/cm²; high power density 1.4×10^4 W/cm²; stable cycling performance 82.7% capacity retention	[70]

Table 2. *Cont.*

Gel Electrolyte	Yarn Composition	Electrodes	Preparation Method	Conductivity Tests and Results	Ref
PVA/KOH	Co wire + CO yarn	NiMnO3-rGO@CO-Cu	$NiMnO_3$-rGO nanocomposites via hydrothermal reaction; CO yarns dip-coating in graphite and PVDF; Cu wire and CO yarns are woven; Co-Cu fibers dip-coating in $NiMnO_3$-rGO; electrode dip-coating in gel electrolyte	Maximum specific capacitance 4.0×10^2 mF/cm^2; maximum specific capacitance 8.3×10 mF/cm^2; specific energy density 17.5 µWh/cm^2	[71]
PVA/KOH	Carbon-based yarns (CBY)	CoNi-layered double hydroxides@CBY and S-dopes carbon nanoparticles@CBY	CoNi-layered double hydroxide nanosheets; in situ growth via hydrothermal reactions; yarn anchoring assembly	Specific surface area 655 m^2/g^1 of CoNi-layered double hydroxides@CBY; voltage operating range 1.4 V; capacitance 2.3×10^2 mF/cm^2; energy density (6.2×10^{-5} Wh/cm^2); stable cycle performance	[72]
PVA/KOH	CNT-coated CO (CCY)	NiCo-LDH@CCY	CO yars dip-coated in CNT dispersion; CCY dip-coating to produce NiCo-LDH@CCY; supercapacitor assembled by coating two yarns in a PET substrate with PVA/KOH or via PVA/KOH solution dip-coating	Good areal capacitance 1.24×10^2 mF/cm^2, current density 0.2 mA/cm^2 and energy density 3.9×10^{-5} Wh/cm^2	[73]
PVA/KOHNa_2SO_4	Transition metal oxide (TMO)-Ni-TMO trilayer nanoribbon yarn	(TMO)-Ni-TMO trilayer nanoribbon	TMO-Ni-TMO nanoimprinting in the mold; nanoribbon array delamination; twisting the nanoribbon array to form yarns; production of YSC, coated with gel electrolyte, using two TMO yarns or one TMO yarn + graphene fiber	Maximum energy density 7.6×10^{-2} Wh/cm^3 achieved for graphene–CoNixOy@Ni (G-CNO) YSCs; power density 2.4×10^{-1} Wh/cm^3 achieved with G-CNO YSCs; 94.2% initial capacitance retention after 10,000 cycles	[74]

Table 2. *Cont.*

Gel Electrolyte	Yarn Composition	Electrodes	Preparation Method	Conductivity Tests and Results	Ref
PVA/H_2SO_4, PVA/KOH, PAAm/NaCl	MWCNT	MWCNT	Floating catalyst chemical vapor deposition to obtain MWCNT, followed by twisting; insulators prepared by connecting two Ag wires to the MWCNT yarns; gel electrolyte added to the coils	High-recognition voltage signals (4–15 mV) with low noise (0.024 mV); high voltage amplitude sensitivity to tensile stretches in multiple frequencies (0.1–10 Hz) and strains (1–80%)	[75]
PVA/LiCl	CO	rGO	Polymer-assisted metal deposition (PAMD) of Ni on CO; rGO electrochemical deposition; yarn embroidering	Capacitance 1.6×10 mF/cm^2 for in-plane MWNT/Ni-CO@fabric with PVA/LiCl at 0.8V; cycling performance—97% capacitance retention	[76]
$ZnCl_2/NH_4Cl/CMC$	Cellulose	PANI (cathode) Zn (anode)	Cellulose yarn electrospinning; PANI in situ polymerization; conductive slurry coating to obtain cathodes; Zn electrochemical deposition in conductive slurry-coated cellulose yarn to obtain anodes; cathode CMC coating, followed by twisting with anode; gel electrolyte coating by CMC swelling	Energy density 153.2 and 6.1×10 Wh/Kg; good cyclic stability, specific capacity maintained at 109.7 mAh/g	[77]

Among all gel electrolytes described in the literature, PVA-based gel electrolytes are among those used for a wider range of applications in combination with different acids or bases. This is justified by most authors as being related to performance issues, since in all cases where more than one gel electrolyte is tested, the best electrochemical results are achieved with PVA. For example, the work of Yang et al. [57] involved YSCs prepared using a PVA/salt gel electrolyte. This method combines Ni_3(2,3,6,7,10, 11-hexahydroxytriphenylene)$_2$ (Ni_3HHTP_2), a conductive metal–organic framework (MOF), with Au-plated PET/polydopamine yarns to produce all-solid-state symmetrical YSCs. The supercapacitor produced comprises two identical PET/Au/Ni-MOF@carbon yarn electrodes, twisted together while fully coated with the KCl gel electrolyte. The resulting yarns presented high-length capacitance (1.1×10^4 mF/cm, at a current density of 0.1 mA/cm), an energy density of 3.9×10^{-6} Wh/cm (at 2.5×10^{-5} W/cm), and a power density of 2.5×10^{-4} W/cm (at 5.9×10^{-7} Wh/cm). No performance loss was observed when a tension up to 42.3% was applied. In the work of Tang et al. [58], $ZnSO_4$ was used to coat G/Zn–MnO_2@stainless-steel CO-blend fibers. This was prepared by the combination of a MnO_2-based anode and a Zn-based cathode, itself prepared via electrochemical deposition. The battery presented exceptional durability and washing resistance, with 89.66% capacity retention after 30 hammering cycles and 77.97% capacity retention after a 30 min washing cycle. The material also presented a good electrochemical performance, with a 43.53 mAh/g specific capacity and a 51.94 mWh/g energy density. PVA can also be used in combination with inorganic acids such as H_2SO_4 or H_3PO_4. One example is the work of Uzun et al. [59], who proposed the development of Ti_3C_2-coated cellulose yarns via a simple two-step dip-coating process, which was used for the preparation of supercapacitors using a H_2SO_4 gel electrolyte. The yarn production was achieved through the dip-coating of cellulose yarns in a homogeneous dispersion of Ti_3C_2-based small- and/or large-size flakes. A good MXene cellulose loading capacity was reported (up to 77% wt, around 2.2 mg/cm), resulting in highly conductive yarns (up to 440.3 ± 0.9 S/cm), which could be knitted into full fabrics using different stitch patterns (half-gauge and interlock) making use of industrial knitting machines. In 2020, Du et al. [60] developed a self-powered and self-sensing energy textile system using a H_2SO_4 gel electrolyte and cellulose/stainless-steel composite yarns. The yarns were coated with PEDOT via in situ polymerization, followed by sorbitol secondary doping, and the YSCs were produced using two parallel electrodes separated by the gel electrolyte. The fabric prepared by weaving demonstrated a solid areal energy density (1.0×10^{-5} Wh/cm^2) at a power density of 1.7×10^{-4} W/cm^2. Another yarn supercapacitor was produced by Zhang et al. [61], where stainless-steel/CO-blended yarns were coated with polypyrrole (PPy) to produce a highly flexible material. In this work, H_2SO_4 was used as the gel electrolyte, and the supercapacitor, produced by electrochemical deposition, presented high areal capacitance (up to 3.4×10^2 mF/cm^2 at a current density of 0.6 mA/cm^2) and high cycle stability (approximately 93% capacitance retention after 1000 cycles). As for the H_3PO_4 combination, Sun et al. [62] proposed a method to prepare a flexible PEDOT:PSS-poly acrylonitrile/Ni CO (PNF/NiC) capacitor using this specific gel electrolyte (Figure 3). The yarns were first coated by dip-coating in a Ni-containing electroless solution, and then PAN nanofibers were wrapped to produce PAN core-spun yarns. Finally, the yarns were dip-coated in a PEDOT:PSS chloroform solution while dripping on an ethanolic solution of ferric chloride. They were then combined with the H_3PO_4 gel electrolyte, resulting in yarns with a high volumetric capacitance (2.6×10^4 mF/cm^3 at 0.08 A/cm^3), and good energy (9.56 mWh/cm^3) and power (830 mW/cm^3) densities. Carvalho et al. [63] also used H_3PO_4 gel electrolytes, paired with in situ regenerated cellulose-based ionic hydrogel, to produce CO/carbon fiber–yarn-based supercapacitors. The yarns were produced via the twisting of CO and carbon fiber yarns. Following this, we coated with electrolytic and gel electrolyte solutions after fixation in an acrylonitrile–butadiene–styrene (ABS) mold. This sustainable approach led to the preparation of stretch-broken carbon fiber yarns, where both the yarns and the cellulose could be recovered after washing with deionized water and drying. The cellulose could then be reused to produce new YSCs without compromising the good electrochemical properties of the material (specific energy density of 1.7×10^{-8} Wh/cm^2, specific power density of 5.3×10^{-4} W/cm^2 at 1 mA/cm^2, and 92% capacitance retention after 10,000 consecutive cycles). Rani and Sharma [64] also presented a protocol for the preparation of a high-energy-density yarn supercapacitor using H_3PO_4 as gel electrolyte. Here, the YSCs were fabricated by the electrophoretic deposition of TiO_2 nanofibers onto multi-walled carbon nanotube (MWCNTs) and then dip-coated in the gel electrolyte solution. After producing fibers by twisting two electrode yarns together, the resulting device revealed good capacitance (3.7×10 mF/cm at 0.1 mA/cm), capacitance retention (90% retention after 10,000 cycles), energy (1.2×10^{-5} Wh/cm), and power (3.7×10^{-4} W/cm) densities. Another work, presented by Park et al. [65], described the use of KI as a redox mediator to improve the electrochemical performance of gel electrolytes in YSCs. For this, a H_3PO_4-KI-PVA gel electrolyte was used to coat carbon fibers, which were then twisted and physically woven together and used to produce the YSCs. The capacitors produced through this methodology presented an enhanced energy storage ability compared to the control, without KI (volumetric capacitance of 363.9 F/L, energy density of 5.1×10^{-2} Wh/L). Improved ion diffusion, charge transfer kinetics, and limited moisture conditions were attributed to the presence of KI. In addition, the YSCs also demonstrated a specific capacitance of 13.9 mF (at a current density of 10 μA) when a 7.5 mM solution of KI was used. In 2020, Levitt et al. [66] described the preparation of CO and Nylon®-based YSCs using H_3PO_4 gel electrolytes in combination with $Ti_3C_2T_X$ MXene. The yarns were coated with an MXene/water solution using an automated coating setup and then knitted into an interlock rib fashion. This was followed by coating the knitted fabric with the gel electrolyte solution. The resulting YSCs demonstrated high capacitance (up to 7.1×10^2 mF/cm^2 and 5.2×10^2 mF/cm^2 at 2 mV/s, in 1 M H_3PO_4 and H_3PO_4 gel electrolyte, and delivery energy and power densities of 2.5×10^{-5} and 4.7×10^{-7} Wh/cm^2, respectively), as well as excellent cycling stability over 10,000 cycles in a gel electrolyte.

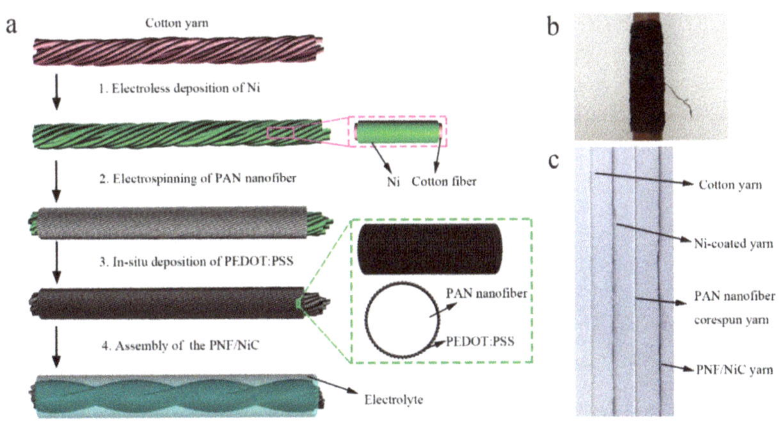

Figure 3. (**a**) Scheme showing fabrication process of PNF/NiC capacitor yarn. (**b**) Image of 500-meter-long Ni-coated CO yarn wound around spinning cone. (**c**) Images of different yarns at various stages of processing [62].

One other example can be found in the work of Pal and Subhedar [67], where YSCs were prepared using PEDOT:PSS-coated carbon nanotubes in combination with

PVA/H_3PO_4 as a gel electrolyte. The authors proposed a biscrolling technique to assemble the hydrophilic carbon nanotubes, where NMP is used to hydrophilize the carbon nanotube forest. In this method, the yarns are assembled by attaching one end of the composite PEDOT:PSS/CNTs-NMP sheet to a motor using sticky double-sided tape, and then twisted at about 2000 turns/min, followed by coating with gel electrolyte. This supercapacitor showed a promising electrochemical performance, having a capacitance of 112.76 F/g, maximum power and energy densities of 9.8×10^2 W/kg (1.2 Wh/kg) and 3.8 Wh/kg (1.5×10^2 W/kg), respectively, and cycling stability performance with the minimum loss after 5000 repeated cycles of charge–discharge. Another possible combination of PVA in gel electrolytes is the inorganic base–gel electrolyte. For example, Wang et al. [68] used a KOH gel electrolyte to produce 1D symmetrical yarn-shaped supercapacitors. The NiCo-layer double hydroxide (NiCo-LDH) was grown in situ on the yarns by using a hydrothermal method. Then, PPy was deposited by electrochemical deposition, forming the PPy@NiCo-layered double hydroxide@stainless-steel yarns. Finally, the KOH gel electrolyte was coated after assembling two parallel yarn electrodes on the PET fabric, while a PAN yarn was electrospun on the yarn electrode to prevent short circuiting. The YSCs demonstrated good specific capacitance (1196 F/g at 1 A/g), energy (6.6×10^{-3} Wh/cm^3), and power (1.6×10^{-1} mW/cm^3) densities. In another work from Wang et al. [69], a 3D metallic conductive-coated Au/CO yarn was prepared for the development of flexible energy storage devices using a KOH gel electrolyte. The $NiCo_2S_4$ nanotube array electrodes were prepared by hydrothermal synthesis, followed by the electrodeposition of Ni-CO LDH to produce 3D core–shell structured $NiCo_2S_4$@Ni-Co LDH yarns. The yarns were then used to assemble a two-ply supercapacitor, which presented high capacitance (5.7×10^3 mF/cm) and good areal energy density values (3.5×10^{-6} Wh/cm^2, at a real power density of 1.6×10^{-4} Wh/cm^2), as well as a stable cycling performance, with only a small capacitance retention decrease (9%) after 3000 cycles. Another example of the use of KOH gel electrolyte can be found in the work of Liu et al. [70], where triboelectric nanogenerators (TENGs) and asymmetric YSCs were prepared. In this publication, yarn-based TENGs and YSCs are produced from Ni and Cu film-coated PET yarns, where the negative electrode is prepared by the hydrothermal reaction of self-assembled rGO/CNT and the positive electrode is prepared by electroplating PET yarns with Ni-Co bimetallic oxyhydroxide. The YSCs and TENGs produced from these yarns exhibited high areal energy density (around 7.8×10^{-5} Wh/cm^2), power density (1.4×10^4 W/cm^2), flexibility (1000 cycles run while bending at 180°), and a stable cycling performance (82.7% capacity retention after 5000 cycles). In 2024, a work by Al-khaykanee et al. [71] used a KOH gel electrolyte to prepare NiMnO3-rGO@CO-Cu YSCs. Here, CO yarns were dip-coated with a graphite and PVDF solution and then woven with the CO wires. After that, the electrodes were prepared by dip-coating with the $NiMnO_3$-rGO nanocomposite (prepared by hydrothermal reaction). The YSCs prepared via this procedure revealed good electrochemical performance, with a maximum specific capacitance of 4.0×10^2 mF/cm^2 (at scan rate of 5 mV/s in a 1 M KOH solution electrolyte), a maximum specific capacitance of 8.3×10 mF/cm^2, and a specific energy density of 1.8×10^{-5} Wh/cm^2 (at a specific power density of 3.8×10^{-4} W/cm^2 in a potential range of 1.7 V). Moreover, the authors also describe how the yarn supercapacitor's capacitance does not suffer any performance issues when bent, having tested a bending angle up to 210°. A novel battery–capacitor hybrid device was prepared by Li and coworkers [72], using carbon-based yarns (CBYs) in combination with KOH gel electrolytes. Here, an in situ growth procedure was used to prepare CoNi-layered double hydroxide (CoNi-LDH) nanosheets, achieved by hydrothermal reaction at 180 °C. The battery–capacitor apparatus was made by combining two yarns, CoNi-LDH@CBY (cathode) and S-doped carbon nanoparticles on CBY (S-

CNP@CBY, anode), coated with a PVA/KOH gel electrolyte. The assembled device presented good electrochemical properties, having a wide voltage operating window of 1.4 V, a large area capacitance (2.3×10^2 mF/cm^2 at 1.8×10^{-3} W/cm^2), and a high energy density (6.2×10^{-5} Wh/cm^2), while also being stable after 5000 cycles of charge–discharge. The CoNi-LDH@CBY electrodes, highlighted as a novelty by the authors, also presented a large specific surface area (655 m^2/g) and abundant active sites, contributing to an improved electrochemical performance. Another example of NiCO-LDH-based electrodes that are used in combination with PVA/KOH gel electrolytes to assemble YSCs can be found in the work of Yu et al. [73]. Here, CO yarns were coated with CNT, and were then dip-coated in a solution of ($Co(NO_3)_2 \cdot 6H_2O$, $Ni(NO_3)_2 \cdot 6\ H_2O$ and urea, while we used variable amounts of NH_4F to tune the NiCo-LDHs morphology into either nanoneedles, nanosheets, needle–sheet composites, or nanoflowers. Then, the supercapacitor is assembled by combining two yarns and either dip-coating them in the gel electrolyte or by placing them in a PET substrate and subsequently coating them with PVA/KOH. The devices prepared in this study presented good areal capacitance (1.2×10^2 mF/cm^2), good current density (0.2 mA/cm^2), and high energy density (3.9×10^{-5} Wh/cm^2 at a current density of 0.2 mA/cm^2), with the best results being achieved with the nanoneedle-shaped NiCO-LDHs. Ahn et al. [74] also developed a yarn supercapacitor using PVA/KOH and PVA/Na_2SO_4 gel electrolytes. In their work, the authors prepared transition metal oxide (TMO)-Ni-TMO trilayer nanoribbon yarns by nanoimprinting the TMO-Ni-TMO sequentially in the mold. Then, after the delamination of the nanoribbon array, the yarn was produced by twisting. Finally, the YSCs were produced using two TMO yarns or one TMO yarn and a graphene fiber coated with gel electrolyte. Among the YSCs prepared using different TMOs, the most promising results were obtained with the graphene–CoNixOy@Ni (G-CNO) YSCs, having maximum energy and power densities of 7.6×10^{-2} and 2.4×10^{-1} Wh/cm^3, respectively, and displaying 94.2% retention of the initial capacitance after 10,000 cycles. In addition, good flexibility, reported as a 0–0.9% resistance change, was observed at a curvature of 5.21 1/cm; we also saw a stable electrical resistance with <1.00% variations after 1000 bending cycles. Li et al. [75] achieved a combination of different gel electrolytes in the preparation of textile sensor devices. In this work, PVA/H_2SO_4, PVA/KOH, and PAAm-NaCl gel electrolytes were used to coat stretchable piezoionic MWCNT-based yarns, which were prepared by simple dip-coating or dip-coating followed by UV curing (for PAAm-NaCl gel electrolyte). The yarns were previously produced via floating catalyst chemical vapor deposition, followed by twisting. The yarns were able to generate sensitive voltage signals (4–15 mV), with only a small noise percentage (0.024 mV) between their endings and asymmetrically stretching, and they also showed the high sensitivity of voltage amplitude to tensile stretches in a wide range of frequencies (0.1–10 Hz) and strains (1–80%). It is suggested that piezoionic activity is derived from so-called "ion-squeezing", which allows for a current to be generated along the conductive yarn without an external electrical pulse being applied, which is optimal for the application of this material as a smart textile and wearable sensor, as well as for artificial muscle feedback. Another example was presented by Huang et al. [76], where the PVA/LiCl gel electrolyte was used to prepare YSCs via an innovative procedure called additive functionalization and embroidery manufacturing (AFEM). This procedure starts with the polymer-assisted metal deposition (PAMD) of Ni onto CO, followed by the electrochemical deposition of rGO in the yarns. Then, these yarns are embroidered into textile substrates in a variable design fashion. The authors test the yarn supercapacitor capacity when coated with a PVA/LiCl gel electrolyte (voltage window of 0.8 V). The YSCs produced possess good electrical properties, such as an areal capacitance of up to 1.6×10 mF/cm^2 for in-plane MWCNT/Ni-CO@fabric with PVA/LiCl at 0.8V), and a stable cycling performance (97% capacitance retention after 1000 charge/discharge cycles). The authors also highlighted that the AFEM method developed was applicable to several types of yarns, as confirmed by testing using different yarn electrodes produced from MWCNT. CMC-based gel electrolytes were used in the production of a Zn-Ion fiber battery using high-strength cellulose yarns [77]. The cellulose yarns were spun from NaOH/urea aqueous solutions and used for the preparation of the cathode and anode yarns. The cathode was prepared by the in situ polymerization of PANI, followed by coating with a conductive slurry (95 wt.% graphite nanosheets and 5 wt.% cellulose nanofibers), and the anode was prepared by the electrochemical deposition of metallic Zn onto cellulose yarns previously coated with the conductive slurry. The battery was produced by intertwining both electrodes, in combination with a cellulose-based gel electrolyte produced by CMC swelling with an aqueous electrolyte (2 M $ZnCl_2$-3 M NH_4Cl). The Zn-ion battery was electrochemically characterized, where energy densities of 153.2 and 6.1 $\times$ 10 Wh/Kg were measured at power densities of 0.16 and 6.5 $\times$ 10^3 W/kg, respectively. Good cyclic stability was also achieved, with the specific capacity maintained at 109.7 mAh/g after 1000 charge/discharge cycles at a current density of 5 A/g.

Similarly to previously reported results about fiber-based materials, the development of yarn-based conductive or energy storage materials often relies on PVA-based gel electrolytes, which are commonly combined with acids and salts (Table 2). Dip-coating and in situ polymerization are widely used techniques for incorporating gel electrolytes into yarn-based conductive or storage materials. These methods enable uniform and conformal coating along the fiber surface, ensuring strong adhesion and intimate contact between the electrolyte and the yarn substrate. This is essential for preserving mechanical flexibility and maintaining stable electrochemical performance under mechanical deformations such as bending, twisting, or stretching. Additionally, both methods are compatible with a wide range of fiber types and can accommodate large-scale or continuous processing. Their simplicity, low cost, and scalability make them particularly suitable for the development of wearable and flexible energy storage devices.

2.3. Woven Fabrics

Woven e-textiles have gained increasing attention due to their remarkable ability to seamlessly integrate electronic functions while maintaining the key advantages of traditional fabrics, which include comfort, a lightweight structure, flexibility, breathability, hydrophilicity, cost-effectiveness, mechanical strength, and softness. Their structured interlaced architecture also enhances durability, wearability, and adaptability for diverse applications [78,79]. Several studies have emerged in the literature that utilize woven structures as substrates for the development of flexible energy storage materials, highlighting their potential for enhanced durability, mechanical stability, and integration into wearable applications (Table 3). CO fabrics, although highly valued, inherently act as electrical insulators. Therefore, their naturally insulating properties require advanced modifications to enhance their suitability for electrical conductivity applications [78]. The carbonization of CO fabric was initially explored to enhance its conductivity and chemical stability (Figure 4). However, this method did not yield satisfactory outcomes for energy storage applications [78]. Recent advanced strategies involve the incorporation of conductive polymers (e.g., polypyrrole (PPy), PANI, and PEDOT) and metallic ions (e.g., copper (Cu), nickel (Ni), cobalt (Co), and silver (Ag)) into untreated fabrics or those subjected to carbonization [79–82].

Table 3. Comparison of woven energy storage devices using gel electrolytes, highlighting preparation methods and electrochemical properties.

Gel Electrolyte	Fabric Composition	Electrodes	Preparation Method	Conductivity Tests and Results	Ref.
PVA/H_3PO_4-commercial fluorescent pigment of ZnS-Mn	CO	CNT	Dip–pad–dry process with CNT; redox-active additive in a solid-gel electrolyte	100% stability after 8000 cycles; energy density of 1.6 Wh/kg; potency density of 6.4×10^2 W/kg; high specific capacitance of 4.37 F/g	[1]
PVA/LiCl	CO	PPy; TsOH	Carbonization of the fabric; in situ electrochemical deposition with PPy; immersion in electrolyte gel	Specific capacity of 5.0×10^2 mF/cm^2; 73.6% stability after 1000 cycles; energy density range of 1.18–0.68 mWh/cm^3; high power density of 1.8–8.4 $\times 10^{-3}$ W/cm^3; breaking strength of 5.27 N	[78]
Nanofibrillated cellu-lose/polyacrylamide; sodium polyacrylate	CO	Co; Cu; Ag; Ni	CO activated with $PdCl_2$ (loading catalyst); electroless deposition with Co, Ni, Cu, and Ag	Conductivity of ~200 kS/m; capacity of 1.0 mAh/cm^2; high energy density of 1.7 m $\times 10^{-3}$ Wh/cm^2; power density of 8.5×10^{-3} W/cm^2; discharge capacity of 132 mAh	[79]
PVA/H_2SO_4	CO	PPy; PU; water repellent; 5-sulfosalicylate; ferric chloride	Spray-coating of CO fabric with PU and thickener; oxide polymerization with PPy; gel electrolyte coating	Stability of 85% after 3000 cycles; energy density of 9.0×10^{-8} Wh/cm^2; power density of 1.9×10^{-5} W/cm^2	[80]
PVA/KOH	CO	Graphene; $Co(NO_3)_2.6H_2O$; $Ni(NO_3)_2.6H_2O$; PANI; cetyl trimethyl ammonium bromide; PANI nanotubes	Chemical deposition with graphene/PANI nanotubes solution; immersion into NiCo/cetyl trimethyl ammonium bromide solution; gel electrolyte coating	Specific capacitance of 4.3×10^2 mF/cm^2; 84.05% stability after 10,000 cycles; 15.95% deterioration after 10,000 cycles; specific surface area of 30.63 m^2/g and pore volume of 0.176 cm^3/g; surface roughness of 98.7 nm; energy density of 8.0×10^{-5} Wh/cm^3	[81]

Table 3. *Cont.*

Gel Electrolyte	Fabric Composition	Electrodes	Preparation Method	Conductivity Tests and Results	Ref.
PVA/H_2SO_4	CO	PEDOT; $Ti_3C_2T_x$ MXene	Vapor phase polymerization with PEDOT; spray-coating with MXene dispersion; immersion into gel electrolyte	Sheet resistance of 3.6 Ω/sq; specific capacitance of 1.0×10^3 mF/cm^2; areal energy density of 1.3×10^{-5} Wh/cm^2; strong joule heating performance of 193.1 °C; 36.62 dB EMI shielding effectiveness; high sensitivity; 25.7% capacitance retention	[82]
$TEABF_4$-polyacrylamide	CO	Carbon black; activated carbon; PVA; 1,2,4-Trichlorobenzene	Spraying with ink formulation (carbon black, activated carbon, poly(ethylene-co-vinyl alcohol), and 1,2,4-Trichlorobenzene); impregnation with gel electrolyte	Specific capacitance of 3.4×10 mF/cm^2; energy and power densities of 18.9 μW h/cm^2 and 2.4×10^{-4} W/cm^2; specific surface area of 1874.2 m^2/g and pore size of 1.6–2.7 nm; 48% capacitance retention after 2 mouths of aging; 88% Coloumb efficiency for 2 mouths of aging assay	[83]
PVA/H_2SO_4	CO	PPy; PVA-co-ethylene nanofiber suspension	Spray-coated on the fabric with PVA-co-etthylene nanofiber suspension; in situ chemical polymerization with PPy layers; gel electrolyte coating	100% capacitance retention after 10,000 cycles; 98.17% stability over 1000 bending cycles; specific capacitance of 6.7×10^2 mF/cm^2; areal energy density of 6.0×10^{-5} Wh/cm^2; 90–100% Coloumb efficiency	[84]
PVA/KOH	CO	$NiCl_2 \cdot 6H_2O$; $CoCl_2 \cdot 6H_2O$; Fe_3C	Electroless plating process (CO fabric/Ni:Co); sandwiched with gel electrolyte and silk fabric as separator	High specific capacitance of 113.7 C/g; 80% stability after 4000 cycles; energy and power densities of 4.7×10 Wh/kg and 1.5×10^3 W/kg	[85]

Table 3. *Cont.*

Gel Electrolyte	Fabric Composition	Electrodes	Preparation Method	Conductivity Tests and Results	Ref.
PVA/KOH	CO	ZnO nanoparticles; CuS microsphere	Dyeing of CO fabric; atomic layer deposition of ZnO nanoparticles; carbonization process; hydrothermal reaction with CuS; gel electrolyte immersion	Specific capacitance of 1.8×10^3 mF/cm^2; 85.2% stability after 5000 cycles; energy and power densities of 0.27 Wh/cm^2 and 4.3×10 W/cm^2	[86]
PVA/KOH	CO	Procion reactive dye; cellulose and polyethylene terephthalate films	Dyeing and carbonization of CO fabrics; dipping in the electrolyte; supercapacitor sandwiched with cellulose and polyethylene terephthalate films	Specific surface areas of 448.2–622.1 m^2/g and pore volumes of 0.22–0.39 cm^3/g; specific capacitance of 1.1×10^3–1.8×10^3 mF/cm^2; 79.3% capacitance retention; sheet resistance of ~30–45 Ω/sq; 92.2% stability after 5000 cycles; volumetric specific capacitance of 1.8×10^4 mF/cm^3; energy density of 5.7×10^{-3} Wh/cm^3 at the power density of 4.5×10^{-3} W/cm^3	[87]
PVA/KOH	CO	$NiCl_2 \cdot 6H_2O$; $Na_2WO_4 \cdot 2H_2O$	Thermal annealing with $NiCl_2 \cdot 6H_2O$ and $Na_2WO_4 \cdot 2H_2O$; sandwiched with gel electrolyte	High specific capacitance of 60.61 F/g; 95% stability after 3000 cycles; 99.7% stability after bending; energy and power densities of 2.3×10 Wh/kg and 8.3×10^2 W/kg	[88]
LoSalt®/ Grenade Energy®/Shopper isotonic drink + agar and k/carrageenan	CO	Carbon black; activated carbon; ethylene-vinyl acetate; 1,2,4-Trichlorobenzene	Spray-coating with ink formulation; gel electrolyte impregnation	Specific capacitance of 2.3×10 mF/cm^2; energy and power densities of 2.3×10^{-6} W h/cm^2 and 2.0×10^{-4} W/cm^2	[89]

Table 3. *Cont.*

Gel Electrolyte	Fabric Composition	Electrodes	Preparation Method	Conductivity Tests and Results	Ref.
PVA/LiCl	Cellulosic	Dopamine; GO	Dopamine in situ polymerization; immersion on GO solution; carbonization; gel electrolyte immersion	Specific capacitance of 1.2×10^3 mF/cm^2; specific area of 347.6 m^2/g; 6% stability after 4000 bending cycles	[90]
PVA/H_2SO_4	CO	PPy; GO; Ag	Vacuum filtration of GO; addition of PPy and Ag^+ through UV-induced in situ polymerization; gel electrolyte immersion	Specific capacitance of 1.7×10^3 mF/cm^2; capacitance retention of 46.9%; energy potency of 1.1×10^{-3} Wh/cm^2; 90.5% stability after 10,000 cycles; 89.7% stability at 180° after 10,000 bending cycles	[91]
PVA/Zn(OTf)$_2$	CO	VO_2; Zn nanosheets; CNT film; $H_2C_4O_4 \cdot 2H_2O$; H_2O_2; ethanol; PDMS	Carbonization of fabrics; O_2 plasma treatment on fabric; VO_2 mixture with carbonized CO fabric; pressure sensors production; gel electrolyte immersion; sandwich structure containing Zn/CNT film (anode) and VO_2/carbonized CO (cathode)	Flexible pressure sensor: electrical conductivity of ~95.8 Ω/sq; tensile stress of ~6.7 MPa; tensile strain of 155%; sensibility of 0.07–7.12 kPa^{-1}; response/recovery time of 12 and 8 ms Aqueous Zn-ion batteries: specific capacity 301.5 mAh/g; 99.8% Coulombic efficiency; 88.7% stability after 5000 cycles	[92]
PVA/KOH	CO	MWCNT; thermoplastic PU	Gel electrolyte applied through dip–pad–dry method; screen-printing with MWCNT ink; multiple layers of electrodes and fabric separators were sandwiched	High resistance of ~120 Ω; specific capacitance of 4.17 mF/cm^2; high bending strain of 20%/s; ~97.4% stability after 1000 cycles	[93]
PVA/KOH	CO	Ag; CNT; graphene	Screen-printing with coating ink (Ag, CNT/graphene, and textile ink); gel electrolyte immersion	Specific capacitance of 6.8×10^2 mF/cm^2; 80% stability after 3000 cycles	[94]

Table 3. *Cont.*

Gel Electrolyte	Fabric Composition	Electrodes	Preparation Method	Conductivity Tests and Results	Ref.
PVA/KOH	CO	Ag; activated carbon; graphene; CNT	Screen-printing with coating ink (Ag and activated carbon); gel electrolyte immersion	Specific capacitance of 3.3×10^3 mF/cm^2; ~130% stability after 10,000 cycles; high energy and power densities of 5.1×10^{-4} Wh/cm^2 and 1.5×10^{-1} W/cm^2	[95]
PVA/KOH	CO	MWCNT; thermoplastic PU; NMP	Textile electrodes printed using MWCNT ink; gel electrolyte coating	Specific capacitance of 1.4×10 mF/cm^2; high bending strain of 20%/s; 90% stability after 2000 cycles	[96]
PVA/H_2SO_4	CO	Graphene; carboxymethyl celulose; PU	Printing of graphene ink onto textiles; gel electrolyte drop-casting	Specific capacitance of ~3.2 mF/cm^2; resistance of 30 Ω/cm; energy and power densities of 2.8×10^{-4} Wh/cm^2 and 3.0×10^{-3} W/cm^2; stability ~10,000 cycles; durability of 3.5-time higher resistance after 10 washing cycles; device retention of 95% after 10,00 cycles	[97]
PVA/LiCl	CO	N-doped carbon; FeO_4; PPy	Impregnation with PPy; oxidative polimerization with $FeCl_3$; carbonization; gel electrolyte immersion	Specific capacitance of 135 F/g; moderate specific surface areas (700.8 m^2/g); rate capability of 1–10 mV/s (44.4%); 88.4% stability after 1000 cycles; 92.3% Coulombic efficiency after 1000 cycles	[98]
PVA/H_3PO_4	CO	PEDOT:PSS; dimethyl sulfoxide; MXene, graphene nanoscroll; PPy	PEDOT:PSS dip-dry coating; dimethyl sulfoxide immersion; one-step in situ polymerization with MXene, graphene nanoscroll, and PPy; gel electrolyte immersion	Capacitance of 2.7×10^3 mF/cm^2; high energy density of 3.2×10^{-4} Wh/cm^2 and power density of 4.6×10^{-4} W/cm^2; 79% stability after 2500 cycles; 97.8% stability after bending; 92% waterproof property after 2 h	[99]

Table 3. *Cont.*

Gel Electrolyte	Fabric Composition	Electrodes	Preparation Method	Conductivity Tests and Results	Ref.
PVA/H_3PO_4	CO	MWCNT; sodium dodecylbenzenesulfonate	Dip–pad–dry with oxidized MWCNT dispersion; gel electrolyte immersion	High energy density of 3.5 Wh/kg; 98% stability after 5000 cycles; specific capacitance of 3.91 F/g	[100]
PVA/H_3PO_4	CO	MWCNT; sodium dodecylbenzenesulfonate	Dip–pad–dry with MWCNT; gel electrolyte coating	Specific capacitance of 9.2×10^3 mF/cm^2; 96.3% stability after 5000 cycles; energy and power densities of 6.3 Wh/kg and 1.1×10^3 W/kg	[101]
PVA/LiCl	CO	MXene	Exfoliating; delaminating; impregnation with MXene suspensions; carbonization; gel electrolyte immersion	Specific capacitance of 5.0×10^2 mF/cm^2; 74% stability after 10,000 cycles; hydrophilic properties; electric conductivity of 5882 S/m	[102]
PVA/KCl	CO	Cu-MOFs; $HAuCl_4$; polydopamine; polyethylene terephthalate film	Dopamine hydrochloride immersion; immersion in $HAuCl_4$; immersion in Cu-MOF solution; gel electrolyte immersion	Specific capacitance of 258 F/g; energy density of 4.3×10^{-4} Wh/cm^2; 83% stability after 3000 cycles; 94% Coulombic efficiency after 3000 cycles	[103]
PVA/H_3PO_4	CO	PPy; PANI; PEDOT; $CuCl_2$	$CuCl_2$ padding; vapor phase polymerization in situ with monomer (PANI, PEDOT or PPy) in ice bath; gel electrolyte immersion	Specific capacitance of 9.0×10^2 mF/cm^2; 86.5% stability after 12,000 cycles; 90% stability at 180° after 1000 bending cycles	[104]
PVA/KOH	CO	CuS nanosheets	Carbonization and oxidation of the CO fabric; electrodeposition with CuS nanosheets; gel electrolyte immersion	Specific capacity of 1.3×10^3 mF/cm^2; 91.8% stability after 2000 cycles; ultrahigh energy density (0.96 Wh/cm^2); power density of $4. 4 \times 10^3$ W/cm^2	[105]

Table 3. *Cont.*

Gel Electrolyte	Fabric Composition	Electrodes	Preparation Method	Conductivity Tests and Results	Ref.
PVA/H_2SO_4	CO	Au nanoparticles; tetraoctylammonium bromide; PANI; PDMS	Layer-by-layer deposition of tetraoctylammonium bromide; PDMS dip-coating eletrodeposition of PANI; gel electrolyte deposition	Au/PDMS: hydrophobicity (120–140°); washing fastness after 60 cycles; resistance to high-frequency ultrasound (300 s); high breathability; corrosion resistance; self-cleaning capability for 10 cycles; 98.88% cleaning efficiency Au/PANI: areal capacitance 5.0×10^2 mF/cm^2; energy 3.3×10^{-5} Wh/cm^2 and power 1.1×10^{-5} Wh/cm^2 density; 77% stabilty after 1000 cycles	[106]
PVA/KOH	CO	Cu; Ni	Sputtering (Cu); electrochemical deposition (Ni); electroplated; gel electrolyte immersion	Specific capacity of 243.9 µAh/cm^2; 70% stability after 5000 cycles; energy and power densities of 4.9×10 Wh/kg and 3.9×10^2 W/kg	[107]
PVA/KOH	CO	Nickel tungstate; niquel oxide; $Ni(NO_3)_2$	Ultrasonic spray-coating method with $Ni(NO_3)_2$; in situ chemical synthesis of a uniform nickel oxide layer; electrochemical deposition of nickel tungstate; gel electrolyte immersion	Specific capacity of 1.4×10^2 mF/cm^2; high specific of energy 12 µWh/cm^2; specific power of 6.9×10^{-8} W/cm^2; 78% stability after 5500 cycles	[108]
PVA/KOH	CO	$Zn(NO_3)_2$+MOFs+CoNi + layered double hydroxides (cathode); Zn-N (anode)	$Zn(NO_3)_2$ immersion; carbonization; Co-MOFs, $NiSO_4 \cdot 6H_2O$ and layered double hydroxide deposition; gel electrolyte application	Discharge capacity retention (60.11% after 1000 cycles); 98% Coulombic efficiency; bending resistance; high specific capacitance (161.25 F/g); energy (4.7×10 Wh/kg) and power (2.7×10 Wh/kg) density	[109]

Table 3. *Cont.*

Gel Electrolyte	Fabric Composition	Electrodes	Preparation Method	Conductivity Tests and Results	Ref.
PVA/KOH	CO	Co-zeolitic imidazole framework-67 nanoparticles	Co-zeolitic imidazole framework-67 nanoparticles in situ deposition; carbonization; gel electrolyte application	Specific capacitance of 288.62 F/g; energy density of 1.6×10 Wh/kg; power density of 6.5×10 W/kg; bending resistance; 76.4% stability after 2000 cycles	[110]
PVA/H_2SO_4	CO 95%; spandex 5%	PANI;carbon;textile	Tandem procedure by imersing the fabric in acidic aniline solution; reaction with ammonium persulfate and drying the fabric at 60 °C, using H_2SO_4 as dopand and carbonizing assistant	Specific capacitance of 3.9×10^2 mF/cm^2; over 70% capacitance retention after 5000 cycles; energy density of 3.6×10^{-2} Wh/m^2 at 7.5×10^{-1} W/m^2 power density; stable capacitance under bending (0–180°) and stretching (up to 50% elongation)	[111]
PVA/H_2SO_4	CO 95%; spandex 5%	PANI;graphen;textile;HCl	Dipping and drying method, followed by in situ polymerization of aniline	Specific capacitance of 1.6×10^3 mF/cm^2; over 75% capacitance after 10,000 cycles; energy density of 7.6×10^{-1} Wh/m^2; stable capacitive performance under bending from 0 to 180°; 77% retention over 600 bending cycles	[112]
PVA/KCl	Polyamide	Metal-coated textiles	Printing a graphite with ethyl cellulose as binder and terpineol as solvent; gel electrolyte sandwiched between the positive and negative electrode with piece of cellulose PET cloth as the separator	Areal capacitance of 3.2×10 mF/cm^2 and an energy density of 2.8×10^{-6} Wh/cm^2	[113]

Table 3. *Cont.*

Gel Electrolyte	Fabric Composition	Electrodes	Preparation Method	Conductivity Tests and Results	Ref.
PVA/KCl	Metal-coated polyamide;PET	Metal-coated textiles	Commercial metal-coated fabrics compared with a metal-free graphite coated PET/cellulose fabric	Ni/Cu-coated PET fabric: capacitance 9.9×10 mF/cm^2; energy density of 8.8×10^{-6} Wh/cm^2 Ni/Cu/Ag-coated polyamide: capacitance 4.7×10 mF/cm^2 at 5 mV/s; energy density of 4.2×10^{-6} Wh/cm^2; stable performance over 5000 charge–discharge cycles	[114]
KOH	PET; CO blend	NiCoAl-LDH); Ti3C2Tx; MXene; Ag nanowires as positive electrode; active carbon (negative electrode)	Interdigital pattern obtained by printing and electroless deposition; anchoring of the battery-type material onto conductive MXene and hydrothermal treatment; active carbon inks deposited by screen-printing; flexible supercapacitor assembled by covering a layer of PVA/KOH gel electrolyte, and encapsulated by scotch tape	Positive electrode: capacity 592 C/g, excellent rate performance and cycling stability over 10,000 cycles Positive electrode and negative electrode: energy density of 2.2×10^{-5} Wh/cm^2 and a power density of 3.0×10^{-3} W/cm^2	[115]
PEO/KOH	PET	Silver oxide	Stencils of different thicknesses used to print different layers of the battery in polyamide-nylon 6 and PET	Capacity of 0.6 mAh/cm^2 with an active electrode area of 0.5 cm $\times$ 1 cm	[116]
PVA/H_2SO_4	PET	rGO nanosheets; PPY	Dipping and drying method	Capacitance of 2.3×10^3 mF/cm^2; volumetric capacitance of 5.5×10^3 mF/cm^3; energy density of 1.1×10^{-5} Wh/cm^2; power density of 3.0×10^{-5} W/cm^2; retains 76% of its initial capacitance after 6000 cycles and mechanical stability under bending	[117]

Table 3. *Cont.*

Gel Electrolyte	Fabric Composition	Electrodes	Preparation Method	Conductivity Tests and Results	Ref.
PVA/Na_2SO_4	Cu; Ni coated conductive PET	PPy–graphene–PPy-coated fabric	Sandwich configuration with PVA/Na_2SO_4 as a gel electrolyte and filter paper as a separator; electrodes and filter paper immersed in the gel electrolyte before being assembled into a supercapacitor device	Capacitance of 6.8×10^2 mF/cm^2; energy density of 6.4×10^{-5} Wh/cm^3; power density of 0.6×10^{-3} W /cm^3; maintained 94.2% of its capacitance after 4000 cycles	[118]
PVA/H_2SO_4	CO; PET	Graphene; microcircuit encapsulant PE773	Pad–dry–cure method with graphene ink and encapsulation (microcircuit encapsulant PE773); immersion in gel electrolyte	Specific capacitance of ~2.7 mF/cm^2; capacitance retention of 98% after 150 cycles at 180° flexion; high domestic washing fastness (10 cycles)	[119]
PVA/H_3PO_4	Silk	CSF; PPy	Carbonization of silk fabrics; potentiostatic electrodeposition of PPy; immersion of the electrodes in gel electrolyte; SC assembly using cellulose sandwich	Composite: capacitance of 4.0×10^3 mF/cm^2 and cycling stability of 88.6%; capacitance retention after 1500 cycles. SC: areal specific capacitance of 6.7×10^2 mF/cm^2; energy density of 6.9×10^{-3} Wh/cm^3	[120]
PVA/H_3PO_4	Nylon®	CNT; Nylon®	Nylon®/CNT electrodes produced by dip-coating; Nylon®/rubber:Nylon®/CNT laminates production; SC production using a Nylon® sheet as separator	Capacitance of 117 F/g (at 2 mV/s); maximum energy density of 4.0 Wh/kg	[121]
PVA/H_3PO_4	Silk	PANI@GO	Dip-coating of silk in the respective mixture (either GO or PANI). For GO@PANI-coated fibers, the GO-coated silk materials were dipped in the PANI mixture.	Specific capacitance of 450 F/g; capacitance of 71.2 F/g obtained with the symmetrical PANI@GO-SL/PVA/PANI@GO–silk capacitor; 87.4% capacity retention at 5000 cycles; energy and power densities of 2.5×10 Wh/kg and 8.0×10^3 W/kg	[122]

Table 3. *Cont.*

Gel Electrolyte	Fabric Composition	Electrodes	Preparation Method	Conductivity Tests and Results	Ref.
PVA/H_3PO_4	PP; PET; PAN	rGO	Reactive inkjet printing of GO in the fabric with concomitant reduction; dip-casting of the gel electrolyte; all-solid-state SC assembly	PP fabric: specific capacitance of 1.3×10 mF/cm^2; power and energy densities of 4.6×10^{-3} W/cm^2 and 1.2×10^{-3} Wh/cm^2; apx. 100% of its original capacitance after 5000 cycles	[123]
PVA/H_3PO_4	Polyamide; carbon fibers	PEDOT:PSS	Weaving a polyamide warp yarn around the self-designed mode; interwoven weft yarn (polyamide yarn/Ag-coated polyamide yarn); supercapacitor assembled by using PEDOT:PSS-coated carbon fibers, coated with gel electrolyte and separated by cellulose	Capacitance 1.3×10 mF/cm^2 (79.9 F/g) at a current density of 0.1 mA/cm^2; power and energy densities of 4.6×10^{-3} W/cm^2 and 1.2×10^{-3} Wh/cm^2	[124]
PVA/H_2SO_4	Kevlar®	rGO	rGO-coated Kevlar® fibers produced by modified hydrothermal gelation; dip-coating of rGO@Kevlar cloth in gel electrolyte	rGO@Kevlar® fibers (38.1% rGO): specific strength of 1.6 MPa.m^3/kg; specific capacitance of 57 F/g rGO@Kevlar® cloth SC. Withstands impact of 9.1 N and deformation of 90°.	[125]
PVA/KOH-Zn(Ac)$_2$-LiOH- Ca(OH)$_2$	PET; polyamide 6,6	Zn; Cu; NiCo	Kapton applied on fabric Ni and Cu deposited by electroless and electrodeposiition; 2 interdigitated Cu electrodes coated; Zn and NiCo BOH nanosheets were electrodeposited; coating with gel electrolyte	Electroplated Zn anode and a Ni cathode; energy density of 2.6×10^2 Wh/kg; power density of 1.0×10^4 W/kg; stable cycling performance of 82.7% for 1500 cycles; good mechanical reliability (bending, twisting and tailoring)	[126]

Table 3. *Cont.*

Gel Electrolyte	Fabric Composition	Electrodes	Preparation Method	Conductivity Tests and Results	Ref.
PVA/LiCl	Bamboo	MnO_2–$NiCo_2O_4$; rGO	Printing of Ni-Co + printing of $KMnO_4$ for anode; printing with rGO + hydrazine reduction for cathode; device prepared using an anode and cathode sheets, separated by a bamboo fabric sheet, and coated with gel electrolyte	MnO_2–$NiCo_2O_4$/rGO device shows stable performance within a 0–1.6 V range; capacitance of 2.1×10^3 mF/cm^2; energy density of 3.8×10^{-2} W/cm^3; power density of 2.7 W/cm^3; 92% of capacitance retention after 5000 cycles and low charge transfer resistance 3.2 Ω	[127]
PVA/KCl	Ag-coated polyamide (Berlin fabric)	Ag-coated polyamide; graphite	Printing coated electrodes and gel electrolyte	Areal capacitance 1.3×10 mF/cm^2	[128]
PVDF/LiTFSI	PP-based satin	LFP/LTO;carbon black	Thermally drawn fibers comprising anode (LTO), cathode (LFP), gel electrolyte (PVDF/LiTFSI) and conductive polymer (carbon black). The fibers produced are then woven into the satin.	Battery discharge capacity of apx. 123 mAh and discharge energy of apx. 2.2×10^{-1} Wh. Woven: 96% of capacity retention after 1000 bending cycles.	[129]

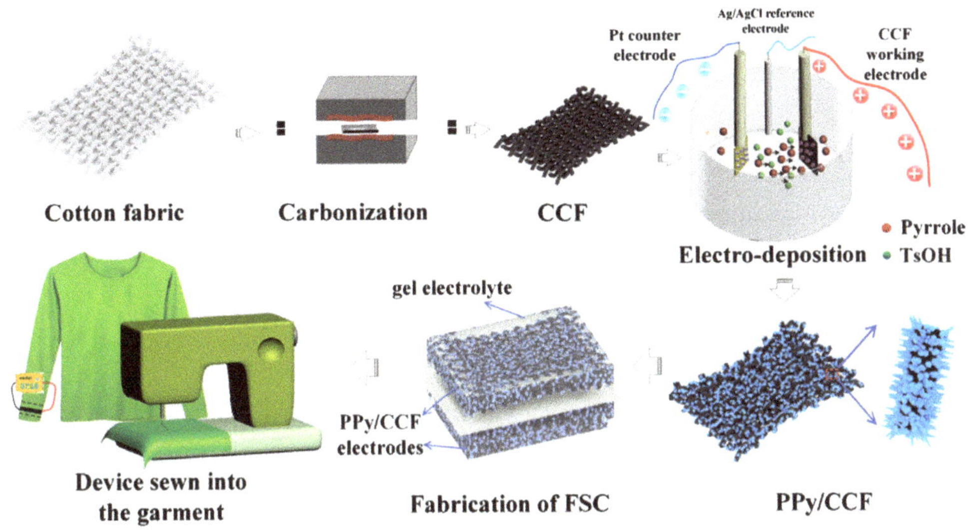

Figure 4. Process flowchart illustrating fabrication of electrode materials and all-solid-state supercapacitor [78].

The incorporation of conductive polymers or metallic ions involves the use of gel electrolytes, which are essential for textile supercapacitors. These materials enhance mechanical stability, safety, and adaptability for wearable applications. The flexibility provided by the gel electrolyte allows the system to withstand deformation without losing functionality, ensuring long-term durability. Furthermore, the solid interface between the electrode and gel electrolyte supports efficient ion transport, preserves indispensable moisture levels, and enhances conductivity, resulting in a superior electrochemical performance. PVA is extensively used to produce gel electrolytes for flexible CO-based supercapacitors. Combined with several acids or bases, these electrolytes enhance dielectric reactions and increase ionic conductivity via H^+ and OH^- ions, promoting electrochemical performance [83]. Thus, Sun et al. [84] used PVA/H_2SO_4 as gel electrolytes in a supercapacitor produced by spraying a PVA-co-ethylene nanofiber suspension onto CO fabric, followed by electropolymerization with PPy. The nanopores in the PPy increased the electrochemically active spots, enhancing Faradaic reactions. The resulting symmetric supercapacitor achieved high specific capacitance (6.7×10^2 mF/cm^2), 100% capacitance retention after 10,000 cycles, and efficient Coulombic efficiency (90–100%). Similarly, Wang et al. [80] coated a PPy-based electrode with a PVA/H_2SO_4 gel electrolyte. A hydrophobic spray and PU coating enhanced the CO fabric's hydrophobicity, promoting oxidative polymerization with 5-sulfosalicylate as a dopant. The microsupercapacitor initially lost some capacitance but stabilized, retaining 85% after 3000 cycles, showing good electrochemical stability. An often-used gel electrolyte in flexible textile supercapacitors is PVA/KOH. Wang et al. [85] used this gel electrolyte in CO-based electrodes modified with Ni:Co via electroless plating (cathode) and carbon/Fe_3C fabric (anode), with silk fabric as the separator. The supercapacitor showed 113.7 C/g at 1 A/g, 80% retention after 4000 cycles, and 4.7×10 Wh/kg energy density, indicating potential for high-performance flexible devices. Zhai et al. [86] applied this gel electrolyte by immersing a dyed and carbonized CO fabric electrode, modified with zinc oxide (ZnO) nanoparticles and copper sulfide (CuS) microspheres. These were prepared via atomic layer deposition and a hydrothermal reaction, respectively. The synergistic effect between the electrochemical and pseudocapacitive properties of ZnO and CuS enhanced ion exchange and redox reactions, increasing the substrate's porosity.

The supercapacitor achieved a capacitance of 1.8×10^3 mF/cm^2 at 2 mA/cm^2 and 85.2% retention after 5000 cycles, showing excellent performance for supercapacitor applications. Shun et al. [87] demonstrated that carbonized CO fabrics exhibit hydrophobicity (140°), rendering them unsuitable for use as electrodes. However, pre-carbonization dyeing with reactive dye (Procion, 10 wt.%) improved hydrophilicity due to heteroatoms (oxygen (O) and nitrogen (N)), allowing use as electrodes. The pyrolysis of the dye created porous structures with high surface areas, enhancing electrochemical properties and ion deposition. Dyeing increased specific capacitance (1.8×10^3 mF/cm^2), slightly reduced resistivity (30 Ω/sq), and maintained capacitance retention (~97.4% after 1000 cycles). Using PVA/KOH electrolytes, cellulose separators, and polyethylene terephthalate (PET) coating, the supercapacitor retained 92.2% capacitance after 5000 cycles. Mallick et al. [88] also employed the same gel electrolyte in their study, developing a hybrid supercapacitor through thermal treatment (annealing) with nickel (II) chloride hexahydrate ($NiCl_2 \cdot 6H_2O$) and sodium tungstate dihydrate ($Na_2WO_4 \cdot 2H_2O$). After 20,000 cycles, the capacitance increased by 34% due to nickel(II) hydroxide ($Ni(OH)_2$) formation. Furthermore, the one-step Ni:Co deposition method proved cost-effective, achieving a specific capacitance of 60.61 F/g, suitable for supercapacitor material applications. In contrast to the studies mentioned above, Hiller et al. introduced two novel gel electrolyte alternatives: one was inspired by culinary energy drinks [89] and the other was developed using tetraethylammonium tetrafluoroborate ($TEABF_4$)/polyacrylamide [83] for supercapacitors produced with CO fabrics functionalized with spray-coated activated carbon. Commercial electrolytes such as Grenade Energy, Shopper Isotonic Drink, and LoSalt® solutions (0.5 and 1 M) were evaluated in the first case. Although energy drinks contained compounds like potassium chloride (KCl), sodium chloride (NaCl), magnesium chloride ($MgCl_2$), and dipotassium hydrogen phosphate (K_2HPO_4), agar–agar and *k*-carrageenan were added to achieve a gel-like consistency. Among the tested options, the LoSalt® (1 M) electrolyte exhibited a promising electrochemical performance, with a capacitance of 2.3×10 mF/cm^2 at 0.5 mA/cm^2 [89]. The application of the organic electrolyte $TEABF_4$/polyacrylamide exhibited electrochemical properties similar to those of PVA-based formulations, improving durability. This system showed capacitance retention of 48% and a Coulombic efficiency of 88% after two months of aging tests. Although further improvements are needed, the study highlighted practical alternatives to overcome the dehydration limitations of PVA-based electrolytes [83]. Conductive materials often compromise softness and porosity, essential for supercapacitor applications. To overcome this limitation, using gels electrolytes, Wang et al. [79] developed a wearable battery. Their method used CO fabric, activated with palladium (II) chloride ($PdCl_2$) and deposited with Co, Cu, Ni, and Ag through electroless deposition. A correlation between conductivity and softness was found in the CO fabrics treated with Co and Ni, revealing conductivities of 1.1×10^3 and 3.3×10^3 S/m and softness values of 8.50 and 8.67 mm, respectively. In contrast, the CO/Cu and CO/Ag fabrics exhibited conductivities of approximately 100×10^3 S/m and softness values exceeding 8.5 mm. The achieved softness addressed common issues in conventional flexible electrodes, enhancing their suitability for wearable battery applications. Chen et al. [90] developed a method using dopamine–cellulose in situ polymerization to overcome low electrode porosity and the application of conductive additives. This was followed by immersion in the GO solution and carbonization. The dopamine multifunctional interface acted as both a binder and conductive additive. After immersion in gel electrolyte, the flexible supercapacitor exhibited nanometric porosity, a high specific surface area (347.6 m^2/g), and a capacitance of 1.2×10^3 mF/cm^2, with only a 6% reduction after 4000 cycles. Moreover, Liang et al. [91] selected PPy as the conducting polymer due to its electrochemical properties and high conductivity. GO was applied via vacuum filtration, followed by the in situ ultraviolet (UV) polymerization of PPy and Ag^+. The π–π interactions between GO and PPy and Ag ions improved electrochemical performance (1.7×10^3 mF/cm^2 at 0.5 mA/cm^2, 90.5% retention after 10,000 cycles). UV polymerization generated PPy/Ag nanoparticles, enhancing ion transport and redox reactions (Ag/Ag^+), with a synergistic effect between PPy, Ag, and GO. Hydrothermal and plasma treatments were evaluated to increase the carbonized CO's electrochemical properties, overcoming this limitation. Via ring annealing, carbonized CO fabric formed an electrically conductive substrate. After plasma treatment with O_2 and a hydrothermal reaction with vanadium dioxide (VO_2) nanosheets, the VO_2/CO electrode was created, containing ~0.96 mg/cm^2 of VO_2. This composite was applied in (i) a flexible piezoresistive sensor with two VO_2/CO layers separated by a polydimethylsiloxane (PDMS) elastomer substrate and in (ii) quasi-solid Zn-ion batteries with a VO_2/CO cathode and Zn/CNT film anode, both in the PVA/Zn trifluoromethanesulfonate (PVA/Zn(OTf)$_2$) gel electrolyte. The sensor showed 95.8 Ω/sq conductivity, high compression sensitivity (7.12 kPa at 0–2 kPa), a fast response (12/8 ms), and superior 160° flexibility due to the VO_2 3D network. Resistance varied with pressure, and energy consumption was low (0.01 V), making the device suitable for lightweight wearable monitoring systems. The battery showed 99.8% Coulombic efficiency, 88.7% capacity retention after 5000 cycles, and a specific capacity of 301.5 mAh/g [92]. The printing process is the most common, straightforward, and cost-effective method of producing flexible CO-based supercapacitors [93]. Keawploy et al. [94] applied conductive ink containing Ag, CNT, and graphene to enhance the conductivity of CO fabric. This enabled the fabrication of an all-solid-state flexible supercapacitor via screen-printing using an Ag/CNT (60/40%) mixture and PVA/KOH as the gel electrolyte. The device achieved a high areal specific capacitance of 6.8×10^2 mF/cm^2 at a current density of 0.0125 mA/cm^2, with around 80% capacitance retention after 3000 cycles. Similarly, these authors [95] applied the same modification technique and gel electrolyte, substituting CNT with activated carbon and removing graphene from the ink formulation. These modifications enhanced the specific capacitance to 3.3×10^3 mF/cm^2 at 5 mV/s and improved the capacitance retention to 130% after 10,000 cycles. The optimization of ink composition is crucial for flexible supercapacitor production. Jiang et al. [96] evaluated four ink formulations based on multi-walled CNT (MWCNT) (5 wt%), thermoplastic PU (2.5, 5, 10, and 15 wt%), and NMP (79.5, 84.5, 89.5, and 92.0 wt%). The optimal formulation contained 5 wt.% thermoplastic PU and 89 wt.% NMP, which produced a porous microstructure, improving electrolyte ion adsorption and increasing capacitance. The supercapacitor achieved a specific capacitance of 26.4 F/g and 1.4×10 mF/cm^2 at 10 mV/s, with stability under a deformation of 20%/s. The same research group [93] also investigated the impact of the structural type used, comparing all-in-one and sandwiched fabric-based supercapacitors. The all-in-one structure exhibited superior performance, achieving a specific capacitance of 4.2 mF/cm^2, high flexibility, approximately 97.4% capacitance retention after 1000 cycles, and good electrochemical stability, offering easier integration than the typically preferred sandwiched design. Furthermore, washing durability presents a challenge for wearable supercapacitors. Islam et al. [97] developed a solid-state supercapacitor. This was spray-coated with graphene (four layers) for healthcare sensors, using a thin PU encapsulant to protect the conductive polymer (Figure 5). This coating preserved the sensor's electrochemical properties, maintaining cyclic stability after 10,000 cycles and a capacitance of 3.2 mF/cm^2 after 10 washing cycles.

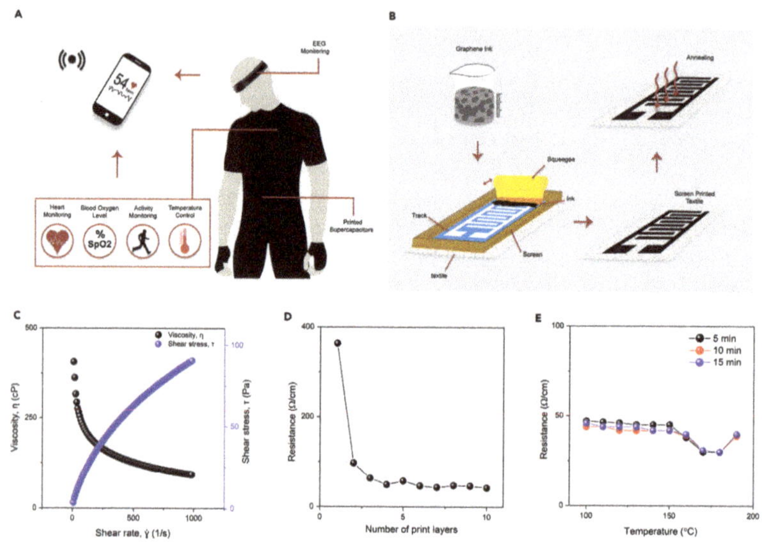

Figure 5. Multifunctional printed wearable e-textiles: (**A**) Illustration highlighting multifunctional features of printed e-textiles. (**B**) Schematic representation of screen-printing technique used to produce graphene-based wearable e-textiles. (**C**) Rheological behavior of graphene ink, showing viscosity and shear stress as functions of shear rate. (**D**) Variation in electrical resistance with increasing print layers on graphene-printed CO fabric. (**E**) Influence of curing time and temperature on electrical resistance of graphene-printed CO fabric [97].

Alongside printing, simple methods, namely dipping and impregnation, can also be applied in the production of textile supercapacitors. Zhou et al. [98] explored an impregnation application method, applying PPy, using iron(III) chloride ($FeCl_3$) as an oxidant. They formed FeOx during polymerization and carbonization to enhance ion transfer within the textile structure. The PVA/LiCl supercapacitor achieved 135 F/g capacitance and 88.4% retention after 1000 cycles, indicating potential for energy storage. Integrating multiple conductive polymers offers a viable strategy for improving supercapacitors' electrochemical efficiency and durability. Thus, Xiaohong et al. [99] developed a flexible symmetric supercapacitor with CO fabric electrodes, functionalized with PEDOT:PSS and in situ polymerized with MXene, graphene nanoscrolls, and PPy. The combination of these conductive polymers—PEDOT:PSS for conductivity and stability, graphene nanoscrolls for mechanical strength, MXene for hydrophilicity, and PPy for flexibility and pseudocapacitance—resulted in a high specific capacitance (4.9×10^3 mF/cm^2 at 1 mA/cm^2), with 90% retention after 3000 cycles. The device also exhibited high energy density (3.2×10^{-4} Wh/cm^2) and flexibility, as well as strong waterproof performance (92% capacitance retention after 2 hours in water), demonstrating its suitability for wearable textiles. Costa et al. [100] studied the impact of oxidizing agents on CNT, applied to CO fabrics via the dip–pad–dry method. Nitric acid (HNO_3), H_2SO_4, and an HNO_3:H_2SO_4 (1:3 v/v) mixture were tested. HNO_3 increased the number of oxygen-containing groups, enhancing redox reactivity, specific surface area (269 m^2/g), and pore volume (0.618 cm^3/g), and thereby improving electrochemical properties. The nitric acid pre-oxidized supercapacitor showed 57, 14, and 106% improvements in capacitance (3.91 F/g), voltage (2.53 V), and energy density (3.5 Wh/kg), respectively, outperforming traditional methods with carbon, Ni, or graphene. Another study by Costa et al. [101] showed that applying MWCNTs to CO fabrics via the dip–pad–dry method, combined with a PVA/H_3PO_4 gel electrolyte, resulted in a flexible textile capacitor. High conductivity, thermal, and chemical stability and large surface area ensure excellent electrochemical performance. The superior cyclic stability was attributed to the solid-gel electrolyte, which enhances the capacitor's durability. Teixeira et al. [1] also evaluated the effect of gel electrolyte composition on a CNT-based supercapacitor's electrochemical properties (energy storage and fluorescent optical properties). Adding a fluorescent ZnS-Mn pigment enhanced redox reactions, pseudocapacitive charge storage, and UV light response. This resulted in a 4% increase in capacitance (4.37 F/g) and 100% retention after 8000 cycles, showing excellent stability. Moreover, Li et al. [102] developed a flexible supercapacitor using exfoliated and delaminated CO fabric immersed in Ti_3C_2Tx (2–8 wt.%), followed by carbonization (800, 1000, and 1200 °C) and immersion in PVA/H_3PO_4. The rough CO surface facilitated the adhesion of MXene more than synthetic fabrics. Furthermore, carbonization temperature and MXene concentration influenced electrochemical properties. The optimal performance was achieved with 6 wt.% $Ti_3C_2T_x$ carbonized at 1000 °C, yielding a specific capacitance of 7.9×10^2 mF/cm^2 (2 mV/s) and 74% retention after 10,000 cycles. To explore the electrochemical properties of developed materials as flexible wearable energy storage devices, the flexible supercapacitors were assembled by placing two electrodes in parallel and coating them with the PVA/H_3PO_4 gel electrolyte. Galvanostatic charge–discharge (GCD) results were displayed near isosceles triangles with no significant voltage drops and IR drops at current densities ranging from 0.5 to 6 mA/cm^2. The calculated capacitance was found to be up to 5.0×10^2 mF/cm^2 at 0.5 mA/cm^2, and 74% capacitance was retained after 10,000 charge and discharge cycles, revealing stable performance at 0.5 mA/cm^2. In contrast to a single flexible supercapacitor with an operating voltage of 0.8 V, the two flexible supercapacitors devices exhibit a 1.6 V charge and discharge voltage window with similar charge/discharge times. To warrant application, flexible wearable energy storage devices must prove to be flexible enough under different mechanical deformations. Thus, different bending angles from 0° to 180° were tested, showing a slight decrease in capacitance, which was related to the weaker interfacial bonding between the conductive material and textile substrate under these deformation modes. Despite this, wrist-worn supercapacitors can power an light-emitting diode (LED) light source (Figure 6).

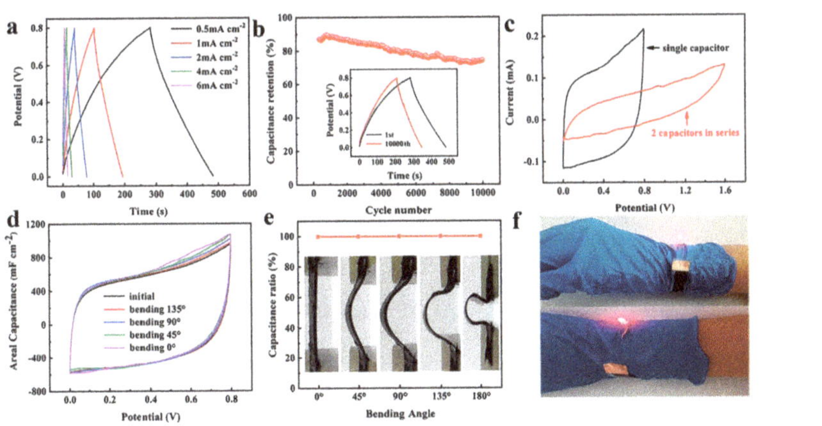

Figure 6. (**a**) Galvanostatic charge–discharge (GCD) test of flexible supercapacitor with 6 wt.% of $Ti_3C_2T_x$ and corresponding (**b**) cycling performance at current density of 0.5 mA/cm^2. (**c**) Cyclic voltammetry (CV) curves of single supercapacitor and two supercapacitors connected in series (the single device operates up to 0.8 V), while series configuration reaches 1.6 V. (**d**,**e**) CV curves at different bending angles recorded at a scan rate of 2 mV/s^1. (**f**) Photograph of flexible assembled device wrapped around wrist, powering red LED [102].

Several studies report the influence of gel electrolytes on porosity, facilitating redox reactions and enhancing electrochemical properties. An asymmetric all-solid-state supercapacitor was fabricated by Ramandi et al. [81] by chemically depositing graphene/PANI nanotubes, followed by immersion in NiCo/cetyl trimethyl ammonium bromide and KOH. The NiCo-layered double hydroxide improved electrochemical performance by enhancing redox reactivity and forming a mesoporous structure. Adding potassium persulfate ($K_2S_2O_8$) to the gel electrolyte promoted ion diffusion, increasing specific capacitance (4.3×10^2 mF/cm^2 at 0.075 mA/cm^2) and reducing degradation (15.95% after 10,000 cycles). Liu et al. [103] demonstrated that ion diffusion in the electrolyte, porosity, and surface area are crucial for supercapacitors with strong electrochemical properties. An asymmetrical all-solid-state textile supercapacitor was produced using dopamine hydrochloride, chloroauric acid ($HAuCl_4$), a Cu-metal–organic framework (MOF) solution, and a gel electrolyte (PVA/KCl) with PET-sealed electrodes. The Cu-MOF and gold (Au) layers improved ion diffusion and redox reactions, leading to a high specific capacitance (258 F/g at 0.075 mA/cm^2) and energy density (4.3×10^{-4} Wh/cm^2). Similarly, Wan et al. [104] found that CO fabric improves electrochemical performance due to its bidimensional structure, which enhances ionic conductivity. The flexible supercapacitor, prepared by padding with copper chloride ($CuCl_2$), followed by in situ vapor-phase polymerization with PPy in an ice bath and immersion in a gel electrolyte, achieved a specific capacitance retention of 86.5% after 12,000 cycles. Zheng et al. [82] explored vapor-phase polymerization with PEDOT and spray-coated MXene, using PVA/H_2SO_4 as the electrolyte. Five layers of PEDOT provided superior performance (1.0×10^3 mF/cm^2), outperforming values from the literature. Moreover, the synergistic interaction between PEDOT and MXene formed an interconnected network that enhanced electron transport and conductivity. The supercapacitor showed strong electrochemical stability, high thermal performance (193.1 °C at 12 V), effective electromagnetic (EMI) shielding (36.62 dB), and high detection, making it ideal for portable energy storage. Electrodeposition and electrochemical deposition are commonly employed as techniques to enhance the conductivity of CO fabrics through electrochemical reactions. In electrodeposition, the fabric is immersed in an electrolytic solution containing metal ions, and the application of an electric current promotes the deposition of a thin metallic layer onto the insulating substrate [105]. Jin et al. [105] investigated the electrodeposition of CuS nanosheets onto carbonized CO fabrics through two techniques: potential (−1.1 V, 400 s) and galvanostatic (9 mA, 2400 s, 60 °C). The galvanostatic method (g-CuS) provided superior electrochemical performance (4676 vs. 3.5×10^3 mF/cm^2) due to its higher crystallinity, surface area, and pore size, leading to lower charge transfer resistance (2.2 Ω) and more efficient redox reactions. Upon the immersion of the electrode in a PVA/KOH solution and the formation of a subsequent sandwich structure between cellulose films, polyethylene terephthalate membranes, and Cu wire, a decrease in specific capacity was observed (4675 vs. 1.3×10^3 mF/cm^2). Despite this decline, the performance remained suitable for wearable energy storage device applications, demonstrating stable cyclic performance.

Conversely, electrochemical deposition is commonly used with conductive polymers, wherein polymer chains are formed through the oxidation and reduction of monomers, which are subsequently deposited onto the electrode surface. Flexible supercapacitor electrodes were produced by Sun et al. [78] using carbonized CO fabric, the in situ electrodeposition of PPy, and immersion in PVA/LiCl. Fabric carbonization improved electron transfer (hydrophobicity), while *p*-toluenesulfonic acid dopant enhanced redox reactions and conductivity. The optimized electrode (PPy/carbonized CO, 1:2) exhibited a high specific capacitance of 3.6×10^3 mF/cm^2 at 2 mA/cm^2. PPy deposition improved conductivity, contact area, and hydrophilicity, reducing charge transfer and series resistance.

Despite a decrease in capacitance to 5.0×10^2 mF/cm^2 after immersion in PVA/LiCl and sandwiching with cellulose, the performance remained stable. The electrochemical and comfort properties of CO fabrics, with Au nanoparticles deposited via layer-by-layer, were evaluated. Modification with PDMS resulted in conductive textiles with self-cleaning capabilities (98.88%), durability (60 washing cycles), hydrophobicity (120–140°), breathability, and corrosion resistance. These properties are crucial for supercapacitors, yet further electrochemical studies are needed to assess their impact. The electrodeposition of PANI onto these fabrics produced a solid-state device with an energy capacitance of 3.3×10^{-5} Wh/cm^2 [106]. In contrast, Wang et al. [107] applied electrochemical deposition via CV with Ni onto CO fabrics previously sputtered with Cu and subsequently electroplated. The CV curves revealed that the initial Cu sputtering did not affect the efficiency of the Ni deposition, as Cu was not involved in the redox reactions. However, the Ni-Co-S structure promoted ion transport during the electrochemical process, enhancing electrochemical performance and demonstrating potential for wearable supercapacitor applications. Similarly, Hekmat et al. [108] developed flexible supercapacitors using ultrasonic spray with nickel(II) nitrate ($Ni(NO_3)_2$), followed by in situ chemical synthesis of nickel oxide, the electrochemical deposition of nickel tungstate, and immersion in PVA/KOH. The electrolyte gel provided flexibility and prevented short circuiting, maintaining electrochemical stability even after multiple cycles. This made it a promising solution for wearable devices. Hybrid materials, including MOFs and their derivatives, offer advantages over inorganic particles, enhancing electrochemical energy storage. An asymmetric supercapacitor with a carbonized CO cathode, modified with Zn-N-MOFs-CoNi layered double hydroxides and a Zn-N-modified carbonized CO anode, achieved high energy (4.7×10 Wh/kg at 1.6×10^3 W/kg) and power density (2.7×10 Wh/kg at 8.0×10^3 W/kg). The high surface area of the layered double hydroxides and the metallic components' enhanced pseudocapacitance improved electrochemical performance, yielding Coulombic efficiency (>98%), a long cycle life, superior bending resistance, and high specific capacitance (161.25 F/g) [109]. The exceptional electrode performance stems from integrating carbonized CO fabric and Co/zeolitic imidazole framework-67 (MOF-derived), which enhances the supercapacitor's electrochemical properties. The spinel cobalt oxide–nitrogen-doped structure enables redox processes and charge storage, while CO-derived carbon fibers improve conductivity and flexibility. Nitrogen doping via the zeolitic imidazole framework−67 further enhances performance, achieving 288.62 F/g (0.6 A/g) capacitance, high energy/power density, and 76.4% retention after 2000 cycles [110].

In addition to single-material CO fabrics, various textile substrates have been explored to develop flexible materials with conductive and/or energy storage properties. A notable study was reported by Song et al. [111], in which a H_2SO_4-assisted carbonization strategy for synthesizing PANI–carbon–textile electrodes was studied through a facile tandem fabrication procedure. This electrode was prepared using H_2SO_4 as an acid dopant and carbonizing assistant. First, a fabric of 95% CO and 5% spandex was immersed in the acidic aniline solution; this was followed by a reaction with ammonium persulfate ($(NH_4)_2S_2O_8$). Lastly, the ternary electrode was synthesized by simply drying the fabric in the vacuum-drying oven. The all-solid-state symmetric supercapacitor was assembled by using two PANI–carbon–textile pieces and PVA/H_2SO_4 gel as the electrolyte and separator, simultaneously. The optimized electrode, PANI/carbon/textile, achieved an impressive areal specific capacitance of 3.9×10^2 mF/cm and maintained over 70% capacitance retention after 5000 cycles. Furthermore, a symmetrical supercapacitor fabricated from these electrodes delivers a high energy density of 3.6×10^{-2} Wh/m at 7.5×10^{-1} W/m power density and the capacitance of the device was stable under bending (0–180°) and stretching (up to 50% elongation). In another study, Song et al. [112] developed PANI–graphene–textile electrodes using a straightforward dipping and drying method, followed by in situ polymerization of aniline. This research systematically examines the influence of acidic dopants (HCl, HNO_3, *D*-tartaric acid or citric acid) on the morphological, structural, and capacitive properties of the electrodes. Among them, the sample prepared with hydrochloric acid (HCl) demonstrates an impressive areal specific capacitance of 1.6×10^2 mF/cm at 1 mA/cm^2. It also exhibits excellent cycling stability, retaining over 75% capacitance after 10,000 cycles at 10 mA/cm^2. The all-solid-state supercapacitor assembled from these electrodes achieves a high energy density of 7.6×10^{-1} Wh/m^2 at 1.4 W/m^2 power density. Additionally, the device maintains stable capacitive performance under bending from 0 to 180°, with 77% retention over 600 bending cycles, confirming its outstanding mechanical flexibility and durability in potential wearable energy storage applications. Pullanchiyodan et al. [113] developed a polyamide fabric-based asymmetric supercapacitor, utilizing metal-coated textiles as both electrodes and current collectors. The device features graphite paste printed on Ag-coated textile (Berlin) as the negative electrode and Ni/Cu/Ag-plated fabric (Nora Dell) as the positive electrode, with a PVA/KCl gel electrolyte ensuring biocompatibility. During electrochemical operation, metal oxides formed on the Nora Dell fabric contribute to pseudocapacitance, while the graphite electrode supports electric double-layer capacitance. The asymmetric supercapacitor exhibited improved performance, achieving an areal capacitance of 3.2×10 mF/cm^2 and an energy density of 2.8×10^{-6} Wh/cm^2 at 25 mV/s. In addition, the device demonstrated strong cyclic stability and self-charging capability when integrated with flexible solar cells, highlighting its potential as a safe and sustainable energy storage solution for wearable electronics. Other substrates have been used for the development of electronic textiles based on gel electrolytes, with CO/spandex mixtures, CO/PET blends, polyamide, PET, silk, aramids, PET/polyamide 6,6, and polypropylene (PP) being some examples. In one study by Pullanchiyodan et al. [114], the researchers investigated the development of fabric-based supercapacitors, using metal-coated textiles (Berlin RS (silk, polyamide), Bremen RS (Parachute silk, polyamide), Nora Dell (polyamide), and Armor FR (PET)) as both active materials and current collectors, combined with a non-toxic PVA/KCl gel electrolyte for wearable energy storage applications. Electrochemical and capacitive performance studies were conducted and compared with a metal-free graphite-printed textile supercapacitor (cellulose-PET fabric) to assess the impact of metal coatings. Graphite electrodes printed on Armor FR (Ni/Cu-coated PET fabric) and Nora Dell (Ni/Cu/Ag-coated polyamide) exhibited considerably enhanced capacitance, achieving 9.9×10 mF/cm^2 and 4.7×10 mF/cm^2, respectively, at 5 mV/s. Notably, this was 24 and 52 times greater than that of the metal-free graphite printed textile. Under the same conditions, the energy densities of Armor FR and Nora Dell supercapacitors reached 8.8×10^{-6} Wh/cm^2 and 4.2×10^{-6} Wh/cm^2. Moreover, the Nora Dell-based supercapacitor demonstrated a stable performance over 5000 charge–discharge cycles, confirming its durability. Furthermore, in vitro cytocompatibility tests with adult human dermal fibroblast cells validated the biocompatibility of the PVA/KCl electrolyte, reinforcing its suitability for wearable applications. In another work, Zhang et al. [115] focused on the development of coplanar supercapacitors on textiles, which offered improved mechanical flexibility and seamless integration with other devices compared to conventional stacked configurations. However, their practical application is often limited by low energy density. To address this, this study proposed an in-plane hybrid supercapacitor on textiles, featuring a specially designed battery-type positive electrode to enhance energy storage performance. This electrode integrates NiCoAl-LDH (for high capacity), $Ti_3C_2T_x$ MXene (for high conductivity and structural stability), and Ag nanowires, achieving an impressive capacity of 592 C/g (at 1 A/g), along with an excellent rate performance and cycling stability over 10,000 cycles. Utilizing this composite as the positive electrode and activated carbon as the negative electrode, the screen-printed supercapacitor demonstrated a high areal energy density of 22.18 μWh/cm^2 and a power density of 3.0×10^{-3} W/cm^2, considerably outperforming traditional carbon-based in-plane textile supercapacitors. Furthermore, this device exhibits exceptional bending capability, reinforcing its potential as a printable and efficient power source for flexible and wearable electronics. Kota et al. [116] developed batteries that were useful for body-conformal wearable sensors in a 2D planar textile-based primary silver (I) oxide (Ag_2O)–Zn battery fabricated using the stencil printing method. The functionalized PET fabric achieved an areal capacity of 0.6 mAh/cm^2 with an active electrode area of 0.5 cm × 1 cm. The discharge duration of the textile-based Ag_2O–Zn battery was enhanced by optimizing the PEO-based gel electrolyte. In particular, the authors investigated how the drying time, electrolyte concentration, and PEO weight percentage of the gel electrolyte influence the battery's discharge capacity. The 5 wt.% PEO gel served as a reliable separator material, exhibiting no signs of delamination after the addition o the KOH electrolyte. To enhance the discharge duration of the textile-based Ag_2O–Zn battery, the drying time of the PEO-based gel electrolyte was carefully optimized. The optimal drying condition for the 5 wt.% PEO-based gel electrolyte was identified as 10 minutes at 40 °C. Barakzehi et al. [117] published another notable study focused on the development of a textile-based electrode by modifying PET fabric. The fabric was modified with rGO nanosheets and PPy nanospherical particles. These conductive composites enable the fabrication of flexible supercapacitors utilizing a gel electrolyte of PVA/H_2SO_4. The optimized device demonstrates an impressive areal capacitance of 2.3×10^2 mF/cm^2 at 1 mV/s, a volumetric capacitance of 5.5×10^3 mF/cm^3 at 1.6 mA/cm^3, an energy density of 1.1×10^{-5} Wh/cm^2, and a power density of 3.0×10^{-2} m^2. Moreover, it retains approximately 76% of its initial capacitance after 6000 cyclic voltametric cycles and exhibits outstanding mechanical stability under bending. In another study [118], the same authors explored the modification of PET fabric with aluminum-based MOFs (MIL-53(Al)) using a layer-by-layer assembly method at room temperature. Subsequently, rGO was deposited via dip-coating in a GO suspension, followed by chemical reduction. Finally, pyrrole was polymerized in situ on the PET/MOF/rGO surface. The resulting composite electrode achieved an impressive areal capacitance of 5.1×10^2 mF/cm^2 at a scanning rate of 1 mV/s in an aqueous H_2SO_4 electrolyte. The optimized electrode was further integrated into a symmetrical solid-state supercapacitor, which delivered a volumetric capacitance of 3.5×10^3 mF/cm^3, an energy density of 6.4×10^{-5} Wh/cm^3, and a power density of 0.6×10^{-3} W /cm^3. Notably, the device exhibited excellent long-term stability, retaining 85% of its initial capacitance after 1000 cyclic voltametric cycles and maintaining performance even after 12 months of ambient storage. Bhargava et al. studied the electrochemical performance of flexible supercapacitors using three commonly used substrates, namely, commercially available conductive PET fabric (Cu/Ni-PET), carbon cloth, and stainless-steel wire mesh. This study specifically examined how substrate properties, including conductivity, surface morphology, and wettability, affected the performance of PPy–graphene–PPy sandwich structure electrodes. The results revealed that the Cu/Ni-PET-based device exhibited the highest electrochemical performance, achieving an areal capacitance of 6.8×10^2 mF/cm^2 at a current density of 2 mA/cm^2 and maintaining 94.2% of its capacitance after 4000 cycles. The superior performance of the Cu/Ni-PET-based supercapacitor was primarily attributed to its high conductivity, favorable wettability, distinctive surface morphology, and excellent flexibility. Afroj et al. [119] applied the pad–dry–cure method to CO-PET (35/65%) fabrics. This group developed a wearable e-textile using graphene ink, an encapsulant, and a PVA/H_2SO_4 gel electrolyte. The material achieved 2.7 mF/cm^2 capacitance and 98% retention after 150 bending cycles at 180°. It also retained conductivity after 10 washes, addressing the durability issues of textile supercapacitors. Several examples of woven textile-based power sources that use PVA/H_3PO_4 as gel electrolyte can be found in the literature. In their work, Li et al. [120] proposed a method to produce all-solid-state supercapacitors using silk fabrics and employing PVA/H_3PO_4 as the gel electrolyte. This procedure proposes the use of silk fabric, carbonized (CSF) under a nitrogen atmosphere at 900 °C, which is then coated with PPy via potentiostatic electrodeposition. The supercapacitor is assembled using a 2-electrode system with a cellulose separator sandwich and sealed with a PET film. The resulting composite has a mass loading of 16.02 mg/cm^2 (CSF/PPy), as well as good areal capacitance (4.0 × 10^3 mF/cm^2 at 2 mA/cm^2) and cycling stability (88.6% capacitance retention after 1500 cycles). Regarding the supercapacitor produced, outstanding areal specific (6.7 × 10^2 mF/cm^2) and volumetric (1.5 × 10^4 mF/cm^3 at the current density of 2 mA/cm^2) capacitances were achieved, as was an energy density of 6.88 mWh/cm^3 (at a power density of 4.0 × 10^{-3} W/cm^3). Sim et al. [121] also proposed another method to produce supercapacitors using PVA/H_3PO_4 as gel electrolyte. In this work, the authors prepared Nylon®/CNT electrodes by dip-coating Nylon® fibers, which were then used to prepare supercapacitors via lamination. The laminates comprised two Nylon®/CNT electrodes, separated by a Nylon® sheet and covered with the gel electrolyte, which were in turn outlined by two Nylon®/rubber composite sheets. This material showed promising performance, with a maximum capacitance of 117 F/g (at 2 mV/s) and a maximum energy density of 4.0 Wh/kg (at a power density of 3.5 × 10^2 W/kg), while losing only a fraction of its performance while under strain up to 200%. Another example can be found in the work of Khairi et al. [122], where woven conductive fibers were prepared using silk and different coatings (namely GO, PANI and GO@PANI composites). These were then tested to assess their electrochemical performance. The capacitors were prepared by dip-coating, with the silk fabric being dipped in either the GO or the PANI solution. The PANI solution, used for both the PANI-SL and the GO@PANI–silk materials, comprised the appropriate reagents to synthesize PANI (namely aniline, ammonium persulphate and nitric acid), and not an aqueous PANI suspension. The authors highlight the GO@PANI–silk fibers as a promising combination that can be used in the production of supercapacitors, presenting the highest specific capacitance of all the coated silk electrodes (450 F/g at 10 mV/s). In addition, a capacitance of 71.2 F/g (at a current density of one A/g) was obtained with the symmetric PANI@GO-SL/PVA/PANI@GO–silk capacitor (using PVA/H_3PO_4 as gel electrolyte), displaying 87.4% retention at 5000 cycles. The Ragone plot of the symmetric cell, produced using this material also showed the highest energy (2.5 × 10 Wh/kg) and power (8.0 × 10^3 W/kg) densities. An all-solid-state supercapacitor was constructed by Stempien et al. [123] using two rGO-coated textile layers combined with PVA/H_3PO_4. For this study, PP, PAN, and PET were used as textile substrates, which were coated with GO by reactive inkjet printing in parallel with ascorbic acid, which acted as a reducing agent. The coated textile layers were then coated with the gel electrolyte by drop casting; they were then combined in a two-layer system to assemble the SC. Promising results were obtained with PP, having achieved a maximum specific capacitance of 1.3 × 10 mF/cm^2 (79.9 F/g) at a current density of 0.1 mA/cm^2, power and energy densities of 4.6 × 10^{-3} W/cm^2 and 1.2 × 10^{-3} Wh/cm^2, respectively, and an increased electrochemical stability that allowed the supercapacitor to maintain around 100% of its original capacitance after 5000 cycles. It also demonstrated good and stable behavior when submitted to bending forces. A direct current triboelectric nanogenerator (DC-TENG) and a supercapacitor were also assembled, using polyamide and carbon fibers as textile substrates to harvest biomotion energy from the electrostatic breakdown of clothes [124]. The TENG was prepared by weaving a polyamide warp yarn around the self-designed mode in 20 rows, followed the production of an interwoven weft yarn (polyamide yarn/Ag-coated polyamide yarn). The results showed that a small-sized DC-TENG (1.5 × 3.5 cm) was able to power 416 LEDs, while the larger DC-TENGs (6.8 × 7 cm) were able to achieve an open-circuit voltage of 4500 V, a short-circuit current of 40 μA, and a short-circuit charge transfer of 4.47 μC per motion cycle. A supercapacitor was also assembled by using PEDOT:PSS-coated carbon fibers, coated with PVA/H_3PO_4 gel electrolyte and separated by a cellulose sheet, proving to be able to power a hygrothermograph or a calculator to proper function after 1.5 min of charging. Another example of a gel electrolyte that can be used in the production of woven textile-based power sources is PVA/H_2SO_4. This type of gel electrolyte is found in the work of Sun et al. [125], where rGO-coated Kevlar®-based woven textiles were used to design supercapacitors, using PVA/H_2SO_4 as gel electrolytes. In this paper, the authors described the gelation process through which they prepared the rGO@Kevlar® materials, describing the optimization process that allowed them to find the optimal solution for the gelation step. The optimized procedure allowed the preparation of rGO@Kevlar® fibers with 38.1% rGO, possessing a specific strength of 1.6 MPa.m^3/kg (comparable to the as-received Kevlar®, 2.0 MPa.m^3/kg) and a specific capacitance of 57 F/g. This procedure was then applied to Kevlar® cloth to produce SC, starting by gelation in an autoclave, in the presence of a GO solution containing NH_4OH. The supercapacitors were assembled by adding the gel electrolyte to two separate sheets of the cloth prepared under optimal conditions (rGO–Kevlar®-90C72h–N_2H_4), which were then pressed together under a pressure of apx. 1 MPa for 20 minutes. The supercapacitors were tested for their bending and impact resistance, proving able to withstand impacts of up to 9.1 N and deformations of 90°. In addition to these acid-based gel electrolytes, base-based gel electrolytes can also be used for textile-centered power source applications. One of these cases is presented by Liu et al. [126], where rechargeable textile alkaline Zn microbatteries (micro-AZBs) were assembled by using a multi-component gel electrolyte (KOH, zinc acetate ($Zn(Ac)_2$), lithium hydroxide (LiOH), calcium hydroxide ($Ca(OH)_2$)) in their composition. The micro-AZBs we assembled by using a Kapton mask as a resistor during Ni electroless deposition and Cu electrodeposition. Then, two interdigitated Cu electrodes were coated with Zn and NiCo BOH nanosheets by electrodeposition, and finally the gel electrolyte solution was drop-coating. The promising results were achieved using an electroplated Zn anode and a Ni0.7Co0.3OOH cathode, which show good energy density (2.6 × 10^2 Wh/kg), power density (1.0 × 10^4 W/kg), and stable cycling performance (82.7% for 1500 cycles), as well as satisfactory mechanical reliability (bending, twisting, tailoring, among others). Another possibility is the use of PVA-based gel electrolytes combined with different inorganic or organic salts. In their report, Sundriyal and Bhattacharya [127] stated that an electronic textile can be assembled using PVA/LiCl as the gel electrolyte. Here, the authors reported a device prepared from bamboo fabric, in which the cathode (rGO) and anode (MnO_2–$NiCo_2O_4$) electrodes were prepared by inkjet printing, using hydrazine as a reducing agent for the former and $KMnO_4$ as an oxidant for the latter. The device was then prepared using two layers, i.e., one anode and one cathode, separated by a bamboo fabric sheet and coated with the PVA/LiCl gel electrolyte. The electrochemical tests demonstrated that the MnO_2–$NiCo_2O_4$//rGO device had a stable performance (within a 0–1.6 V range), as well as a high capacitance (2.1 × 10^3 mF/cm^2) and energy density (3.8 × 10^{-2} W/cm^3), with a maximum power density of 2.7 × 10 W/cm^3. Additionally, the device also presented a reasonable cycle life, with a 92% capacitance retention after 5000 cycles and low charge transfer resistance (3.2 Ω). Another textile-based supercapacitor was prepared by Pullanchiyodan et al. [128] by using a PVA/KCl gel electrolyte in combination with Ag-coated polyamide Berlin fabric. In this work, two types of flexible supercapacitor were prepared, B-flexible supercapacitors and BGr-flexible supercapacitors, which differed only in the incorporation of a graphite paste (containing also ethyl cellulose and Triton X-100 as dispersant) as a coating, applied to the latter, which acted as the electrode instead of the Berlin fabric. Then, the gel electrolyte was coated on top of either the Berlin fabric (for a B-flexible supercapacitor) or the graphite coating (for a BGr-flexible supercapacitor), after which a separator is laminated on top of the device. In both cases, a Ag wire was applied as an external connection, connected to the fabric via a Ag paste. The electrochemical evaluation of the material demonstrated its potential, possessing an areal capacitance of 1.3×10 mF/cm^2, which was almost 4× higher than the capacitance (3.53 mF/cm^2) of B-flexible supercapacitor flexible supercapacitors (without the graphite paste). Lastly, an example of a polyvynilidene/bis trifluoromethane sulfonimide lithium salt (PVDS/LiTMDS) gel electrolyte being used in the preparation of textile-based batteries can be found in the work of Khudiyev et al. [129]. In this work, a fiber battery was produced by thermal drawing, combining the anode (LTO), cathode (LFP), gel electrolyte (PVDF/LiTFSI), and conductive polymer (carbon black) in gel forms. A correlation was established between the fiber battery length and its capacity using a 140 m long battery with a discharge capacity and energy of apx. 123 mAh and 217 mWh, respectively. The fibers produced were then woven into the satin fabric using an automated weaving machine to produce a 2D electronic battery. The 2D device was submitted to electrochemical and mechanical tests, demonstrating 96% retention in capacity (after 1000 bending cycles), as well as the ability to sustain a maximum tensile stress of apx. 33 MPa (with a failure strain of apx. 34%). In fact, it could retain its capacity with minimal losses after 10 washing cycles. Table 3 presents strategies for producing supercapacitors using conductive polymers (e.g., PPy, PANI, PEDOT), metallic ions (e.g., Cu, Ni, Co, Ag), and hybrid structures (e.g., MOFs) on untreated or carbonized fabrics, with the dominance of PVA-based gel electrolytes. Methods include electrochemical deposition, electroless deposition, spray-coating, screen-printing, dipping, polymerization, and sputtering. Key performance metrics such as specific capacitance, capacitance retention (max. of 100% after 10,000 cycles), energy density, and power density are used to evaluate electrochemical performance and determine effective designs for energy storage. Despite the focus on electrochemical and comfort properties, research should prioritize biodegradability for a more sustainable, non-toxic alternative to conventional devices. Biocompatible gel electrolytes, such as PVA-based gels, have already shown excellent performance in wearable applications.

Carbon cloth, specifically carbon woven textiles, emerges as a viable type of textile structure due to its durability, flexibility, porosity, and simplicity of integration into various energy storage systems. The usage of this material to produce electrodes (cathodes and anodes) in energy devices is extremely visible. However, despite its growing popularity, there is a major paucity of information regarding the construction parameters of these structures. As a result, many published studies do not provide the full characterization of the structural properties of the carbon fabrics utilized, inhibiting the direct comparison of the influence of structural parameters on device efficiency. According to the scientific literature, carbon textiles are widely applied as structural components in a variety of energy storage devices, including batteries and supercapacitors, as shown in the summary table (Table 4). Researchers justify this fact by citing its three-dimensional structure, which combines carbon's high conductivity with the ability to provide mechanical support and the necessary porosity to optimize interaction with the electrolyte, resulting in ease of ion and electron transport, making it ideal for improving the performance and durability of these devices. Ping Li et al. [130] experimentally showed the first prototype of a highly flexible quasi-solid-state Zn–polyaniline-*co*-azure C (PANAC) battery at the level of battery applications. Providing a rich protonated nitrogen in the charged state and a special ability of "proton self-supply", a considerable increase in the electroactivity of the polymer in nearly neutral aqueous electrolytes occurred while permitting rapid electron/ion transport. The experiment's challenge was to conformally assemble the PANAC copolymer onto a carbon cloth wrapped in porous carbon, which was used as a cathode. The researchers revealed the peculiarity of developing a bicontinuous porous carbon-sheathed carbon cloth (CC-PC) cathode with a hierarchical and porous structure to increase the active polymer load (PANAC) and provide rapid ion and electron transport. Comparing the CC-PC@PANAC cathode to the Pt-plate@PANAC cathode (produced by electropolymerization on a flat platinum (Pt) plate under equivalent conditions), the CV curves of the electrodes (at various scan speeds) show that Pt-plate@PANAC cathode has a much lower capacity and rate performance. This demonstrates the critical role of the CC–PC@PANAC's hierarchical 3D electrodynamic architecture, which can enhance electron transfer and decrease ion transfer distance while increasing the percentage of near-surface/surface reactions. Applications for this kind of hierarchical architecture in supercapacitors are also being investigated, employing carbon cloth as a support framework. The research of Po-Yuan Cheng et al. [131] studied high-performance flexible gel-type symmetric supercapacitors. The authors fabricated a nitrogen-doped hierarchical porous carbon nanostructure (NH-PCN) onto a carbon cloth (CC), utilizing self-organized mesoporous silica spheres as a template. For this effect to be replicated, the porosity properties of the CC support structure proved to be highly significant. Because of this property, mesoporous silica spheres were utilized to form a mold in the original CC support structure, which allowed the formation of a thin hollow hierarchical porous carbon nanostructure with large pore volumes and high specific surface areas. Thus, a layer of N-doped carbon was created by etching away the silica mold and subsequently forming thin layers of carbon (polydopamine carbonization). This layer was able to provide electric double-layer capacitance and pseudocapacitance. This was made possible by the adsorption/desorption of electrolyte ions on the surfaces of the high-specific-surface-area electrodes as well as by surface redox reactions, made possible by their numerous nitrogen-doped active sites. This fact was confirmed in the results by comparing two symmetric supercapacitors, using LiCl/PVA as the electrolyte. The electrodes created without MSS templating exhibited lower energy density, power density, and specific capacitance. Along with functionalization procedures that guaranteed optimal attachment and the interaction of functional particles with carbon structures, the study by Xiao-Man Cao et al. [132] evidenced the need to functionalize electrodes made with the CC support structure to promote improved electrochemical reactions. By using a technique known as "root-etch-wrap," scientists have shown that it is possible to combine the benefits of a stable structural design with electrodes and components to obtain a synergistic effect. Thus, to produce the electrodes, the authors deposited hollow ZnO spheres onto the activated CC through an in situ growth process. Next, the hollow core and shell structure ZnO@ZIF-8 were synthesized in a controlled manner through a controlled synthesis process of ZnO@ZIF-8-CC, forming a ZIF-8 MOF shell composed of interconnected nanoscale crystals. After that, PANI was electropolymerized on the inner and outer surfaces of the ZnO@ZIF-8 structure, forming a uniform and dense layer that resulted in the overall PANI/ZnO@ZIF-8-CC structure. Analyzing the various structures produced by each functionalization step, this research revealed that the final structure had a lower porosity but a higher specific capacitance level, which was explained by the ZnO, ZIF-8, and PANI particles. This indicates that the PANI/ZnO@ZIF-8-CC structure has a substantially lower equivalent series resistance (Rs = 1.22 Ω) than ZnO@ZIF-8-CC (Rs = 6.92 Ω). These features make these structures effective for building energy storage devices. In this study, a high energy density of 1.4×10^{-4}–8.9×10^{-5} Wh/cm^3 (at a power density of 1.4–2.4 × 10 W/cm^3) and good long-term cycling capacity (87% after 10,000 cycles at 5 mA/cm^2) were demonstrated after a supercapacitor system was assembled using a PVA/KCl gel electrolyte intercalated between the PANI/ZnO@ZIF-8-CC electrodes. Functionalizing CC-based electrodes boosts energy storage efficiency by enhancing conductivity, surface area, and porosity. However, the study conducted by Xia et al. [133] highlights the importance of not only optimizing the structures that compose the electrodes of electronic systems but also exploring new electrolyte formulations to enhance the redox reactions essential for improving the electrochemical performance of the supercapacitors produced. In order to enhance electrical conductivity and provide pseudocapacitive qualities, the researchers first created electrodes by depositing them onto a CC structure using a PANI/CNTs core–shell structure via a layer-by-layer growth process. Therefore, in contrast to the CC hierarchical structures, the H_2SO_4/0.02M Fe^{3+}/Fe^{2+} electrolyte gel was created by adding iron(II) sulfate heptahydrate ($FeSO_4 \cdot 7H_2O$) and iron(III) sulfate ($Fe_2(SO_4)_3$) to an aqueous solution of H_2SO_4-PVA to enhance the redox reactions between the electrolyte and the electrodes. Comparing electrochemical properties using a H_2SO_4 electrolyte, it was possible to verify the presence of new redox peaks centered between 0.30 and 0.55 V, resulting in a higher current density. Also, the increase in the capacitance properties of the electrodes favored by the Fe^{3+}/Fe^{2+} ion pair was very notable, proving that interventions at the level of the electrolyte can directly influence the properties of the materials that functionalize the CC structures. In another study, led by Han and coworkers [134], sodium perchlorate ($NaClO_4$)/PVA gel electrolyte was combined with CC to achieve a biologically compatible quasi-solid-state supercapacitor. In the study, the MnO_2 nanowires/CC electrodes were prepared by dip-coating the carbon cloth in a $KMnO_4$ solution, after which the mixture was submitted to hydrothermal treatment at 180 °C in an autoclave for variable periods of time. Then, the fibers were disintegrated by a hydrothermal reaction and used in combination with activated carbon fibers and a $NaClO_4$/PVA gel electrolyte previously prepared by the freeze–thaw method to produce the flexible supercapacitors. The devices assembled through this procedure presented good capacitance retention, with only 19% performance loss after 25,000 cycles while exposed to external environmental conditions. Additionally, the device also presented stable and consistent charge storage, being able to operate under a −40 to 40 °C range, with only a 40% loss in specific capacitance after 7000 cycles. This brings novelty to the field, since it constitutes a device that, mainly due to the hydrogel used, affords good water retention and frost resistance, while maintaining good electrochemical performance. Mohd Shoeb et al. [135] also proposed the use of woven carbon fibers (WCFs) to assemble high-performance supercapacitors, with the aid of PVA/Na_2SO_4 and PVA/$EMIBF_4$ gel electrolytes. This study presents a recent methodology to assemble the supercapacitor, namely Vacuum-Assisted Resin Transfer Molding, where the device is assembled in a chamber containing the electrodes and the gel electrolyte (PVA/Na_2SO_4), to which a PET resin containing an additional gel electrolyte (PVA/$EMIBF_4$) is then added, with both functioning as electrode separators. After adding a curing agent, methyl ethyl ketone peroxide, while decreasing pressure, the chamber is then enclosed to ensure constant temperature (room temperature, in this case). The electrodes used in this study were prepared by a hydrothermal reaction at 180 °C for 12h, using a solution of samarium (III) nitrate hexahydrate ($Sm(NO_3)_3 \cdot 6H_2O$) and ammonium metavanadate (NH_4VO_3). Then, the carbon nanotube–$SmVO_4$, MoS_2 nanocomposites was also developed via hydrothermal treatment at 180 °C, for 48h, using the $SmVO_4$-MoS_2 matrix previously prepared. The authors report an excellent electrochemical performance, with a specific capacitance of 1.0×10^3 mF/cm^2 being achieved at a current density of 2.187 mA/cm^2 for Sm-Mo-C5 in a three-electrode system. The authors highlight that performance improvement is related to a synergistic effect between the SmVO4,MoS2, and CNTs in the composite, which significantly enhances conductivity and active site availability. Regarding the supercapacitor, the scientists highlight its exceptional performance, with the setup achieving a specific capacitance of 2.9×10^5 mF/cm^2 at a current density of 2 A/cm^2, 72.5% capacitance retention after 50,000 charge–discharge cycles, and a maximum energy density of 8.9×10 Wh/Kg at a power density of 1.0×10^3 W/Kg. This brings a new approach to the field of automotive and aerospatial engineering, as this device can operate in conditions of controlled space and weight while maintaining efficiency and durability. Lin et al. [136] proposed and studied another intervention at the electrolyte layer level, aiming to develop an electrolyte that ensures strong electrochemical interactions with electrodes fabricated on a CC support substrate functionalized through the electrodeposition of PANI nanowires. Additionally, this electrolyte was designed to incorporate self-healing capabilities, allowing it to recover after deformations in the overall electronic system. The self-healing hydrogel electrolyte was developed from an Fe dual physically cross-linked polyelectrolyte, with acrylic acid (AA) as a hydrophilic monomer and stearyl methacrylate (C18) as a hydrophobic monomer in an aqueous solution of iron nitrate, along with the cationic surfactant cetyltrimethylammonium bromide (CTAB). The resulting hydrogel, containing a mild quantity of H_2SO_4, exhibited high ionic conductivity (>30 mS/cm), excellent self-healing efficiency, and high extensibility. The copolymerization was carried out, simply by providing an electrolyte with mobile protons that generated capacitance equivalent to that of conventional systems, such as PVA/H_2SO_4. With a self-repair capacity of roughly 86% following the seventh healing cycle, the scientists demonstrated that self-repairable supercapacitors demonstrated exceptional electrochemical performance. Additionally, it demonstrated outstanding cycle stability and good resistance to aging, with a minor decline in performance under environmental conditions after a month. To summarize, CC structures are highly valued because of their intrinsic material qualities, which include electrical and mechanical capabilities as well as porosity structural features. As a result, the necessity of the activation/functionalization of CC structures is the focus of ongoing research and experimentation aimed at improving specific features for energy devices. Concurrently, the benefits of operation are demonstrated, not only at the structural level (with usually hierarchical structures), but also through the enhancement of the electrolytes used to increase electrochemical interactions and synergistic relationships, thereby improving the efficiencies of the energy devices developed.

Table 4. Comparison of CC-based energy storage devices using gel electrolytes, highlighting preparation methods and electrochemical properties.

Gel Electrolyte	Textile Composition	Electrodes	Preparation Method	Conductivity Tests and Results	Ref.
PVA/NH_4Cl-ZnCl	3D bicontinuous porous carbon-sheathed carbon cloth (CC–PC)	CCPC@PANAC (cathode); CC@Zn NP (anode)	PANAC cathode produced by the copolymerization of PANI with a redox-active phenothiazine derivative. The anode constructed by depositing Zn nanoplate arrays onto CC by the electrochemical method. The gel electrolyte is placed between the electrodes.	Energy density of 3.5×10^2 Wh/kg; specific capacity of 306.3 mAh/g; capacitance retention of 86.6% after 2000 bending cycles	[130]
LiCl/PVA	CC	NHPCN@CC (electrodes)	NHPCN@CC was obtained by self-assembly of a sol–gel MSS template onto CC, followed by PDA coating and subsequent carbonization. The electrodes were immersed in a LiCl/PVA gel electrolyte.	Energy density of 1.0×10 Wh/kg (at 1.0×10^4 W/kg) and 2.4×10 Wh/kg (at 5.0×10^2 W/kg); capacitance retention of 85% (for 8000 cycle)	[131]
PVA/KCl	CC	Flexible conductive porous electrode; PANI/ZnO@ZIF-8-CC	PANI/ZnO@ZIF-8-CC electrode: in situ growth of hollow ZnO spheres on activated CC; core–shell structure, created by coating the ZnO core with a ZIF-8 shell; aniline electropolymerization used to deposit a homogeneous PANI coating on both the inner and outer surfaces of ZnO@ZIF-8-CC	Areal capacitance of 4.8×10^3–4.0×10^3 mF/cm^2 (at 5–30 mA/cm^2); energy density of 1.4×10^{-4}–8.9×10^{-5} Wh/cm^3; equivalent series resistance of 1.22 Ω; specific capacitance of 4.3×10^3 mF/cm^2; capacitance retention of 7% for 10,000 cycles	[132]
PVA/H_2SO_4-Fe^{3+}-Fe^{2+}	CC	PANI; CNTs core–shell (hybrid electrode)	PANI/CNTs@CC electrodes were obtained by the deposition of CNTs onto CC via LPCVD, followed by PANI coating through electropolymerization. The gel electrolyte was prepared using H_2SO_4, PVA, $FeSO_4 \cdot 7H_2O$, and $Fe_2(SO_4)_3$. The SCs were assembled by pressing.	Diffusion resistance of 0.236 Ω (PANI/CNTs); specific capacitance of 6.0×10^2 mF/cm^2 (at 5 mV/s); energy density of 2.3×10 Wh/kg (at a power density of 7.0×10^2 W/kg); capacitance retention of 97% (after 2000 cycles)	[133]

Table 4. *Cont.*

Gel Electrolyte	Textile Composition	Electrodes	Preparation Method	Conductivity Tests and Results	Ref.
$NaClO_4$/PVA	CC	MnO_2 nanowires/CC and activated carbon	MnO_2 nanowires/CC electrode prepared on CC by dip-coating and autoclave; the supercapacitor is assembled using both MnO_2 nanowires/CC and activated carbon fibers, coated with the gel electrolyte.	Capacitance retention of 81% after 25,000 cycles while exposed directly to axternal enrivronment; consistent and stable charge storage between −40 and 40 °C.	[134]
PVA/Na_2SO_4; PVA: $EMIBF_4$	CC	Sm-Mo-C5 woven carbon fibers (WCFs)	$SmVO_4$ nanoparticles were synthesized via hydrothermal techniques, using $Sm(NO_3)_3{\cdot}6H_2O$ and NH_4VO_3. The $SmVO_4$-MoS_2 and CNT-$SmVO_4$-MoS_2 nanocomposites were synthesized by the same method. The supercapacitor was assembled by Vacuum-Assisted Resin Transfer Molding.	Specific capacitance of 1.0×10^3 mF/cm^2 (current density of 2.187 mA/cm^2) in a three-electrode system. The SCs show specific capacitance of 2.9×10^2 mF/cm^2 (at a current density of 2 A/cm^2). Capacitance retention: 72.5% to 50.000 cycles; maximum energy density of 8.0×10 Wh/Kg (at a power density of 1.0×10^3 W/Kg).	[135]
FeDPCL (self-healing hydrogel electrolyte)	CC	PANI-CC	PANI@CWF electrodes were obtained by in situ electropolymerization of aniline onto carbon fabric. Hydrogel was prepared using AA, CTAB, C18, $Fe(NO_3)_3{\cdot}6H_2O$, and potassium persulfate. The gels were immersed in H_2SO_4. The eletrodes were directly paved on hydrogel eletrolyte without separator.	Energy density of 1.9×10 Wh/kg (at power densities of 6.7×10^2 W/kg); capacitance retention of 86% (after healing behavior).	[136]

2.4. Knitted Fabrics

The scientific literature highlights the advancement of electronics as a key factor in the progress of flexible energy devices. In this context, the use of knitted fabrics has emerged as a promising approach for these devices (Table 5). These structures are in high demand and have been extensively researched due to their versatility. They are particularly valued for their ability to integrate with wearable technologies and other flexible devices. When functionalized, these textile substrates enable devices to exhibit high power density, long lifespan, fast charge and discharge capabilities, and safety. They also offer high flexibility and adaptability in terms of application. Thus, in this research field, there is growing interest in the use of textile substrates, particularly in knitted fabric-type structures, which, as justified in material selection for the development of flexible electronic devices, are typically chosen for being highly stretchable, flexible, and breathable, as well as comfortable for use [137,138]. Studies such as those conducted by Zengqing Li et al. [139] demonstrated that the surface properties of knitted fabrics are crucial for improving the efficiency of these energy storage devices. In this work, bionic fiber microarrays could use the active materials such as graphene nanosheets (GNS) or PEDOT:PSS) as hedgehog spines. The suspension comprising GNS and PEDOT:PSS was successively deposited onto the hierarchical fabric through the spraying strategy. The knitted fabric structure produced played a pivotal role and provides an advantage due to its versatility in the possible structures that can be produced. The study revealed that the three-dimensional configuration of the knitted fabric allows for the optimization of energy storage and charging properties, providing a larger contact area between the electrode and the electrolyte. Thus, a multidimensional knitted fabric with a microarray structure of bionic fibers was developed, specifically for applications in supercapacitors. Due to the "capture effect" of the bionic microarrays, the textile electrode can accommodate more electrochemically active materials, while the significantly high-accessibility interfacial area allows the active materials to be fully exposed to the electrolyte, significantly facilitating ion diffusion during the electrochemical process. In this way, the knitted fabric structure not only increases the storage capacity but also improves the performance and efficiency of energy storage devices such as supercapacitors (Figure 7). In this way, the scientific research conducted by Chuanli Sue et al. [140] demonstrated that the use of a novel diamond-shaped stainless-steel mesh (SSM) in an adapted knitting process, combined with a coating of a three-dimensional graphene network, enabled the construction of a textile substrate with a configuration that considerably increased the available surface area. This allows for the deposition of more conductive material, which can improve the charge capacity and overall efficiency of these devices.

Table 5. Comparison of knitted fabric-based energy storage devices using gel electrolytes, highlighting preparation methods and electrochemical properties.

Gel Electrolyte	Structure and Composition	Electrodes	Preparation Method	Conductivity Tests and Results	Ref.
PVA/H_2SO4	CO;CNT	CNT; PPy	Dip-coating pyrrole; supercapacitor assembled with two fabrics as electrodes, a CO fabric as a separator, and PVA/H_2SO_4 as the gel electrolyte	Areal capacitance of electrode of 4.1×10^3 mF/cm^2; quasi-rectangular curve cyclic voltametry; 93% Coulombic efficiency; 89% under bending	[137]
PAM/KOH	Spandex knitted fabric	Ni@NiCoP; SWCNT ST	Ni@NiCoP@SWCNT ST dip-coated in Ni@NiCoP ST into SWCNT ink; PAM-based hydrogel immersed in KOH; a layer of Ni@NiCoP@SWCNT ST and a layer of Ni@NiCoP ST attached to asymmetric electrodes in the supercapacitor; in between PVA/H_2SO_4 gel electrolyte	Eletrode conductivity 532 S/cm; areal capacitance 8.8×10^2 mF/cm^2; gravimetric capacitance 713 F/g; 101% retention after 6000 cycles	[138]
PVA/H_2SO_4	Cellulose	GNS; PEDOT:PSS	Sprayed GNS and PEDOT; gel electolyte coating	Specific areal capacitance of 2.4×10^2 mF/cm^2; energy density of 2.2×10^{-5} Wh/cm^2; 83.9% capacitance after 10,000 cycles	[139]
PVA/H_3PO_4	Diamond-shaped warp knitting grid structure (stainless-steel mesh-SSM)	PEDOT; RGO; SSM	Two-step electrodeposition of PEDOT/RGO@SSM process in; PVA/H_3PO_4 gel electrolyte;	Areal capacitance of 5.3×10 mF/cm^2; ~73% capacitance after 5000 cycles; rate capability of 3.6×10 mF/cm^2; ~78% capacitance retention at 10% strain for 500 stretching cycles	[140]
PVA/LiCl	PET 90%; spandex 10%	Ni/rGO	Ni-coated textile was immersed in a GO dispersion for a hydrothermal reaction; PVA/LiCl gel electrolyte coating	Areal capacitance of 5.1×10 mF/cm^2; no degradation at 50% of tensile strain; capacitance retention of 85.3% retention after charging/discharging for 5000 cycles; slightly slopped rectangular shape ciclic voltametry	[141]

Table 5. *Cont.*

Gel Electrolyte	Structure and Composition	Electrodes	Preparation Method	Conductivity Tests and Results	Ref.
ACN/PC/PMMA/$LiClO_4$	Knitted Fabric (Nylon® 82%; spandex 18%)	MWCNT; MoO_3	MWCNT/MoO_3 nanocomposite spray-coated over knitted textile; gel electrolyte	100% Coulombic efficiency; 86 and 76% capacitance after 5000 and 10,000 charge/discharge cycles; quasi-rectangular-shaped cyclic voltametry; areal capacitance of 3.4×10 mF/cm^2	[142]
PVA/ H_2SO_4	CO	PPy	Dip-coating for in situ polymerization of PPy with, iron nitrate, and 5-sulfosalicylic acid e; gel electrolyte coating; device assembled simetrically	Areal capacitance of 4.5×10^2 mF/cm^2; energy density of 0.4 Wh; 30% capacitance after 500 cycles; ~84–160% retention rate during 1000 stretching procedures	[143]
KOH/Na_2SO_4	Knitted graphite fabric	Graphite and Mn-Cu alloys	Graphite fibers coated with Mn–Cu alloys by electrodeposition in an electrolyte solution	Specific capacitance of 9.3×10^4 mF/cm^2; areal capacitance of 9000 mF/cm^2	[144]

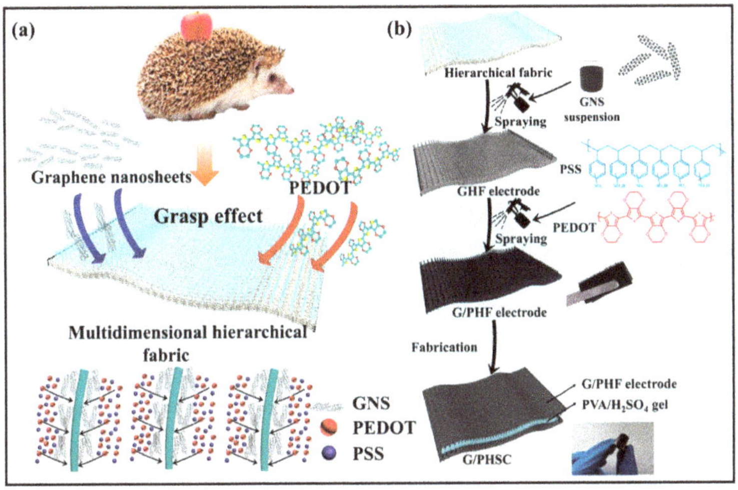

Figure 7. (**a**) Scheme showing deposition of active materials onto hierarchical fabric structure. (**b**) Illustration outlining preparation of graphene nanosheets (GNS)/PEDOT-coated fabric electrodes and assembly process of fabric-based supercapacitors [139].

Another advantage of using these textile structures is the ability to integrate multiple electronic devices, referred to as all-in-one textiles. In terms of wearable applications, these not only prioritize the intrinsic properties of knitted fabric structures but also indicate the growing demand for the development of self-sufficient technical structures. They can be manufactured using existing textile substrate production lines, ensuring commercial viability [141,142]. This fact was proven by the construction of a self-sustaining coplanar and stretchable energy textile (SCPT) by Zifeng Cong et al. [141]. These knitted fabrics integrated TENG through a method similar to resistance dyeing, with applications in energy harvesting devices, microsupercapacitors (MSCs), and energy storage devices. In this article, the researchers demonstrated the functionalization of knitted fabric by printing with conductive material on the textile structure. The authors tested rGO in order to produce textile electrodes. Regarding the reproducibility of the MSCs, the application of the PVA/LiCl electrolyte coating by immersion was necessary to achieve proper functionality and electrochemical reactions. In the study performed by Heun Park et al. [142], the authors produced a high-performance, dynamically stretchable supercapacitor to power an integrated sensor of the same structure in order to detect various biosignals using spray techniques. In this context, the supercapacitor and the deformation sensor were fabricated in two distinct stretching directions of the knitted fabric. MWCNT and molybdenum trioxide (MoO_3) nanowires (NWs) were manufactured for both the stretchable supercapacitor electrodes and the deformation sensor. These pseudocapacitive materials presented high sensitivity, stable performance, and fast response times. PVA-based electrolytes are also commonly applied in knitted fabrics due to their elastomeric hydrogel properties. They can be used as adhesives to improve the bonding forces at the interfaces of the electrodes, transforming into an elastomeric composite film, which adapts to tensile deformation without mechanical degradation [141]. This was also corroborated by the study conducted by Bo Wang et al. [143], who investigated the capacitance stability of electrodes with 0 and 40% strain, initially using a liquid electrolyte (NaCl) and a gel-type electrolyte (PVA/H_2SO_4). Although capacitance values were lower with the gel-type electrolyte, justifiable by the relatively poorer diffusion process of the PVA/H_2SO_4 gel compared to the liquid electrolyte, the adjusted capacitance values of the electrodes (15.2, 41.2, and 95.5 mF for 0, 20, and

40% strain) showed a higher increase rate due to the deeper diffusion process of the gel. Ongoing research in this field aims not only to enhance the properties of materials but also to ensure that the transition from rigid to flexible devices occurs efficiently, presenting the use of knitted fabric structures for these purposes as a viable solution. In more recent studies, Yavuz et al. [144] highlighted the benefits of electrodepositing a manganese–copper (Mn-Cu) alloy onto a knitted fabric substrate made of graphite filaments. This approach, along with the construction of electrodes for the development of flexible energy storage devices, demonstrated the advantage of combining two compounds that are typically used individually in the production of electrolytes. Specifically, the mixture of KOH and Na_2SO_4 combines the benefits of both electrolytes, yielding the excellent ionic conductivity of KOH and the electrochemical stability of Na_2SO_4. This combination results in improved overall performance, emphasizing the synergistic effect of the metal alloy in enhancing charge storage capacity.

2.5. Non-Wovens

Applications comprising non-woven structures correspond to the least found applications in the literature (Table 6). This may be mainly due to the importance of fiber orientation in conductivity. A positive correlation between fiber orientation and electrical conductivity was observed [145,146]. The differences between oriented and non-oriented fiber reached nearly 2-fold S/cm [146]. However, fiber orientation is not always feasible or simple to achieve, as seen with wet-spinning and electrospinning synthesis processes. Notwithstanding the randomly distributed fibers, wet spinning has been proven to be successfully applied to develop fiber-based supercapacitors, batteries, gel polymer electrodes, and battery separators [147]. Several works can be found in the literature using non-woven and gel electrolytes prepared by wet spinning. Using a two-phase microchannel conical reactor, porous GO was injected and underwent interphase with ethanol, creating a porous GO filament enveloped by ethanol that led to its coagulation. The generated filament was drawn by a rotor in ethanol and acetic acid solution at 60 °C for 8 hours. The obtained filaments were collected through filtration and consolidated using a hot press. Two porous GO non-woven meshes were used to produce a laminated structure, consisting of a sandwich of the two meshes enveloping a PVA/H_2SO_4 electrolyte gel. The laminated structure was further consolidated through hot pressing. The obtained non-woven laminated superconductor displayed an impressive performance, with a high capacitance of nearly 1.4×10^3 mF/cm^2, high energy density of approximately 1.2×10^{-4} Wh/cm^2, and ultrahigh stability within 60,000 cycles. These results were correlated with the high specific surface area of approximately 220 g/m^2, notwithstanding adequate mechanical properties (1.2 MPa). As an interesting remark, the authors successfully applied the non-woven laminated supercapacitor as a power source for a powered boat, a self-powered fan, and a smartwatch, demonstrating its validity and eclectic applicability [147]. Qiu and coworkers [148] also developed a non-weaving-based supercapacitor through wet spinning in an attempt to develop a wearable supercapacitor with adequate mechanical properties and high energy density. In a $MgSO_4$ coagulation bath, microfluidic wet-fusing spinning chemistry was used to produce spin-orientated delaminated $Ti_3C_2T_x$-MXene graphene-containing quantum dots. After consolidation through filtration, a radical polymerization reaction of PANI was performed on the non-woven surface. The energy density was 21.19 mWh/cm^3, the power density was 9.8×10^{-1} W/cm^3, and the materials able to withstand 5000 cycles. The high specific surface area, which reached nearly 64 g/m^2, was also mentioned as a key factor for both ion accumulation and transfer. The developed non-woven supercapacitor was connected to a solar cell for charging and converted the energy to power a wearable health monitor on the wrist and an LED.

Table 6. Comparison of non-weaving-based energy storage devices using gel electrolytes, highlighting preparation methods and electrochemical properties.

Gel Electrolyte	Non-Woven Composition	Electrodes	Preparation Method	Conductivity Tests and Results	Ref.
PVA/H_2SO_4	Graphene	Porous graphene fiber-assembled fabric	Micro wet spinning	Capacitance of 1.2×10^3 mF/cm^2; 100% for 60,000 cycles; energy density of 1.2×10^{-4} Wh/cm^2	[147]
PVA/H_2SO_4	MXene-based fiber fabrics	MXene (Ti_3C_2Tx) graphene quantum dots	Micro wet spinning	Capacitance of 1.8×10^6 mF/cm^3; 100% stable for 5000 cycles; energy density of 2.1×10^{-2} Wh/cm^3	[148]
PVA/H_2SO_4	rGO enveloped with PANI	rGO enveloped with PANI	Wet spinning	Capacitance of 3.8×10^3 mF/cm^2; 100% for 10,000 cycles; energy density of 4.9×10^{-3} W/cm^2	[149]
Lithium perchlorate-PCL	PCL nanofibers containg silicon dioxide nanoparticles	PEDOT:PSS screen-printing on indium tin oxide coated polyethylene terephthalate	Electrospinning	Ionic conductivity of 5.2×10^{-3} S/cm; 100% for 100 cycles	[150]
Polyvinylidene fluoride/MgClO4/propylene carbonate	PVDF nanofibers	PVDF nanofibers	Electrospinning	Capacitance of 3832 mAh/cm^3; ionic conductivity of 1×10^{-3} S/cm	[151]
Poly (ethylene oxide)	Poly (ethylene oxide)	Li-ions	Solvent casting and dipping	Run stably > 750 h at 0.5 mA/cm^2; good security; 300 cycles with 1 mA/cm^2	[152]
PVA/H_2SO_4/KOH	Pure cellulose	Carbon paper	Dipping	Capacitance of 1.4×10^2 mF/cm^2; 100% stable for 1000 cycles at 180° bending	[153]
PEGDA; ETPTA; liquid electrolyte; AIBN	PEGDA; ETPTA; liquid electrolyte; AIBN	Li (anode); $LiPF_6$ (cathode)	Injection	Ionic conductivity 0.87 mS/cm; 100% for 200 cycles; resistance of 275 Ω	[154]

Another study reported by Shao et al. [149] used a novel coagulation bath fabrication technique to prepare reduced graphene oxide fiber fabrics for areal-energy-dense supercapacitors. Scanning electron microscopy images revealed that the produced fibers possess a three-dimensional, non-woven fabric-like structure composed of overlapping reduced graphene oxide fibers. Each individual fiber presented a diameter of approximately 20 μm, with clearly visible wrinkled microstructures on their surfaces. The non-woven supercapacitor CV curve was approximately rectangular at 300 mV/s, denoting a high-rate performance. For practical use, flexible and symmetrical all-solid-state supercapacitors were constructed by placing the PVA/H_2SO_4 gel electrolyte between two layers of reduced graphene oxide fiber fabrics (60 μm, 2.8 mg/cm^2). The current density was not impaired after bending the non-woven supercapacitor at 90 and 180°. It exhibited a specific gravimetric capacitance of 285 F/g at 0.1 A/g and 220 F/g at 1 A/g. Its area capacitance reached 8.2×10^2 mF/cm^2 at 1 mA/cm^2 and 3.2×10^2 mF/cm^2 at 20 mA/cm^2. Finally, its gravimetric energy density ranged from 28 μWh/cm^2 to 1.1×10^{-5} Wh/cm^2, and could provide energy for 3 minutes to a 3 V LED (Figure 8).

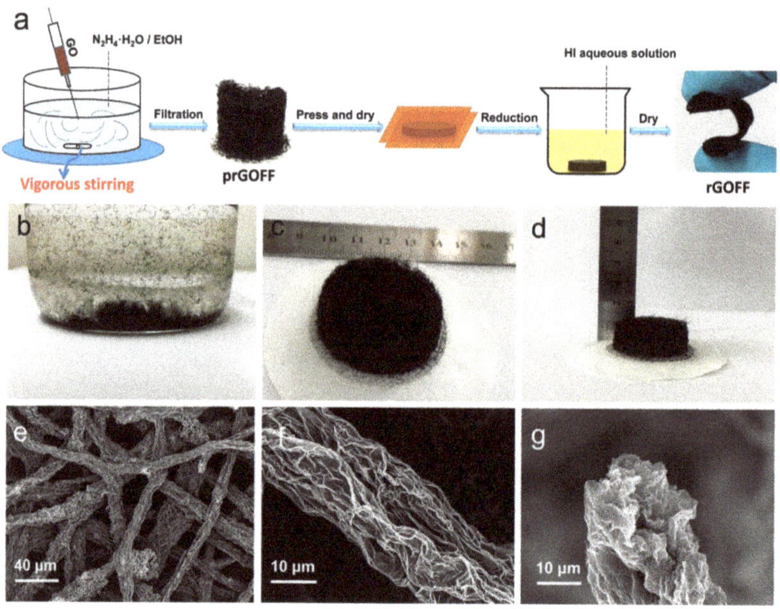

Figure 8. (**a**) Procedure to produce three-dimensional reduced graphene oxide fiber fabric electrodes. (**b**) Digital photo of short partially reduced graphene oxide gel fibers in ethanol. (**c**,**d**) Digital photos of free-standing fabrics. (**e**–**g**) SEM images of reduced graphene oxide fiber fabrics [149].

Electrospinning is a popular strategy for the development of both conductive nonwovens and separation membranes. In one study from Liu et al. [150], poly(ε-caprolactone) was dissolved in poly-ε-caprolactone (PCL) hexafluoroisopropanol. Afterwards, silicon dioxide nanoparticles were added in different ratios. The electrospinning setup encompassed a distance between the needle and the collector of 12 cm, a flow rate of 1 mL/h, and a voltage of 18 kV. Then, the electrospun mat was dipped in 1 M of lithium perchlorate. Subsequently, electrochromic devices were produced by PEDOT:PSS screen-printing on indium tin oxide-coated PET electrodes. The porosity, electrolyte uptake, and mechanical properties peaked at 3 wt.% of silicone dioxide (SiO_2) nanoparticles. The electrospun mats exhibited negligible losses of efficiency after 100 cycles at different bending angles (20 and 40°). They exhibited good conductivity 5.2 mS/cm, and an impressive electrolyte uptake (higher than 800%). Another example was the study developed by Singh et al. [151], where polyvinylidene fluoride was dissolved in a mixture of *N,N*–dimethylformamide and acetone (85:15 v/v). The electrospinning process unfolded at a flow rate of 0.6 mL/h, 16 kV, and the distance between the needle and the rotating drum (at the speed of 1400 rpm) was 15 cm. After a drying step, the electrospun membrane was dipped in the electrolyte magnesium perchlorate ($MgClO_4$) at 0.3 M, achieving the preparation of a gel polymer electrolyte comprising fibers averaging 335 nm in size. An ionic conductivity of nearly 1 mS/cm was observed, with a voltage range of up to 5.0 V. However, the thermostability of the gel polymer electrolyte was relatively poor at temperatures higher than 90 °C, which can considerably limit their application.

Dipping or injection was the last strategy employed for the development of e-textiles with gel electrolytes using non-woven structures. Yang et al. [152] prepared a high-performance flexible gel electrolyte by adding poly(ethylene oxide) powder to a mixture containing graphene aqueous solution, *N,N*-Dimethylformamide and poly(vinylidene fluoride-hexafluoro pentaene (PVDF)) in acetone, and a brown slurry was obtained. An undisclosed non-woven was dipped in the slurry and dried. Once dried, it was immersed in a liquid electrolyte composed of bis(trifluoromethanesulfonyl)imide in tetraethylene glycol dimethyl ether, thus producing a gel polymer electrolyte. The gel polymer electrolyte exhibited a hydrophobic nature up to 15 minutes and a battery was subsequently assembled using a disk made of gel polymer electrolyte, coated with a conductivity of nearly 0.3×10^{-3} S/cm. The poly(ethylene oxide) formed hydrogen bonds with graphene oxide, generating high interface stability. This allowed the battery to power an LED panel for over 750 h without observable dendrite formation at 0.5 mA/cm^2. Furthermore, it was stable for 300 cycles of charge–discharge using 1 mA/cm^2. In another work [153], pure cellulose non-woven fabric with a thickness of 0.38 mm and a weight per area of 48 g/m^2 was immersed in a graphene suspension. Afterwards, it was dried and immersed in an aqueous solution of potassium permanganate. After a drying step, ethyl alcohol was dropwise added onto the textile substrate. A solid-state supercapacitor was assembled using two non-woven substrates that sandwiched an H_2SO_4-PVA gel. Finally, both outer layers of the textile substrates were coated with carbon paper as electrodes. A supercapacitor was obtained, displaying a high specific capacitance of approximately 1.4×10^2 mF/cm^2 at 0.5 mA/cm^2 and exhibiting relevant flexibility and cycle stability up to 2500 cycles. Interestingly, the same methodology was applied to a woven textile of pure cellulose, exhibiting a thickness of 0.2 mm and a weight per area of 130 g/m^2. However, it exhibited poorer performance in comparison to the non-woven. A non-woven textile, of an undisclosed nature, was used as a scaffold for in situ polymerization by Shao et al. [154]. Three main components were mixed: poly(ethylene glycol) diacrylate (PEGDA), ethoxylated trimethylolpropane triacrylate (ETPTA), and liquid electrolyte. The liquid electrolyte encompassed ethylene carbonate, diethyl carbonatem and lithium hexafluorophosphate ($LiPF_6$). Azobisisobutyronitrile (AIBN) was added as the initiator, and the solution was injected into a micro- to nanofiber-sized non-woven. After complete cross-linking, a gel polymer was achieved that was subsequently assembled into a battery, stacking a lithium anode with lithium hexafluorophosphate as the cathode. The battery was able to withstand 200 cycles without considerable Coulombic efficiency loss and exhibited an ionic conductivity of approximately 9×10^{-4} S/cm. The authors used the same combination of chemicals to produce a gel between wafers of the same cathode and anode, and differences were observed. The latter battery exhibited a poorer performance in all performed characterizations. This was most likely due to the nano- and microporous architecture of the non-woven; however, this feature was not described or explored. Nevertheless, the battery successfully powered an LED setup.

3. Perspectives

The growing interest in sustainable energy solutions has driven a significant shift toward the development of ecofriendly and scalable gel polymer electrolytes for use in flexible and wearable energy storage devices. Electrolytes play a pivotal role in electrochemical systems by transporting ions and determining the electrochemical stability window, directly influencing the energy and power densities of devices like supercapacitors, batteries, and conductive materials. Biopolymer-based gel electrolytes have emerged as a promising alternative to conventional petroleum-based materials due to their inherent properties, specifically, hydrophilicity, biocompatibility, mechanical robustness, thermal stability, and environmental friendliness. Recent studies have demonstrated the advantages of using biopolymers such as cellulose, chitosan, alginate, and gelatin, which not only enable efficient ion transport through their functional groups but also allow the integration of multifunctional properties like self-healing, stretchability, and temperature tolerance [155]. Among these, alginate stands out for its capacity to form complexes with a wide range of mono- to tetravalent cations, tuning the physicochemical behavior of the electrolyte and enabling its use across diverse fields. Furthermore, these biopolymers can mitigate key limitations in Zn batteries, such as hydrogen evolution, Zn corrosion, and dendrite formation. The fabrication of biopolymer-based electrolytes often employs environmentally friendly techniques. Examples include cross-linking, which can be established through either the external or internal gelation of the biopolymers, the use of composite matrices for the incorporation of micro- and/or nanostructures, and advanced manufacturing methods such as 2D or 3D printing technologies. These methods support scalability and reduce the ecological impact of gel electrolyte use. Performance evaluation metrics specific to the mechanisms of different types of supercapacitors have also been proposed, helping to standardize and advance the field [156,157]. However, significant efforts are still needed to enhance the key properties of biopolymer-based gel electrolytes, including their relatively low ionic conductivity at room temperature, electrochemical and thermal stability, and durability, and to address issues such as mechanical flexibility without compromising conductivity in order to aid the development of low-cost, scalable, and environmentally friendly methods for biopolymer extraction and electrolyte production [158]. In conclusion, the transition toward sustainable gel electrolytes is not only vital from an environmental perspective but also opens new opportunities for the design of high-performance, safe, and multifunctional flexible energy storage systems. Supported by recent publications, this measure underscores the need for continued innovation in bio-based materials and green fabrication strategies.

4. Conclusions

The integration of gel electrolytes into textile-based power sources presents a promising pathway for advancing flexible and wearable electronics. Compared to traditional liquid electrolytes, gel electrolytes offer significant advantages, including enhanced safety, leak resistance, mechanical flexibility, and improved interface compatibility. These attributes make them particularly well-suited for applications in sectors such as healthcare, sports, and fashion, where reliable and adaptable energy solutions are essential. Despite these advantages, challenges such as lower ionic conductivity and long-term stability must be addressed to fully optimize their performance. Regarding these restrictions, doped and hybrid gel electrolytes offer a promising approach to transforming conventional textiles into conductive and flexible materials for use in wearable electronics with improved properties. Recent advancements in gel electrolytes have related to enhancing ionic conductivity through the incorporation of various additives, namely inorganic nanoparticles, ionic liquids, carbon-based materials and polymer blends. Notably, MXene and PEDOT:PSS are two advanced materials that, when combined, have shown significant promise in enhancing the conductivity of gel electrolytes for textile applications. Despite these improvements, there are still a few challenges. In future research, we recommend the uniformization of units to ensure consistency and facilitate comparison across studies. The variability in textile substrates, fabrication methods, and testing conditions makes it difficult to directly compare performance metrics. As a result, there is an urgent need for unified test protocols and consistent reporting standards in the field. Continued research and innovation in material formulations, structural designs, and fabrication techniques will be crucial to overcoming these limitations. This review underscored the potential of gel electrolytes in various textile forms, including fibers, yarns, and woven or non-woven fabrics, and highlighted strategies for their effective integration. As advancements continue, the development of high-performance, durable, washable, and scalable gel electrolytes will be key to unlocking new possibilities in electronic textiles, paving the way for the design of next-generation wearable energy storage systems.

Author Contributions: Conceptualization, A.I.R.; methodology, A.I.R.; formal analysis, A.I.R., C.A., M.F., J.A., F.P.d.L. and J.P.; investigation, A.I.R., C.A., M.F., J.A., F.P.d.L. and J.P.; resources, A.Z.; writing—original draft preparation, A.I.R., C.A., M.F., J.A., F.P.d.L. and J.P.; writing—review and editing, A.I.R., C.A., M.F., J.A., F.P.d.L. and J.P.; visualization, A.I.R., C.A., M.F., J.A., F.P.d.L. and J.P.; supervision, A.Z.; funding acquisition, A.Z. All authors have read and agreed to the published version of the manuscript.

Funding: This work was funded by the European Regional Development Fund through the Operational Competitiveness Program and the National Foundation for Science and Technology of Portugal (FCT) under the projects UID/CTM/00264/2020 of Centre for Textile Science and Technology (2C2T) on its components Base (https://doi.org/10.54499/UIDB/00264/2020) and programmatic (https://doi.org/10.54499/UIDP/00264/2020). C.A. acknowledges FCT for Ph.D. funding via scholarship 2022.10454.BD.

Data Availability Statement: No new data were created or analyzed in this study.

Conflicts of Interest: The authors declare no conflicts of interest.

References

1. Teixeira, J.S.; Pereira, A.M.; Pereira, C. Smart dual-functional energy storage/fluorescent textile device based on a new redox-active Mn-doped ZnS solid-gel electrolyte. *Chem. Eng. J.* **2021**, *426*, 131274. [CrossRef]
2. Rafique, A.; Ferreira, I.; Abbas, G.; Baptista, A.C. Recent Advances and Challenges Toward Application of Fibers and Textiles in Integrated Photovoltaic Energy Storage Devices. *Nano-Micro Lett.* **2023**, *15*, 40. [CrossRef] [PubMed]
3. Sha, D.; Ding, D.; Tang, S.; Ma, Z.; Liu, C.; Yuan, Y. Solvent-Triggered, Ultra-Adhesive, Conductive, and Biocompatible Transition Gels for Wearable Devices. *Small* **2024**, *20*, e2310731. [CrossRef] [PubMed]
4. Lu, X.; Mo, Z.; Liu, Z.; Hu, Y.; Du, C.; Liang, L.; Liu, Z.; Chen, G. Robust, Efficient, and Recoverable Thermocells with Zwitterion-Boosted Hydrogel Electrolytes for Energy-Autonomous and Wearable Sensing. *Angew. Chem. Int. Ed.* **2024**, *63*, e202405357. [CrossRef]
5. Tordi, P.; Tamayo, A.; Jeong, Y.; Bonini, M.; Samorì, P. Multiresponsive Ionic Conductive Alginate/Gelatin Organohydrogels with Tunable Functions. *Adv. Funct. Mater.* **2024**, *34*, 2410663. [CrossRef]
6. Gao, X.; Guo, C.; Xu, S.; Song, H. Stretchable ionic conductive gels for wearable human-activity detection. *Chem. Eng. J.* **2024**, *489*, 151231. [CrossRef]
7. Zhu, C.; Wu, J.; Yan, J.; Liu, X. Advanced Fiber Materials for Wearable Electronics. *Adv. Fiber Mater.* **2022**, *5*, 12–35. [CrossRef]
8. Nie, S.; Cai, M.; Yang, H.; Shen, L.; Wang, S.; Zhu, Y.; Song, J. Soft, stretchable thermal protective substrates for wearable electronics. *npj Flex. Electron.* **2022**, *6*, 36. [CrossRef]
9. Hou, Z.; Liu, X.; Tian, M.; Zhang, X.; Qu, L.; Fan, T.; Miao, J. Smart fibers and textiles for emerging clothe-based wearable electronics: Materials, fabrications and applications. *J. Mater. Chem. A* **2023**, *11*, 17336–17372. [CrossRef]
10. Islam, M.R.; Afroj, S.; Novoselov, K.S.; Karim, N. Smart Electronic Textile-Based Wearable Supercapacitors. *Adv. Sci.* **2022**, *9*, e2203856. [CrossRef]
11. Guan, L.; Yu, L.; Chen, G.Z. Capacitive and non-capacitive faradaic charge storage. *Electrochim. Acta* **2016**, *206*, 464–478. [CrossRef]

12. Sharma, S.; Chand, P. Supercapacitor and electrochemical techniques: A brief review. *Results Chem.* **2023**, *5*, 100885. [CrossRef]
13. Newby, S.; Mirihanage, W.; Fernando, A. Modern Developments for Textile-Based Supercapacitors. *ACS Omega* **2023**, *8*, 12613–12629. [CrossRef]
14. Grube, A.; Shaban, M.M.; Hilger, L.; Firouzjaei, M.D.; Shamsabadi, A.A.; Demirel, Y.; Elliott, M.; Nejati, S.; Bavarian, M. Wearable Textile Supercapacitors: Material Advancements and Applications. *J. Energy Storage* **2024**, *99*, 113228. [CrossRef]
15. Pal, B.; Yang, S.; Ramesh, S.; Thangadurai, V.; Jose, R. Electrolyte selection for supercapacitive devices: A critical review. *Nanoscale Adv.* **2019**, *1*, 3807–3835. [CrossRef]
16. Hao, Q.; Ma, X.; Gao, Y.; Chen, F.; Chen, X.; Qi, Y.; Li, N. Commercial carbonate based gel polymer electrolytes enable safe and stable high-voltage Li-metal batteries. *Energy Storage Mater.* **2024**, *70*, 103509. [CrossRef]
17. Lu, J.; Chen, Y.; Lei, Y.; Jaumaux, P.; Tian, H.; Wang, G. Quasi-Solid Gel Electrolytes for Alkali Metal Battery Applications. *Nano-Micro Lett.* **2025**, *17*, 194. [CrossRef]
18. Duay, J.; Gillette, E.; Liu, R.; Lee, S.B. Highly flexible pseudocapacitor based on freestanding heterogeneous MnO_2/conductive polymer nanowire arrays. *Phys. Chem. Chem. Phys.* **2012**, *14*, 3329–3337. [CrossRef]
19. Lu, C.; Jiang, H.; Cheng, X.; He, J.; Long, Y.; Chang, Y.; Gong, X.; Zhang, K.; Li, J.; Zhu, Z.; et al. High-performance fibre battery with polymer gel electrolyte. *Nature* **2024**, *629*, 86–91. [CrossRef]
20. Zhou, Y.; Zhang, D.; Zhang, S.; Liu, Y.; Ma, R.; Wallace, G.; Chen, J. Highly stretchable double-network gel electrolytes integrated with textile electrodes for wearable thermo-electrochemical cells. *SusMat* **2024**, *4*, e225. [CrossRef]
21. Adusei, P.K.; Kanakaraj, S.N.; Gbordzoe, S.; Johnson, K.; DeArmond, D.; Hsieh, Y.-Y.; Fang, Y.; Mishra, S.; Phan, N.; Alvarez, N.T.; et al. A scalable nano-engineering method to synthesize 3D-graphene-carbon nanotube hybrid fibers for supercapacitor applications. *Electrochim. Acta* **2019**, *312*, 411–423. [CrossRef]
22. Yu, Y.; Guo, J.; Ma, B.; Zhang, D.; Zhao, Y. Liquid metal-integrated ultra-elastic conductive microfibers from microfluidics for wearable electronics. *Sci. Bull.* **2020**, *65*, 1752–1759. [CrossRef] [PubMed]
23. Hunpratub, S.; Chullaphan, T.; Chumpolkulwong, S.; Chanlek, N.; Phokha, S. Characterization and electrochemical properties of Carbon/CeO2 composites prepared using a hydrothermal method. *Mater. Chem. Phys.* **2023**, *303*, 127820. [CrossRef]
24. Chen, C.; Feng, J.; Li, J.; Guo, Y.; Shi, X.; Peng, H. Functional Fiber Materials to Smart Fiber Devices. *Chem. Rev.* **2022**, *123*, 613–662. [CrossRef]
25. Shalini, S.; Naveen, T.B.; Durgalakshmi, D.; Balakumar, S.; Rakkesh, R.A. Progress in flexible supercapacitors for wearable electronics using graphene-based organic frameworks. *J. Energy Storage* **2024**, *86*, 111260. [CrossRef]
26. Luo, Z.; Quan, J.; Ding, T.; Xu, B.; Li, W.; Mao, Q.; Ma, W.; Li, M.; Xiang, H.; Zhu, M. Recent advances in iron oxide/graphene composites for flexible supercapacitors. *J. Alloys Compd.* **2024**, *980*, 173614. [CrossRef]
27. Li, M.; Luo, Z.; Quan, J.; Ding, T.; Xu, B.; Li, W.; Mao, Q.; Ma, W.; Xiang, H.; Zhu, M. Oxygen defect enriched hematite nanorods @ reduced graphene oxide core-sheath fiber for superior flexible asymmetric supercapacitor. *J. Colloid Interface Sci.* **2024**, *653*, 77–84. [CrossRef]
28. Gao, K.; Wang, S.; Liu, W.; Yue, Y.; Rao, J.; Su, J.; Li, L.; Zhang, Z.; Liu, N.; Xiong, L.; et al. All Fiber Based Electrochemical Capacitor towards Wearable AC Line Filters with Outstanding Rate Capability. *ChemElectroChem* **2019**, *6*, 1450–1457. [CrossRef]
29. Park, H.; Ambade, R.B.; Noh, S.H.; Eom, W.; Koh, K.H.; Ambade, S.B.; Lee, W.J.; Kim, S.H.; Han, T.H. Porous Graphene-Carbon Nanotube Scaffolds for Fiber Supercapacitors. *ACS Appl. Mater. Interfaces* **2019**, *11*, 9011–9022. [CrossRef]
30. Zhang, X.; Pei, Z.; Wang, C.; Yuan, Z.; Wei, L.; Pan, Y.; Mahmood, A.; Shao, Q.; Chen, Y. Flexible Zinc-Ion Hybrid Fiber Capacitors with Ultrahigh Energy Density and Long Cycling Life for Wearable Electronics. *Small* **2019**, *15*, e1903817. [CrossRef]
31. Li, J.; Shao, Y.; Jiang, P.; Zhang, Q.; Hou, C.; Li, Y.; Wang, H. 1T-Molybdenum disulfide/reduced graphene oxide hybrid fibers as high strength fibrous electrodes for wearable energy storage. *J. Mater. Chem. A* **2019**, *7*, 3143–3149. [CrossRef]
32. Chen, Q.-L.; Wu, X.; Cheng, H.; Li, Q.; Chen, S. Facile synthesis of carbon nanobranches towards cobalt ion sensing and high-performance micro-supercapacitors. *Nanoscale Adv.* **2019**, *1*, 3614–3620. [CrossRef] [PubMed]
33. Nayan, R.; Sinha, S.; Dixit, V.; Satnami, M.L.; Ghosh, K.K.; Pervez, S.; Deb, M.K.; Shrivas, K.; Rai, M.K.; Yenchalwar, S.G.; et al. PANI-grafted boron, nitrogen co-doped carbon fiber: An outstanding, high-performance supercapacitor electrode. *J. Energy Storage* **2024**, *96*, 112668. [CrossRef]
34. Huang, H.; Abbas, S.C.; Deng, Q.; Ni, Y.; Cao, S.; Ma, X. An all-paper, scalable and flexible supercapacitor based on vertically aligned polyaniline (PANI) nano-dendrites@fibers. *J. Power Sources* **2021**, *498*, 229886. [CrossRef]
35. Li, J.; Qiu, S.; Liu, B.; Chen, H.; Xiao, D.; Li, H. Strong interaction between polyaniline and carbon fibers for flexible supercapacitor electrode materials. *J. Power Sources* **2021**, *483*, 229219. [CrossRef]
36. Jiang, Q.; Shang, Y.; Sun, Y.; Yang, Y.; Hou, S.; Zhang, Y.; Xu, J.; Cao, A. Flexible and multi-form solid-state supercapacitors based on polyaniline/graphene oxide/CNT composite films and fibers. *Diam. Relat. Mater.* **2019**, *92*, 198–207. [CrossRef]
37. Lai, H.; Li, W.; Zhou, Y.; He, T.; Xu, L.; Tian, S.; Wang, X.; Fan, Z.; Lei, Z.; Jiao, H. Hydrophilically engineered polyacrylonitrile nanofiber aerogel as a soft template for large mass loading of mesoporous poly(3,4-ethylenedioxythiophene) network on a bare metal wire for high-rate wire-shaped supercapacitors. *J. Power Sources* **2019**, *441*, 227212. [CrossRef]

38. He, C.; Cheng, J.; Liu, Y.; Zhang, X.; Wang, B. Thin-walled hollow fibers for flexible high energy density fiber-shaped supercapacitors. *Energy Mater.* **2022**, *1*, 100010. [CrossRef]
39. Gibertini, E.; Magagnin, L. PEDOTS:PSS@KNF Wire-Shaped Electrodes for Textile Symmetrical Capacitor. *Adv. Mater. Interfaces* **2022**, *9*, 2200513. [CrossRef]
40. Hong, Y.; Cheng, X.-L.; Liu, G.-J.; Hong, D.-S.; He, S.-S.; Wang, B.-J.; Sun, X.-M.; Peng, H.-S. One-step Production of Continuous Supercapacitor Fibers for a Flexible Power Textile. *Chin. J. Polym. Sci.* **2019**, *37*, 737–743. [CrossRef]
41. Xu, Q.; Lu, C.; Sun, S.; Zhang, K. Electrochemical properties of PEDOT: PSS /V2O5 hybrid fiber based supercapacitors. *J. Phys. Chem. Solids* **2019**, *129*, 234–241. [CrossRef]
42. Wang, R.; Young Jang, W.; Zhang, W.; Venkata Reddy, C.; Kakarla, R.R.; Li, C.; Gupta, V.K.; Shim, J.; Aminabhavi, T.M. Emerging two-dimensional (2D) MXene-based nanostructured materials: Synthesis strategies, properties, and applications as efficient pseudo-supercapacitors. *Chem. Eng. J.* **2023**, *472*, 144913. [CrossRef]
43. Qin, L.; Jiang, J.; Hou, L.; Zhang, F.; Rosen, J. MXene-based multifunctional smart fibers for wearable and portable electronics. *J. Mater. Chem. A* **2022**, *10*, 12544–12550. [CrossRef]
44. Lu, M.; Zhang, Z.; Kang, L.; He, X.; Li, Q.; Sun, J.; Jiang, R.; Xu, H.; Shi, F.; Lei, Z.; et al. Intercalation and delamination behavior of Ti3C2Txand MnO2/Ti3C2Tx/RGO flexible fibers with high volumetric capacitance. *J. Mater. Chem. A* **2019**, *7*, 12582–12592. [CrossRef]
45. Wei, S.; Wan, C.; Huang, Q.; Chai, H.; Chai, Y.; Li, X.; Wu, Y. 3D cellulose network confining MXene/MnO_2 enables flexible wet spinning microfibers for high-performance fiber-shaped Zn-ion capacitors. *Int. J. Biol. Macromol.* **2024**, *276*, 134152. [CrossRef]
46. Ovhal, M.M.; Lee, H.B.; Satale, V.V.; Tyagi, B.; Chowdhury, S.; Kang, J.W. One-Meter-Long, All-3D-Printed Supercapacitor Fibers Based on Structurally Engineered Electrode for Wearable Energy Storage. *Adv. Energy Mater.* **2023**, *14*, 2303053. [CrossRef]
47. Xiao, X.; Xiao, X.; Zhou, Y.; Zhao, X.; Chen, G.; Liu, Z.; Wang, Z.; Lu, C.; Hu, M.; Nashalian, A.; et al. An ultrathin rechargeable solid-state zinc ion fiber battery for electronic textiles. *Sci. Adv.* **2021**, *7*, eabl3742. [CrossRef]
48. Zhang, Q.; Li, C.; Li, Q.; Pan, Z.; Sun, J.; Zhou, Z.; He, B.; Man, P.; Xie, L.; Kang, L.; et al. Flexible and High-Voltage Coaxial-Fiber Aqueous Rechargeable Zinc-Ion Battery. *Nano Lett.* **2019**, *19*, 4035–4042. [CrossRef]
49. Pan, Z.; Yang, J.; Yang, J.; Zhang, Q.; Zhang, H.; Li, X.; Kou, Z.; Zhang, Y.; Chen, H.; Yan, C.; et al. Stitching of Zn3(OH)2V2O7·2H2O 2D Nanosheets by 1D Carbon Nanotubes Boosts Ultrahigh Rate for Wearable Quasi-Solid-State Zinc-Ion Batteries. *ACS Nano* **2019**, *14*, 842–853. [CrossRef]
50. Li, C.; Zhang, Q.; Li, T.; Zhu, Z.; He, B.; Zhou, Z.; Man, P.; Li, Q.; Yao, Y. An ultra-high endurance and high-performance quasi-solid-state fiber-shaped Zn–Ag2O battery to harvest wind energy. *J. Mater. Chem. A* **2019**, *7*, 2034–2040. [CrossRef]
51. Li, M.; Li, Z.; Ye, X.; Zhang, X.; Qu, L.; Tian, M. Tendril-Inspired 900% Ultrastretching Fiber-Based Zn-Ion Batteries for Wearable Energy Textiles. *ACS Appl. Mater. Interfaces* **2021**, *13*, 17110–17117. [CrossRef] [PubMed]
52. Xiong, T.; He, B.; Zhou, T.; Wang, Z.; Wang, Z.; Xin, J.; Zhang, H.; Zhou, X.; Liu, Y.; Wei, L. Stretchable fiber-shaped aqueous aluminum ion batteries. *EcoMat* **2022**, *4*, e12218. [CrossRef]
53. Zhao, C.; Wang, R.; Liang, H.; Fang, B.; Li, R.; Wu, R.; Mo, R. Autonomous self-healing strategy for flexible fiber lithium-ion battery with ultra-high mechanical properties and volumetric energy densities. *Chem. Eng. J.* **2024**, *496*, 154153. [CrossRef]
54. Liu, S.; Ban, J.; Shi, H.; Wu, Z.; Shao, G.; Cao, G.; Hu, J. Near solution-level conductivity of polyvinyl alcohol based electrolyte and the application for fully compliant Al-air battery. *Chem. Eng. J.* **2022**, *431*, 134283. [CrossRef]
55. Alipoori, S.; Torkzadeh, M.M.; Mazinani, S.; Aboutalebi, S.H.; Sharif, F. Performance-tuning of PVA-based gel electrolytes by acid/PVA ratio and PVA molecular weight. *SN Appl. Sci.* **2021**, *3*, 310. [CrossRef]
56. Wu, Y.; Li, Y.; Wang, Y.; Liu, Q.; Chen, Q.; Chen, M. Advances and prospects of PVDF based polymer electrolytes. *J. Energy Chem.* **2022**, *64*, 62–84. [CrossRef]
57. Yang, S.-Y.; Wang, Y.-F.; Yue, Y.; Bian, S.-W. Flexible polyester yarn/Au/conductive metal-organic framework composites for yarn-shaped supercapacitors. *J. Electroanal. Chem.* **2019**, *847*, 113218. [CrossRef]
58. Tang, Y.; Li, M.; Liu, Y.; Guo, X.; Liu, Y.; Wang, C.; Li, Y.; Tian, M. Washable and anti-impact conductive cellulose yarn-based energy meta-textile against harsh environments. *Appl. Mater. Today* **2024**, *36*, 102055. [CrossRef]
59. Uzun, S.; Seyedin, S.; Stoltzfus, A.L.; Levitt, A.S.; Alhabeb, M.; Anayee, M.; Strobel, C.J.; Razal, J.M.; Dion, G.; Gogotsi, Y. Knittable and Washable Multifunctional MXene-Coated Cellulose Yarns. *Adv. Funct. Mater.* **2019**, *29*, 1905015. [CrossRef]
60. Du, X.; Tian, M.; Sun, G.; Li, Z.; Qi, X.; Zhao, H.; Zhu, S.; Qu, L. Self-Powered and Self-Sensing Energy Textile System for Flexible Wearable Applications. *ACS Appl. Mater. Interfaces* **2020**, *12*, 55876–55883. [CrossRef]
61. Zhang, C.; Chen, Z.; Rao, W.; Fan, L.; Xia, Z.; Xu, W.; Xu, J. A high-performance all-solid-state yarn supercapacitor based on polypyrrole-coated stainless steel/cotton blended yarns. *Cellulose* **2018**, *26*, 1169–1181. [CrossRef]
62. Sun, X.; He, J.; Qiang, R.; Nan, N.; You, X.; Zhou, Y.; Shao, W.; Liu, F.; Liu, R. Electrospun Conductive Nanofiber Yarn for a Wearable Yarn Supercapacitor with High Volumetric Energy Density. *Materials* **2019**, *12*, 273. [CrossRef] [PubMed]
63. Carvalho, J.T.; Cunha, I.; Coelho, J.; Fortunato, E.; Martins, R.; Pereira, L. Carbon-Yarn-Based Supercapacitors with In Situ Regenerated Cellulose Hydrogel for Sustainable Wearable Electronics. *ACS Appl. Energy Mater.* **2022**, *5*, 11987–11996. [CrossRef]

64. Rani, S.; Sharma, Y. Fabrication of Binder-Free and High Energy Density Yarn Supercapacitor for Wearable Electronics. *IEEE Trans. Power Electron.* **2022**, *37*, 13022–13029. [CrossRef]
65. Park, Y.; Choi, H.; Kim, M.-C.; Tran, N.A.T.; Cho, Y.; Sohn, J.I.; Hong, J.; Lee, Y.-W. Effect of ionic conductivity in polymer-gel electrolytes containing iodine-based redox mediators for efficient, flexible energy storage systems. *J. Ind. Eng. Chem.* **2021**, *94*, 384–389. [CrossRef]
66. Levitt, A.; Hegh, D.; Phillips, P.; Uzun, S.; Anayee, M.; Razal, J.M.; Gogotsi, Y.; Dion, G. 3D knitted energy storage textiles using MXene-coated yarns. *Mater. Today* **2020**, *34*, 17–29. [CrossRef]
67. Pal, M.; Subhedar, K.M. Biscrolled hierarchical hybrid structure of PEDOT:PSS/CNTs-NMP yarn based solid state linear supercapacitor for wearable e-textile. *Carbon* **2024**, *229*, 119552. [CrossRef]
68. Wang, S.; Wang, J.; Liao, S.; Chen, J.; Wei, Q. Hierarchical core-shell polypyrrole@NiCo layered double hydroxide arrays grown on stainless steel yarn with high flexibility for 1D symmetric yarn-shaped supercapacitors. *J. Alloys Compd.* **2022**, *926*, 166811. [CrossRef]
69. Wang, Y.-F.; Wang, H.-T.; Yang, S.-Y.; Yue, Y.; Bian, S.-W. Hierarchical NiCo2S4@Nickel–Cobalt Layered Double Hydroxide Nanotube Arrays on Metallic Cotton Yarns for Flexible Supercapacitors. *ACS Appl. Mater. Interfaces* **2019**, *11*, 30384–30390. [CrossRef] [PubMed]
70. Liu, M.; Cong, Z.; Pu, X.; Guo, W.; Liu, T.; Li, M.; Zhang, Y.; Hu, W.; Wang, Z.L. High-Energy Asymmetric Supercapacitor Yarns for Self-Charging Power Textiles. *Adv. Funct. Mater.* **2019**, *29*, 1806298. [CrossRef]
71. Al-khaykanee, A.H.; Ghorbani, S.R.; Arabi, H.; Ghanbari, R. $NiMnO_3$-rGO nanocomposites in a cotton-based flexible yarn supercapacitor. *J. Mater. Sci. Mater. Electron.* **2024**, *35*, 141. [CrossRef]
72. Li, Y.; Liu, Y.; Chen, Y.; Liu, T.; Liu, X.; Ni, S.; Ru, J.; Meng, F.; Li, M.; Cao, Z. A battery–capacitor hybrid yarn device with excellent flexibility and high electrochemical performance. *J. Mater. Chem. A* **2025**, *13*, 12374–12382. [CrossRef]
73. Yu, Y.; Liao, Y.; Fan, J.; Ding, Y.; Fan, Y.; Cao, J.; Zhou, X.; Wang, Y.; Yan, J.; Li, H.; et al. Tip effect of NiCo-LDH with low crystallinity for enhanced energy storage performance of yarn-shaped supercapacitors. *J. Colloid Interface Sci.* **2025**, *679*, 1242–1252. [CrossRef] [PubMed]
74. Ahn, J.; Padmajan Sasikala, S.; Jeong, Y.; Kim, J.G.; Ha, J.-H.; Hwang, S.H.; Jeon, S.; Choi, J.; Kang, B.-H.; Ahn, J.; et al. High-Energy–Density Fiber Supercapacitors Based on Transition Metal Oxide Nanoribbon Yarns for Comprehensive Wearable Electronics. *Adv. Fiber Mater.* **2024**, *6*, 1927–1941. [CrossRef]
75. Li, M.; Qiao, J.; Zhu, C.; Hu, Y.; Wu, K.; Zeng, S.; Yang, W.; Zhang, H.; Wang, Y.; Wu, Y.; et al. Gel-Electrolyte-Coated Carbon Nanotube Yarns for Self-Powered and Knittable Piezoionic Sensors. *ACS Appl. Electron. Mater.* **2021**, *3*, 944–954. [CrossRef]
76. Huang, Q.; Wang, D.; Hu, H.; Shang, J.; Chang, J.; Xie, C.; Yang, Y.; Lepró, X.; Baughman, R.H.; Zheng, Z. Additive Functionalization and Embroidery for Manufacturing Wearable and Washable Textile Supercapacitors. *Adv. Funct. Mater.* **2020**, *30*, 1910541. [CrossRef]
77. Yi, H.; Ma, Y.; Zhang, S.; Na, B.; Zeng, R.; Zhang, Y.; Lin, C. Robust Aqueous Zn-Ion Fiber Battery Based on High-Strength Cellulose Yarns. *ACS Sustain. Chem. Eng.* **2019**, *7*, 18894–18900. [CrossRef]
78. Sun, C.; Li, X.; Cai, Z.; Ge, F. Carbonized cotton fabric in-situ electrodeposition polypyrrole as high-performance flexible electrode for wearable supercapacitor. *Electrochim. Acta* **2019**, *296*, 617–626. [CrossRef]
79. Wang, D.; Sun, J.; Xue, Q.; Li, Q.; Guo, Y.; Zhao, Y.; Chen, Z.; Huang, Z.; Yang, Q.; Liang, G.; et al. A universal method towards conductive textile for flexible batteries with superior softness. *Energy Storage Mater.* **2021**, *36*, 272–278. [CrossRef]
80. Lu, J.; Zhang, L.; Xing, C.; Jia, G.; Lu, Z.; Tian, Q.; Zhang, S.; Lv, J. Polypyrrole and cotton fabric-based flexible micro-supercapacitors. *J. Appl. Polym. Sci.* **2022**, *139*, e52801. [CrossRef]
81. Ramandi, S.; Entezari, M.H. Self-supporting electrode for flexible supercapacitors: NiCo-layered double hydroxide derived from metal organic frameworks wrapped on graphene/polyaniline nanotubes@cotton cloth. *J. Energy Storage* **2022**, *56*, 106106. [CrossRef]
82. Zheng, X.; Shen, J.; Hu, Q.; Nie, W.; Wang, Z.; Zou, L.; Li, C. Vapor phase polymerized conducting polymer/MXene textiles for wearable electronics. *Nanoscale* **2021**, *13*, 1832–1841. [CrossRef] [PubMed]
83. Hillier, N.; Yong, S.; Cruden, A.; Beeby, S. Acetonitrile-Free Organic Electrolyte for Textile Supercapacitor Applications. *J. Electrochem. Soc.* **2021**, *168*, 080520. [CrossRef]
84. Sun, D.; Liu, Q.; Yi, C.; Chen, J.; Wang, D.; Wang, Y.; Liu, X.; Li, M.; Liu, K.; Zhou, P.; et al. The construction of sea urchin spines-like polypyrrole arrays on cotton-based fabric electrode via a facile electropolymerization for high performance flexible solid-state supercapacitors. *Electrochim. Acta* **2020**, *354*, 136746. [CrossRef]
85. Wang, C.; Song, Z.; Wan, H.; Chen, X.; Tan, Q.; Gan, Y.; Liang, P.; Zhang, J.; Wang, H.; Wang, Y.; et al. Ni-Co selenide nanowires supported on conductive wearable textile as cathode for flexible battery-supercapacitor hybrid devices. *Chem. Eng. J.* **2020**, *400*, 125955. [CrossRef]

86. Zhai, S.; Jin, K.; Zhou, M.; Fan, Z.; Zhao, H.; Li, X.; Zhao, Y.; Ge, F.; Cai, Z. A novel high performance flexible supercapacitor based on porous carbonized cotton/ZnO nanoparticle/CuS micro-sphere. *Colloids Surf. A Physicochem. Eng. Asp.* **2020**, *584*, 124025. [CrossRef]
87. Sun, C.; Zhao, J.; Guo, Z.; Zhao, Y.; Cai, Z.; Ge, F. A Novel Method to Fabricate Nitrogen and Oxygen Co-Doped Flexible Cotton-Based Electrode for Wearable Supercapacitors. *ChemElectroChem* **2019**, *6*, 4049–4058. [CrossRef]
88. Mallick, S.; Mondal, A.; Raj, C.R. Rationally designed mesoporous carbon-supported Ni–NiWO4@NiS nanostructure for the fabrication of hybrid supercapacitor of long-term cycling stability. *J. Power Sources* **2020**, *477*, 229038. [CrossRef]
89. Hillier, N.; Yong, S.; Beeby, S. Culinary inspired electrolytes for textile supercapacitors. *Energy Rep.* **2021**, *7*, 81–86. [CrossRef]
90. Chen, R.; Yang, Y.; Huang, Q.; Ling, H.; Li, X.; Ren, J.; Zhang, K.; Sun, R.; Wang, X. A multifunctional interface design on cellulose substrate enables high performance flexible all-solid-state supercapacitors. *Energy Storage Mater.* **2020**, *32*, 208–215. [CrossRef]
91. Liang, C.; Zang, L.; Shi, F.; Yang, C.; Qiu, J.; Liu, Q.; Chen, Z. High-performance cotton fabric-based supercapacitors consisting of polypyrrole/Ag/graphene oxide nanocomposite prepared via UV-induced polymerization. *Cellulose* **2022**, *29*, 2525–2537. [CrossRef]
92. Han, F.; Luo, J.; Pan, R.; Wu, J.; Guo, J.; Wang, Y.; Wang, L.; Liu, M.; Wang, Z.; Zhou, D.; et al. Vanadium Dioxide Nanosheets Supported on Carbonized Cotton Fabric as Bifunctional Textiles for Flexible Pressure Sensors and Zinc-Ion Batteries. *ACS Appl. Mater. Interfaces* **2022**, *14*, 41577–41587. [CrossRef] [PubMed]
93. Jiang, L.; Hong, H.; Hu, J.; Yan, X. Fabrication and Seamless Integration of Insensitive-Bending Fully Printed All-in-One Fabric-Based Supercapacitors Based on Cost-Effective MWCNT Electrodes. *ACS Appl. Mater. Interfaces* **2022**, *14*, 12214–12222. [CrossRef] [PubMed]
94. Keawploy, N.; Venkatkarthick, R.; Wangyao, P.; Qin, J. Screen printed textile electrodes using graphene and carbon nanotubes with silver for flexible supercapacitor applications. *J. Met. Mater. Miner.* **2020**, *30*, 39–44. [CrossRef]
95. Keawploy, N.; Venkatkarthick, R.; Wangyao, P.; Zhang, X.; Liu, R.; Qin, J. Eco-Friendly Conductive Cotton-Based Textile Electrodes Using Silver- and Carbon-Coated Fabrics for Advanced Flexible Supercapacitors. *Energy Fuels* **2020**, *34*, 8977–8986. [CrossRef]
96. Jiang, L.; Hong, H.; Hu, J.; Yan, X. Development of flexible supercapacitors with coplanar integrated multi-walled carbon nanotubes/textile electrode and current collectors. *J. Mater. Sci. Mater. Electron.* **2022**, *33*, 5297–5310. [CrossRef]
97. Islam, M.R.; Afroj, S.; Beach, C.; Islam, M.H.; Parraman, C.; Abdelkader, A.; Casson, A.J.; Novoselov, K.S.; Karim, N. Fully printed and multifunctional graphene-based wearable e-textiles for personalized healthcare applications. *iScience* **2022**, *25*, 103945. [CrossRef] [PubMed]
98. Zhou, M.; Yang, B.; Zhao, Y.; Jin, Z.; Li, K.; Tang, L.; Cai, Z. Facile preparation of N-doped carbon/FeOx-decorated carbon cloth for flexible symmetric solid-state supercapacitors. *Cellulose* **2019**, *27*, 1591–1601. [CrossRef]
99. Liu, X.; Chen, Y.; Zhang, H.; Zhuo, L.; Huang, Q.; Zhang, W.; Chen, H.; Ling, Q. Synthesis of MXene-based nanocomposite electrode supported by PEDOT:PSS-modified cotton fabric for high-performance wearable supercapacitor. *J. Colloid Interface Sci.* **2024**, *660*, 735–745. [CrossRef]
100. Costa, R.S.; Soares, O.S.G.P.; Vilarinho, R.; Moreira, J.A.; Pereira, M.F.R.; Pereira, A.; Pereira, C. Unveiling the role of oxidative treatments on the electrochemical performance of carbon nanotube-based cotton textile supercapacitors. *Carbon Trends* **2021**, *5*, 100137. [CrossRef]
101. Costa, R.S.; Guedes, A.; Pereira, A.M.; Pereira, C. Fabrication of all-solid-state textile supercapacitors based on industrial-grade multi-walled carbon nanotubes for enhanced energy storage. *J. Mater. Sci.* **2020**, *55*, 10121–10141. [CrossRef]
102. Li, Y.; Lu, Z.; Xin, B.; Liu, Y.; Cui, Y.; Hu, Y. All-solid-state flexible supercapacitor of Carbonized MXene/Cotton fabric for wearable energy storage. *Appl. Surf. Sci.* **2020**, *528*, 146975. [CrossRef]
103. Liu, Y.-N.; Feng, L.-J.; Bian, S.-W. Surface Structure Construction of Fibers in a Conductive Metal–Organic Framework/Metal/Cotton Electrode for Flexible Textile Supercapacitors. *ACS Appl. Electron. Mater.* **2022**, *4*, 4595–4604. [CrossRef]
104. Wan, J.; Hu, R.; Li, J.; Mi, S.; Xian, J.; Xiao, Z.; Liu, Z.; Mei, A.; Xu, S.; Fan, M.; et al. A universal construction of robust interface between 2D conductive polymer and cellulose for textile supercapacitor. *Carbohydr. Polym.* **2022**, *284*, 119230. [CrossRef] [PubMed]
105. Jin, K.; Zhou, M.; Zhao, H.; Zhai, S.; Ge, F.; Zhao, Y.; Cai, Z. Electrodeposited CuS nanosheets on carbonized cotton fabric as flexible supercapacitor electrode for high energy storage. *Electrochim. Acta* **2019**, *295*, 668–676. [CrossRef]
106. Sneha, N.; Shakthivel, K.R.; Kiruthika, S. Flexible and Durable Conducting Fabric Electrodes for Next-Generation Wearable Supercapacitors. *ACS Appl. Mater. Interfaces* **2025**, *17*, 7568–7580. [CrossRef]
107. Wang, Z.; Wang, H.; Ji, S.; Wang, H.; Brett, D.J.L.; Wang, R. Design and synthesis of tremella-like Ni–Co–S flakes on co-coated cotton textile as high-performance electrode for flexible supercapacitor. *J. Alloys Compd.* **2020**, *814*, 151789. [CrossRef]
108. Hekmat, F.; Tutel, Y.; Unalan, H.E. Wearable supercapacitors based on nickel tungstate decorated commercial cotton fabrics. *Int. J. Energy Res.* **2020**, *44*, 7603–7616. [CrossRef]
109. Xia, Z.; Zhang, X.; Shen, L.; Wang, N.; Wang, T.; Wu, L. Design of freestanding flexible electrodes with hierarchical porous structure from cotton textiles for high-performance supercapacitors. *J. Energy Storage* **2025**, *116*, 115972. [CrossRef]

110. Vashishth, E.; Raulo, A.; Srivastava, R.K.; Nandan, B. Smart Integration of Cobalt-Based Metal Organic Framework with Textile Waste for a Sustainable and Flexible High-Performance Supercapacitor. *ACS Sustain. Resour. Manag.* **2025**, *2*, 108–118. [CrossRef]
111. Song, P.; He, X.; Tao, J.; Shen, X.; Yan, Z.; Ji, Z.; Yuan, A.; Zhu, G.; Kong, L. H_2SO_4-assisted tandem carbonization synthesis of PANI@carbon@textile flexible electrode for high-performance wearable energy storage. *Appl. Surf. Sci.* **2021**, *535*, 147755. [CrossRef]
112. Song, P.; He, X.; Xie, M.; Tao, J.; Shen, X.; Ji, Z.; Yan, Z.; Zhai, L.; Yuan, A. Polyaniline wrapped graphene functionalized textile with ultrahigh areal capacitance and energy density for high-performance all-solid-state supercapacitors for wearable electronics. *Compos. Sci. Technol.* **2020**, *198*, 108305. [CrossRef]
113. Pullanchiyodan, A.; Manjakkal, L.; Dahiya, R. Metal Coated Fabric Based Asymmetric Supercapacitor for Wearable Applications. *IEEE Sens. J.* **2021**, *21*, 26208–26214. [CrossRef]
114. Pullanchiyodan, A.; Manjakkal, L.; Dervin, S.; Shakthivel, D.; Dahiya, R. Metal Coated Conductive Fabrics with Graphite Electrodes and Biocompatible Gel Electrolyte for Wearable Supercapacitors. *Adv. Mater. Technol.* **2020**, *5*, 1901107. [CrossRef]
115. Zhang, C.; Guo, R.; Wang, H.; Xie, X.; Du, C. Composite electrodes with NiCoAl-LDH coated Ti3C2Tx MXene and incorporated Ag nanowires for screen-printable in-plane hybrid supercapacitors on textiles. *Appl. Surf. Sci.* **2022**, *598*, 153796. [CrossRef]
116. Kota, A.; Gogia, A.; Neidhard-Doll, A.T.; Chodavarapu, V.P. Printed Textile-Based Ag2O–Zn Battery for Body Conformal Wearable Sensors. *Sensors* **2021**, *21*, 2178. [CrossRef]
117. Barakzehi, M.; Montazer, M.; Sharif, F.; Norby, T.; Chatzitakis, A. A textile-based wearable supercapacitor using reduced graphene oxide/polypyrrole composite. *Electrochim. Acta* **2019**, *305*, 187–196. [CrossRef]
118. Bhargava, P.; Liu, W.; Pope, M.; Tsui, T.; Yu, A. Substrate comparison for polypyrrole-graphene based high-performance flexible supercapacitors. *Electrochim. Acta* **2020**, *358*, 136846. [CrossRef]
119. Afroj, S.; Tan, S.; Abdelkader, A.M.; Novoselov, K.S.; Karim, N. Highly Conductive, Scalable, and Machine Washable Graphene-Based E-Textiles for Multifunctional Wearable Electronic Applications. *Adv. Funct. Mater.* **2020**, *30*, 2000293. [CrossRef]
120. Li, X.; Sun, C.; Cai, Z.; Ge, F. High-performance all-solid-state supercapacitor derived from PPy coated carbonized silk fabric. *Appl. Surf. Sci.* **2019**, *473*, 967–975. [CrossRef]
121. Sim, K.; Gao, Y.; Chen, Z.; Song, J.; Yu, C. Nylon Fabric Enabled Tough and Flaw Insensitive Stretchable Electronics. *Adv. Mater. Technol.* **2018**, *4*, 1800466. [CrossRef]
122. Khairy, M.; Kamal, R.; Mousa, M.A. Preparation and physical properties of conductive silk fabrics used in wearable clothes and flexible supercapacitors. *J. Ind. Text.* **2022**, *52*, 15280837221130512. [CrossRef]
123. Stempien, Z.; Khalid, M.; Kozicki, M.; Kozanecki, M.; Varela, H.; Filipczak, P.; Pawlak, R.; Korzeniewska, E.; Sąsiadek, E. In-situ deposition of reduced graphene oxide layers on textile surfaces by the reactive inkjet printing technique and their use in supercapacitor applications. *Synth. Met.* **2019**, *256*, 116144. [CrossRef]
124. Chen, C.; Guo, H.; Chen, L.; Wang, Y.-C.; Pu, X.; Yu, W.; Wang, F.; Du, Z.; Wang, Z.L. Direct Current Fabric Triboelectric Nanogenerator for Biomotion Energy Harvesting. *ACS Nano* **2020**, *14*, 4585–4594. [CrossRef] [PubMed]
125. Sun, W.; Shah, S.A.; Lowery, J.L.; Oh, J.H.; Lutkenhaus, J.L.; Green, M.J. Lightweight Kevlar-Reinforced Graphene Oxide Architectures with High Strength for Energy Storage. *Adv. Mater. Interfaces* **2019**, *6*, 1900786. [CrossRef]
126. Liu, M.; Pu, X.; Cong, Z.; Liu, Z.; Liu, T.; Chen, Y.; Fu, J.; Hu, W.; Wang, Z.L. Resist-Dyed Textile Alkaline Zn Microbatteries with Significantly Suppressed Zn Dendrite Growth. *ACS Appl. Mater. Interfaces* **2019**, *11*, 5095–5106. [CrossRef] [PubMed]
127. Sundriyal, P.; Bhattacharya, S. Textile-based supercapacitors for flexible and wearable electronic applications. *Sci. Rep.* **2020**, *10*, 13259. [CrossRef] [PubMed]
128. Pullanchiyodan, A.; Manjakkal, L.; Dahiya, R. Metal Coated Fabric Based Supercapacitors. In Proceedings of the 2020 IEEE International Conference on Flexible and Printable Sensors and Systems (FLEPS), Manchester, UK, 16–19 August 2020; pp. 1–4.
129. Khudiyev, T.; Grena, B.; Loke, G.; Hou, C.; Jang, H.; Lee, J.; Noel, G.H.; Alain, J.; Joannopoulos, J.; Xu, K.; et al. Thermally drawn rechargeable battery fiber enables pervasive power. *Mater. Today* **2022**, *52*, 80–89. [CrossRef]
130. Li, P.; Fang, Z.; Zhang, Y.; Mo, C.; Hu, X.; Jian, J.; Wang, S.; Yu, D. A high-performance, highly bendable quasi-solid-state zinc–organic battery enabled by intelligent proton-self-buffering copolymer cathodes. *J. Mater. Chem. A* **2019**, *7*, 17292–17298. [CrossRef]
131. Cheng, P.-Y.; Cheng, H.-Y.; Huang, C.-L.; Chen, Y.-A.; Hsieh, C.-T.; Lu, S.-Y. N-Doped Hierarchical Continuous Hollow Thin Porous Carbon Nanostructure for High-Performance Flexible Gel-Type Symmetric Supercapacitors. *ACS Sustain. Chem. Eng.* **2019**, *7*, 17020–17029. [CrossRef]
132. Cao, X.-M.; Han, Z.-B. Hollow core–shell ZnO@ZIF-8 on carbon cloth for flexible supercapacitors with ultrahigh areal capacitance. *Chem. Commun.* **2019**, *55*, 1746–1749. [CrossRef] [PubMed]
133. Xia, C.; Leng, M.; Tao, W.; Wang, Q.; Gao, Y.; Zhang, Q. Polyaniline/carbon nanotube core–shell hybrid and redox active electrolyte for high-performance flexible supercapacitor. *J. Mater. Sci. Mater. Electron.* **2019**, *30*, 4427–4436. [CrossRef]
134. Han, J.; Fan, C.; Li, D.-s.; Jiang, L. A hydrogel electrolyte with dynamically Tunable mechanical properties for wide-temperature flexible supercapacitor. *J. Power Sources* **2025**, *631*, 236217. [CrossRef]

135. Shoeb, M.; Mashkoor, F.; Jeong, H.; Khan, M.N.; Jeong, C. Investigating the Effect of Carbon Nanotubes Decorated $SmVO_4$-MoS_2 Nanocomposite for Energy Storage Enhancement via VARTM-Fabricated Solid-State Structural Supercapacitors Using Woven Carbon Fiber. *Small* **2024**, *21*, e2408283. [CrossRef]
136. Lin, Y.; Zhang, H.; Liao, H.; Zhao, Y.; Li, K. A physically crosslinked, self-healing hydrogel electrolyte for nano-wire PANI flexible supercapacitors. *Chem. Eng. J.* **2019**, *367*, 139–148. [CrossRef]
137. Luo, X.; Liang, Y.; Weng, W.; Hu, Z.; Zhang, Y.; Yang, J.; Yang, L.; Zhu, M. Polypyrrole-coated carbon nanotube/cotton hybrid fabric with high areal capacitance for flexible quasi-solid-state supercapacitors. *Energy Storage Mater.* **2020**, *33*, 11–17. [CrossRef]
138. Sun, P.; Qiu, M.; Li, M.; Mai, W.; Cui, G.; Tong, Y. Stretchable Ni@NiCoP textile for wearable energy storage clothes. *Nano Energy* **2019**, *55*, 506–515. [CrossRef]
139. Li, Z.; Ma, Y.; Wang, L.; Du, X.; Zhu, S.; Zhang, X.; Qu, L.; Tian, M. Multidimensional Hierarchical Fabric-Based Supercapacitor with Bionic Fiber Microarrays for Smart Wearable Electronic Textiles. *ACS Appl. Mater. Interfaces* **2019**, *11*, 46278–46285. [CrossRef]
140. Su, C.; Shao, G.; Yu, Q.; Huang, Y.; Jiang, J.; Shao, H.; Chen, N. A flexible, lightweight and stretchable all-solid-state supercapacitor based on warp-knitted stainless-steel mesh for wearable electronics. *Text. Res. J.* **2022**, *92*, 1807–1819. [CrossRef]
141. Cong, Z.; Guo, W.; Guo, Z.; Chen, Y.; Liu, M.; Hou, T.; Pu, X.; Hu, W.; Wang, Z.L. Stretchable Coplanar Self-Charging Power Textile with Resist-Dyeing Triboelectric Nanogenerators and Microsupercapacitors. *ACS Nano* **2020**, *14*, 5590–5599. [CrossRef]
142. Park, H.; Kim, J.W.; Hong, S.Y.; Lee, G.; Lee, H.; Song, C.; Keum, K.; Jeong, Y.R.; Jin, S.W.; Kim, D.S.; et al. Dynamically Stretchable Supercapacitor for Powering an Integrated Biosensor in an All-in-One Textile System. *ACS Nano* **2019**, *13*, 10469–10480. [CrossRef] [PubMed]
143. Wang, B.; Song, W.; Gu, P.; Fan, L.; Yin, Y.; Wang, C. A stretchable and hydrophobic polypyrrole/knitted cotton fabric electrode for all-solid-state supercapacitor with excellent strain capacitance. *Electrochim. Acta* **2019**, *297*, 794–804. [CrossRef]
144. Yavuz, A.; Faal, H. Alloy coating from an ionic liquid on graphite filaments and knitted graphite fabrics for stretchable supercapacitor applications. *Mater. Sci. Eng. B* **2025**, *317*, 118227. [CrossRef]
145. Schneidmadel, S.; Koch, M.; Bruchmüller, M. Effects of fiber orientation on the electrical conductivity of filled plastic melt. *AIP Conf. Proc.* **2016**, *1779*, 030007.
146. Yu, L.; Tatsumi, D.; Morita, M. Relationship between Viscoelasticity and Electrical Conductivity of Carbonized Cellulose Fiber Networks. *Nihon Reoroji Gakkaishi* **2014**, *41*, 331–336. [CrossRef]
147. Guan, T.; Shen, S.; Cheng, Z.; Wu, G.; Bao, N. Microfluidic-assembled hierarchical macro-microporous graphene fabrics towards high-performance robust supercapacitors. *Chem. Eng. J.* **2022**, *440*, 135878. [CrossRef]
148. Qiu, H.; Qu, X.; Zhang, Y.; Chen, S.; Shen, Y. Robust PANI@MXene/GQDs-Based Fiber Fabric Electrodes via Microfluidic Wet-Fusing Spinning Chemistry. *Adv. Mater.* **2023**, *35*, e2302326. [CrossRef]
149. Shao, F.; Hu, N.; Su, Y.; Yao, L.; Li, B.; Zou, C.; Li, G.; Zhang, C.; Li, H.; Yang, Z.; et al. Non-woven fabric electrodes based on graphene-based fibers for areal-energy-dense flexible solid-state supercapacitors. *Chem. Eng. J.* **2020**, *392*, 123692. [CrossRef]
150. Liu, X.; Li, K.; Hou, C.; Li, H.; Chen, P.; Zhang, Q.; Li, Y.; Wang, H. Poly-ε-caprolactone nanofibrous mats as electrolyte host for tailorable flexible electrochromic devices. *Mater. Sci. Eng. B* **2019**, *241*, 36–41. [CrossRef]
151. Singh, R.; Janakiraman, S.; Khalifa, M.; Anandhan, S.; Ghosh, S.; Venimadhav, A.; Biswas, K. An electroactive β-phase polyvinylidene fluoride as gel polymer electrolyte for magnesium–ion battery application. *J. Electroanal. Chem.* **2019**, *851*, 113417. [CrossRef]
152. Yang, T.; Shu, C.; Hou, Z.; Zheng, R.; Hei, P.; Li, M.; Ran, Z.; Zhang, Q.; Long, J. 3D porous network gel polymer electrolyte with high transference number for dendrite-free LiO_2 batteries. *Solid State Ion.* **2019**, *343*, 115088. [CrossRef]
153. Li, Z.; Tian, M.; Sun, X.; Zhao, H.; Zhu, S.; Zhang, X. Flexible all-solid planar fibrous cellulose nonwoven fabric-based supercapacitor via capillarity-assisted graphene/MnO_2 assembly. *J. Alloys Compd.* **2019**, *782*, 986–994. [CrossRef]
154. Shao, D.; Wang, X.; Li, X.; Luo, K.; Yang, L.; Liu, L.; Liu, H. Internal in situ gel polymer electrolytes for high-performance quasi-solid-state lithium ion batteries. *J. Solid State Electrochem.* **2019**, *23*, 2785–2792. [CrossRef]
155. Ding, J.; Yang, Y.; Poisson, J.; He, Y.; Zhang, H.; Zhang, Y.; Bao, Y.; Chen, S.; Chen, Y.M.; Zhang, K. Recent Advances in Biopolymer-Based Hydrogel Electrolytes for Flexible Supercapacitors. *ACS Energy Lett.* **2024**, *9*, 1803–1825. [CrossRef]
156. Tordi, P.; Ridi, F.; Samorì, P.; Bonini, M. Cation-Alginate Complexes and Their Hydrogels: A Powerful Toolkit for the Development of Next-Generation Sustainable Functional Materials. *Adv. Funct. Mater.* **2024**, *35*, 2416390. [CrossRef]
157. Jia, R.; Wei, C.; Ma, B.; Li, L.; Yang, C.; Wang, B.; Tan, L.; Feng, J. Biopolymer-Based Gel Electrolytes for Advanced Zinc Ion Batteries: Progress and Perspectives. *Adv. Funct. Mater.* **2024**, *35*, 2417498. [CrossRef]
158. Yang, W.; Yang, W.; Zeng, J.; Chen, Y.; Huang, Y.; Liu, J.; Gan, J.; Li, T.; Zhang, H.; Zhong, L.; et al. Biopolymer-based gel electrolytes for electrochemical energy Storage: Advances and prospects. *Prog. Mater. Sci.* **2024**, *144*, 101264. [CrossRef]

gels MDPI

Article

Boosting the Sensitivity and Hysteresis of a Gel Polymer Electrolyte by Embedding SiO_2 Nanoparticles and PVP for Humidity Applications

Michelle Cedeño Mata *, Albert Orpella, Manuel Dominguez-Pumar and Sandra Bermejo *

MNT Group, Electronic Engineering Department, Polytechnic University of Catalonia (UPC), C/Jordi Girona 1-3, 08034 Barcelona, Spain
* Correspondence: kristel.michell.cedeno@upc.edu (M.C.M.); sandra.bermejo@upc.edu (S.B.)

Abstract: Enhancing sensitivity and hysteresis in capacitance humidity sensors is vital for precise, reliable, and consistent humidity control. This study explores this concern by incorporating polyvinylpyrrolidone (PVP) and SiO_2 nanoparticles into a polyvinyl alcohol (PVA)-based ionic liquid gel polymer electrolyte (ILGPE), studying two capacitor types: ILGPE and SiO_2 composite ILGPE (CILGPE) capacitors. These novel electrolytes use ammonium acetate as a plasticiser, 1-butyl-3-methylimidazolium bromide as an ionic liquid, SiO_2 nanoparticles as a composite, and PVA and PVP as host polymers. Capacitors were characterised and modelled using impedance spectroscopy (IS), providing an electrophysical insight into their working principle. Sensitivity and hysteresis were evaluated within a 20–90% relative humidity (RH) range at 25 °C. The SiO_2 CILGPE capacitor with PVP presented superior sensitivity and hysteresis, revealing the beneficial combination of SiO_2 nanoparticles and PVP. These benefits are due to the creation of pathways that facilitate water molecule diffusion and crystallinity reduction in PVA-ILGPE. In particular, at 10 kHz, it demonstrates a calibrated capacitance sensitivity of 2660 pF/%RH and a hysteresis of 3.28 %RH. This optimised capacitor outperforms some previous humidity capacitive sensors in sensitivity while exhibiting low hysteresis.

Keywords: gel polymer electrolytes; ionic liquid; colloidal nanoparticles; impedance spectroscopy

1. Introduction

Accurate control and monitoring of environmental conditions are vital in various domains, including agriculture, healthcare, and industrial processes. Therefore, a lot of research has been focused on developing sensors that exhibit high sensitivity, low hysteresis, and rapid linear response and recovery over a wide range of humidity without compromising the cost [1]. Several studies have highlighted sensors capable of detecting humidity changes through detecting alterations in parameters such as impedance, voltage, capacitance, resistance, or frequency. Among these, impedimetric and capacitive sensors stand out due to their simple and cost-effective manufacturing advantages, easy use and integration, low power consumption, and good performance. Moreover, capacitive devices have been in the limelight of the sensing field due to their long operating life and design flexibility [2]. These factors, coupled with the ease of capacitance amplification via oscillator circuits, have propelled these devices to a dominant position in the market [3].

Particularly in the sensing field, the study of sensitive materials has been considered an important research topic due to the wide range of suitable materials and combinations that lead to an easy driving of the humidity-sensing performance and mechanism. In this context, polymers, electrolytes, ceramics, and carbon-based materials have become common moisture-sensing materials [1,4]. Among these humidity-sensitive materials, polymers have been proven to be excellent materials for humidity sensing due to their compatibility with other sensing materials and sensitivity to %RH variation, mainly exhibited in their dielectric constant and conductivity. Furthermore, polymers offer benefits such as low cost, low weight, resistance to corrosion, easy and flexible fabrication, high stability, and commercial potential, which make them suitable materials for humidity-sensing applications. Polymers used for humidity sensors include polyvinyl alcohol (PVA) [5,6], polyvinylpyrrolidone (PVP) [7,8], polypyrrole (PPy) [9], keratin [10], and Nafion [11]. Among these polymers, PVA has been widely used due to its properties, such as non-toxicity, biocompatibility, highly hygroscopic nature, and hydrophilicity [5,6].

Nevertheless, PVA-based humidity sensors often have low sensitivity and slow transient responses to relative humidity variations, mainly attributed to PVA's low ion conductivity and high intrinsic impedance. Therefore, several studies combine PVA with other materials to address these limitations and enhance humidity-sensing performance [5]. The electrical and humidity-sensing performance of PVA-based films can be improved by incorporating polymers, plasticisers, ionic liquids (ILs), and doping fillers.

In particular, incorporating ceramic nanostructured materials into PVA-based humidity-sensing layers has revealed significant improvements. Hydrophilic ceramics exhibiting excellent thermal and chemical stability are well suited as humidity-sensitive materials. The effectiveness of these nanostructures in humidity sensing can mainly be attributed to their large surface area, which offers numerous sites for water interaction. Oxide ceramics such as titanium oxide (TiO_2) [12,13], alumina (Al_2O_3) [14], barium titanate ($BaTiO_3$) [15], and silicon oxide (SiO_2) [16] demonstrate strong responses to humidity changes due to their excellent water adsorption and desorption properties. Additionally, clay ceramics like palygorskite [17], sepiolite [18,19], and halloysite [20] also attract attention for their cost-effectiveness and environmentally friendly nature. Halloysite nanotubes (HNTs) are widely used due to their excellent biocompatibility. However, their brittleness and heterogeneity in size, surface charge, and formation of hydrogen bonds require additional surface modification processes to overcome these challenges and enhance their performance [21,22]. These concerns, in combination with the maturity of oxide ceramic surface-treatment techniques, their excellent properties, and their reliable long-term performance, make oxide ceramics more suitable for humidity sensing and as effective fillers for PVA-based humidity sensors in terms of stability and durability.

On the other hand, regarding ILs, their integration into polymers leads to flexible materials that possess high ionic conductivity and good mechanical properties and stability. ILs have been recognised as green solvents that serve as excellent plasticisers for polymers [23–25]. Their tunability, thermal stability, intrinsic ionic conduction, nonflammability, and low volatility make them highly effective for sensing applications [23,24,26,27], either as modifier materials for electrodes or as sensitive materials [23,25]. In particular, combining hygroscopic ILs and polymers in humidity sensors leads to materials with enhanced mechanical properties, improved ionic conductivity, and boosted humidity-sensing capabilities. Among different ILs, the 1-butyl-3-methylimidazolium (Bmlm) class, in combination with polymers such as polyvinylidene fluoride (PVDF) and PVA, has shown promising results. For instance, in [25], a highly sensitive humidity sensor based on 1-butyl-3-methylimidazolium tetrachloroferrate ($BmlmFeCl_4$) and PVDF is exhibited. In contrast, 1-butyl-3-methylimidazolium bromide (BmlmBr) is a hygroscopic, environmentally friendly, low-cost IL. Its combination with PVA has been widely studied, demonstrating its crucial beneficial role as a plasticiser [28].

Finally, the strategic approach of combining different polymers to improve electrical and humidity-sensing performance is rooted in the leveraging of the strengths of each polymer and their mutual tunability. In humidity sensors, amalgamating polymers with different affinities for water molecules or porous or amorphous structures can lead to a sensing layer with fine-tuned humidity-sensing performance. In particular, some works exhibit the combination of PVA/PVP sensing films as a good alternative for humidity sensors [29,30]. PVP is an amorphous polymer that exhibits high compatibility with PVA. PVP can improve PVA's physical properties by reducing its crystallinity and protecting the electrolyte during drying, preserving its structural integrity [29,30].

This work presents the synthesis and characterisation of ionic liquid gel polymer electrolytes (ILGPEs) formulated using PVA, PVP, ammonium acetate, and 1-butyl-3-methylimidazolium bromide. Additionally, composite ILGPEs (CILGPEs) have been developed by embedding SiO_2 nanoparticles that act as fillers. PVA has been selected as the host polymer because of its chemical stability, capacity to absorb water, and non-toxicity. PVP has been included as a lyophilised agent that can also boost the humidity-sensing performance. Ammonium acetate has been selected as a plasticiser due to its effective plasticising properties, while 1-butyl-3-methylimidazolium bromide, an ionic liquid, has been blended in to enhance ionic conductivity. Finally, nonporous silica nanobeads have been selected as a composite material due to their high stability, biocompatibility, hydrophilicity, and high surface area, all of which are promising properties for humidity sensing. In previous research, Véliz et al. demonstrated the effectiveness of nonporous SiO_2 nanoparticles in sensing humidity changes [31]. Moreover, Cedeño et al. revealed the constructive impact of incorporating silica nanofillers into ILGPE [32]. In particular, SiO_2 nanoparticles promote structural changes by forming amorphous regions and incorporating additional pathways for ion transportation. Such findings could be beneficial for humidity sensitivity and hysteresis, leading SiO_2 nanofillers to be suitable materials for humidity sensors. By incorporating and combining PVP and SiO_2 nanoparticles, the humidity-sensing performance of the ILGPE- and SiO_2 CILGPE-based capacitors has been driven.

Consequently, this research compares the impedance properties of different ionic liquid gel polymer electrolyte-based capacitors, delving into the subtle effects of their composition. It investigates the influence of PVP and SiO_2 nanoparticles on the humidity sensitivity and hysteresis of a PVA-based ILGPE capacitor across a relative humidity (RH) range of 20 to 90%. This work intends to provide an initial insight into the potential of these ILGPEs and SiO_2 CILGPEs as humidity-sensing layers. Furthermore, by studying the effect of PVP and silica nanofillers on sensitivity and hysteresis, this work aims to identify an optimal electrolyte composition regarding these critical parameters. The novelty of this work lies in three points: (i) the amalgamation of various strategies for tuning and boosting the humidity-sensing performance of a PVA-based ILGPE, (ii) the novel approach of combining PVP and SiO_2 nanoparticles with a PVA-based ILGPE for humidity sensing, and (iii) the introduction of novel formulated ILGPEs that incorporate SiO_2 and PVP, presenting an unexplored composition with optimised humidity-sensing capabilities.

2. Results and Discussion

2.1. Impedance Spectroscopy (IS) Characterisation and Modelling

The IS method has been employed to investigate the electrical behaviour of the fabricated devices. Capacitance and equivalent series resistance (ESR) have been obtained and analysed to assess the impact of the employed materials and their concentrations on the electrical and humidity-sensing performance. Figure 1a,b illustrate the obtained bode plots of the fabricated devices at different relative humidity levels during the adsorption process. Figure 1c,d compare the Nyquist and capacitance plots, respectively. In particular, capacitance (C') curves have been calculated from the following equation:

$$C' = \frac{Z''}{2\pi f |Z|^2} \tag{1}$$

where $|Z|$ is the modulus and Z'' is the imaginary part of the complex impedance.

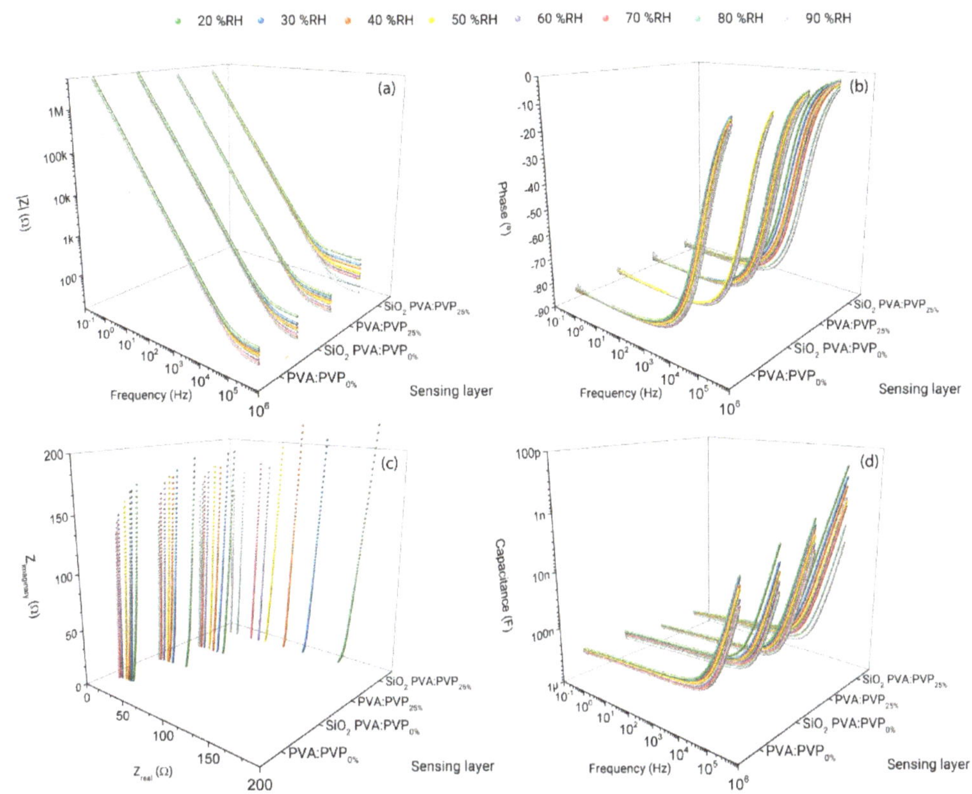

Figure 1. Impedance data comparison of PVA:PVP$_{0\%}$-ILGPE-, SiO_2 PVA:PVP$_{0\%}$-CILGPE-, PVA:PVP$_{25\%}$-ILGPE-, and SiO_2 PVA:PVP$_{25\%}$-CILGPE-based capacitors. (**a**) Impedance module vs. frequency vs. electrolyte content; (**b**) phase responses vs. frequency vs. electrolyte content; (**c**) Nyquist vs. electrolyte content; and (**d**) capacitance vs. frequency vs. electrolyte content. $|Z|$, Z'', and Z' represent the impedance modulus and the imaginary and real parts of the complex impedance, respectively.

Figure 1 depicts the impedance spectrum data obtained from PVA:PVP$_{0\%}$, SiO_2 PVA:PVP$_{0\%}$, PVA:PVP$_{25\%}$, and SiO_2 PVA:PVP$_{25\%}$ capacitors at various relative humidity (RH) levels, thereby exhibiting the trend of the electrical behaviour as a function of %RH and sensing layer composition. Figure 1a compares the impedance modulus of the fabricated devices at different relative humidity levels. As can be noticed, all the capacitors exhibit a similar impedance modulus behaviour, thereby revealing two distinct trends. The first trend is characterised by a drop in the impedance modulus, with a slope of -1 from 0.1 Hz to a specific frequency, indicative of a capacitive behaviour. After this particular frequency, the impedance modulus flattens, exhibiting a slope equal to 0 and tending towards a particular value, revealing a transition from a capacitive to a resistive behaviour. Therefore, the frequency at which the impedance modulus changes its slope acts as a threshold frequency. This behaviour change can also be observed in Figure 1b, where, between a frequency of 0.1 Hz and the specific threshold frequencies, the fabricated devices show phase values smaller than $-45°$, which are attributed to a capacitive behaviour. Beyond these threshold frequencies, the phase responses increase, with their values approaching 0°, indicative that at high frequencies, resistive behaviour predominates. The second trend observed relies on the variation in the impedance modulus in response to changes in humidity levels. This trend is characterised by a reduction in the impedance modulus values with increasing relative humidity. This finding is also evident in Figure 1c, where the Nyquist plots intersect with the real impedance axis at lower values. This phenomenon can be attributed to an enhancement of the charge contribution due to increased charge carrier density. A higher density of charge carriers may result in increased capacitance values. Consequently, the capacitance spectres also reveal increased values in response to a humidity increase (Figure 1d). This behaviour is widely exhibited by capacitive humidity sensors [16,33,34].

The overall trend exhibited in Figure 1 can be fitted according to the Cole–Cole model as an equivalent circuit (Figure 2). In particular, Figure 1c compares the obtained Nyquist plots, wherein part of a characteristic semi-circular shape is exhibited by these curves, revealing a capacitor and resistor connected in parallel. The semi-circular shape represents the capacitive behaviour of the fabricated composite and non-composite ILGPE-based capacitors. Nevertheless, at high frequencies, the semi-circular responses present some shifts from the origin, indicating the existence of a resistance in series with the parallel combination. This deviation is typically ascribed to the bulk resistance of the materials that comprise the capacitors, demonstrating that two sub-circuits connected in series form the equivalent circuit.

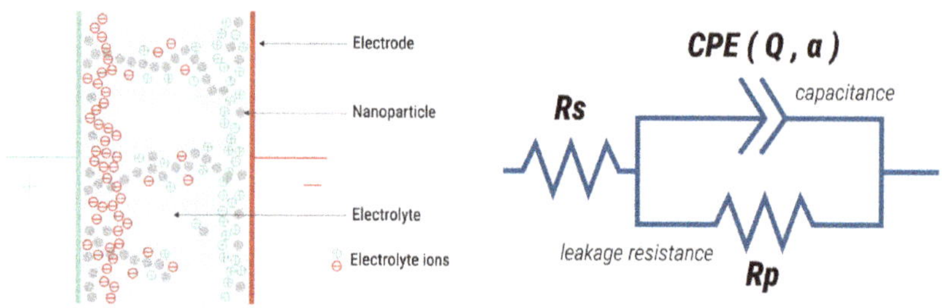

Figure 2. Schematic representation of the SiO_2 CILGPE-based capacitor and its equivalent circuit. R_s, R_p, and CPE represent the series resistance, the leakage resistance, and the constant phase element, respectively.

The first sub-circuit within the equivalent circuit model comprises a series resistor (R_s). Meanwhile, the other circuit involves the parallel combination of a constant phase element (CPE) and a resistor (R_p) related to a leakage resistance. The CPE is a frequency-dependent element that simulates the formation of a double-layer capacitance. The CPE appears due to the charge accumulation at the electrode–electrolyte interfaces, as well as at the boundaries of the nanoparticles. This element can be defined by Equation (2).

$$Z_{CPE} = \frac{1}{Q_{CPE}(j2\pi f)^{\alpha}} \tag{2}$$

Q_{CPE} represents the admittance of an ideal capacitance, and α is the parameter employed to model the frequency dispersion resulting from non-ideal capacitive behaviour owing to factors such as electrode roughness or electrolyte inhomogeneity [35,36].

The R_s represents the ESR, which describes the combination of the internal resistive effects of the materials that compose a capacitor. Therefore, R_s arises from the bulk resistance of the electrolyte and the electrodes internal resistance [36]. Specific frequency ranges have been selected to illustrate a portion of the Nyquist plots (Figure 1c) to appreciate each device's exhibited ESR values at different %RH levels. The R_s values have been estimated by determining the intersections of the Nyquist curves with the x-axis at higher frequencies [35,36].

2.2. Surface Topography

Figure 3 illustrates both plan and side views of the PVA:$PVP_{0\%}$, SiO_2 PVA:$PVP_{0\%}$, PVA:$PVP_{25\%}$, and SiO_2 PVA:$PVP_{25\%}$ capacitors. In particular, Figure 3a shows the top surface and side profile of the PVA:$PVP_{0\%}$-based capacitor. As expected, the PVA:$PVP_{0\%}$-ILGPE composite exhibits a transparent and smooth surface topography that reflects its homogeneity and uniformity.

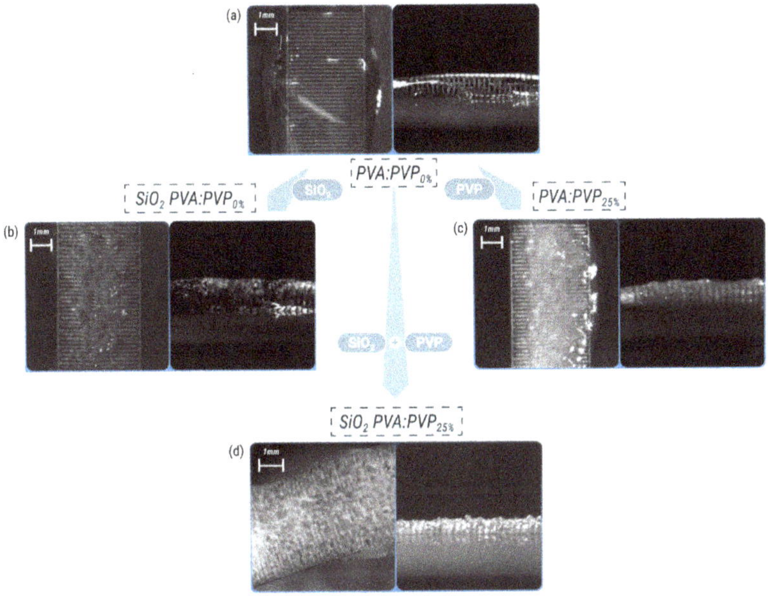

Figure 3. Plan and side views of a (**a**) PVA:$PVP_{0\%}$-ILGPE-, (**b**) SiO_2 PVA:$PVP_{0\%}$-CILGPE-, (**c**) PVA:$PVP_{25\%}$-ILGPE-, and (**d**) SiO_2 PVA:$PVP_{25\%}$-CILGPE-based capacitors.

Incorporating silica nanoparticles or PVP into PVA:$PVP_{0\%}$-ILGPE stimulates topography changes. SiO_2 PVA:$PVP_{0\%}$- and PVA:$PVP_{25\%}$-based electrolytes reveal slight roughness regions. These textured regions suggest a degree of amorphousness within the layers, ascribed to the capability of PVP and SiO_2 nanofillers to disrupt the crystalline structure of PVA. The influence of PVP in diminishing the crystal phase of PVA is also evident in [37–39]. The emergence of this topography transformation leads to an increase in the air-contact area of the sensing layer, which could positively affect the humidity-sensing performance.

The sensing layer integrating SiO_2 nanoparticles and PVP into PVA:$PVP_{0\%}$-ILGPE shows a distinct topography. The SiO_2 PVA:$PVP_{25\%}$-CILGPE exhibits a translucent quality and presents a more pronounced surface roughness than the PVA:$PVP_{0\%}$-, SiO_2 PVA:$PVP_{0\%}$-, and PVA:$PVP_{25\%}$ -based ILGPEs. This effect could be due to the simultaneous presence of PVP and silica fillers, which increases the concentration of agents disrupting the crystalline phase of PVA. The appearance of a rougher topography implies an increase in the air-contact surface area of the sensing layer, suggesting the enhancement of the sensitivities exhibited by SiO_2 PVA:$PVP_{0\%}$- and PVA:$PVP_{25\%}$-based capacitors.

2.3. Humidity Sensor Performance

The humidity impedimetric and capacitive responses of the different fabricated ILGPE- and SiO_2 CILGPE-based capacitors have been evaluated at various frequencies in a relative humidity range from 20 %RH to 90 %RH. Seven different %RH levels have been applied to test the impedance and capacitance responses during the adsorption and desorption processes. The responses of three PVA:$PVP_{0\%}$, SiO_2 PVA:$PVP_{0\%}$, PVA:$PVP_{25\%}$, and SiO_2 PVA:$PVP_{25\%}$ capacitors are displayed in Figure 4.

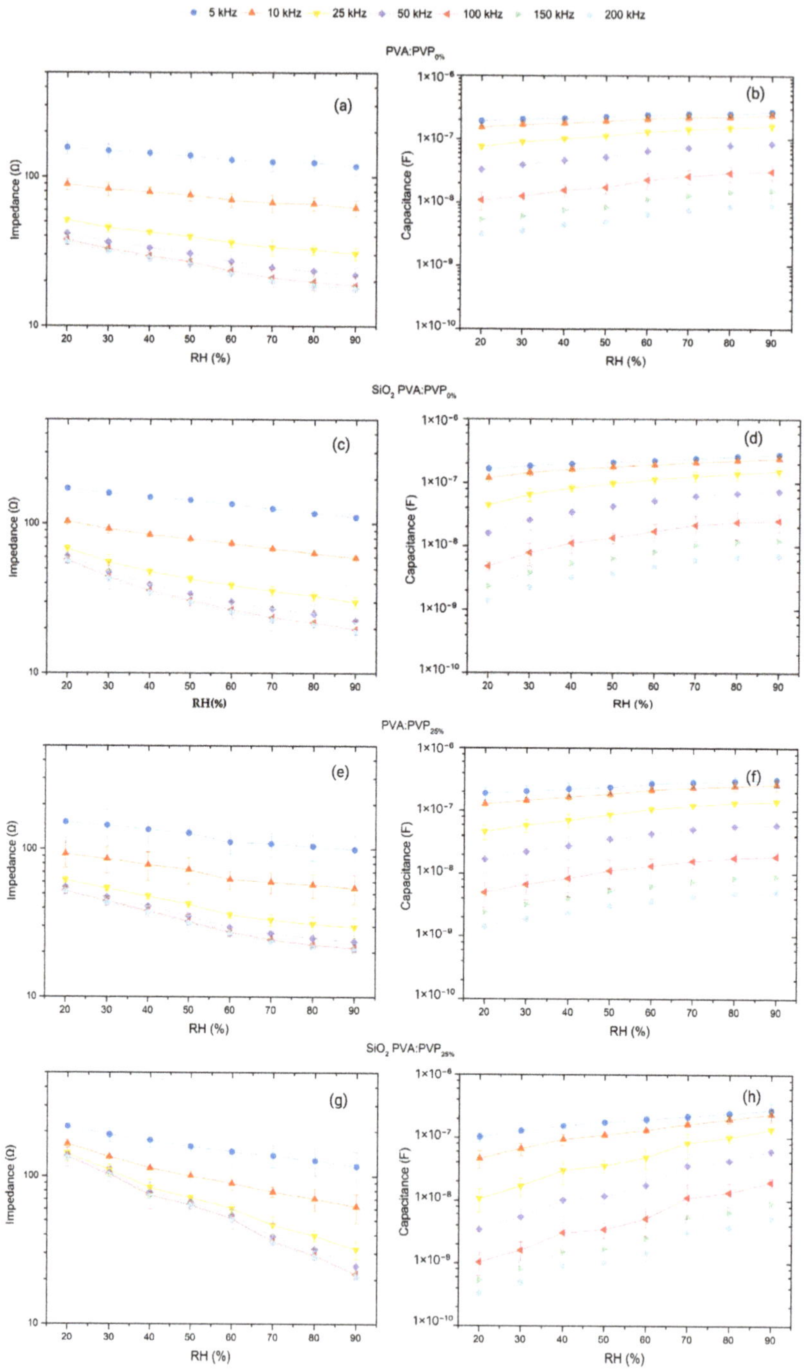

Figure 4. Impedance modulus and capacitance humidity-sensing responses at the evaluated frequencies exhibited by (**a**,**b**) PVA:PVP$_{0\%}$-ILGPE-, (**c**,**d**) SiO_2 PVA:PVP$_{0\%}$-CILGPE-, (**e**,**f**) PVA:PVP$_{25\%}$-ILGPE-, and (**g**,**h**) SiO_2 PVA:PVP$_{25\%}$-CILGPE-based capacitors.

As illustrated in Figure 4, the impedance responses of the fabricated humidity sensors exhibit a decreasing trend with relative humidity; meanwhile, the capacitance responses reveal an increasing trend due to the relative humidity increase. The exhibited impedance and capacitance behaviours could be ascribed to the enhancement of ion conduction resulting from the interaction of the sensing layer with water molecules. Higher relative humidity levels imply an increase in the probability of the adsorption of water molecules and, hence, an increase in the charge carrier density. As a result, the developed capacitors' impedance diminishes, and capacitance rises. Regarding the role of the operating frequency, it can be noticed that the impedance and capacitance responses tend to exhibit lower variations at high frequencies, indicating a decrease in the sensitivity of the ILGPE and SiO_2 CILGPE capacitors with the frequency. This effect can also be ascribed to water molecules' polarisation dynamics. Adsorbed water molecules can follow changes in the electric field at low frequencies. However, they are not able to follow faster electric field changes that occur at high frequencies [10,16].

Furthermore, it is noticeable that the PVA:$PVP_{0\%}$ capacitors reveal linear impedance responses to humidity changes. However, adding SiO_2 nanoparticles, PVP, or both stimulates the interaction with water molecules, leading to impedance responses displaying non-linear behaviours. This observation reduces the viability of using impedance modulus as a humidity-sensing parameter since one of the desired sensor characteristics is linearity. Consequently, evaluations of sensitivity and hysteresis have been exclusively focused on the capacitance responses.

Following the analysis of the exhibited humidity-sensing responses at different frequencies, the capacitance sensitivities of the sensors to changes in relative humidity in the ambient chamber at 25 °C have been calculated using Equation (3), as typically used [1,13,34,40,41]:

$$S_\gamma = \frac{\Delta_y}{\Delta_{RH}} = \frac{\gamma_{90\%} - \gamma_{20\%}}{90 - 20} \quad (3)$$

where, $\gamma_{90\%}$ and $\gamma_{20\%}$ are the capacitance values at 90% and 20% of relative humidity, respectively. Figure 5 compares the sensitivity values vs. the evaluated frequencies and shows an evident impact on capacitance sensitivities because of the addition of PVP and SiO_2 nanofillers to the PVA-based ILGPE. The exhibited capacitance responses indicate that using PVP and SiO_2 nanoparticles is supportive in driving the sensitivity at the evaluated frequencies. Different hydrophilic properties mean distinct ways to interact and react with water molecules. In particular, in Figure 5, SiO_2 PVA:$PVP_{0\%}$ and SiO_2 PVA:$PVP_{25\%}$ capacitors exhibit higher capacitance sensitivities than the PVA:$PVP_{0\%}$ capacitor, thereby revealing the positive impact of the charge pathways provided by the nanoparticles on the sensitivity.

Figure 5 also reveals that incorporating PVP and SiO_2 into the PVA-ILGPE increases the capacitance sensitivity within the frequency range of 1 kHz to 100 kHz. These findings could be ascribed to the number of active sites per volume available in the sensing layer. Firstly, the presence of PVP in the sensing layer composition implies a relevant change in the available active sites per volume in the sensing layer because of its amorphous nature. Secondly, the presence of SiO_2 nanoparticles affects the sensitivity since SiO_2 fillers can increase the available surface area and number of active sites per volume [1]. This effect could arise from the reduction in the crystallinity of the ILGPEs, which is supported by the findings revealed in Figure 3, and the high area-to-volume ratio of the nanoparticles available for ion exchanges. Finally, it is noteworthy that the device that exhibits the highest capacitance sensitivities is the SiO_2 PVA:$PVP_{25\%}$ capacitor, revealing the positive contribution of the combination of the SiO_2 and PVP benefits. This result validates the hypothesis that incorporating SiO_2 nanobeads and PVP can stimulate a reduction in the degree of crystallinity of PVA. Therefore, the resultant electrolyte exhibits a more amorphous layer with a rougher surface, increasing the sensing area and enhancing the sensitivities.

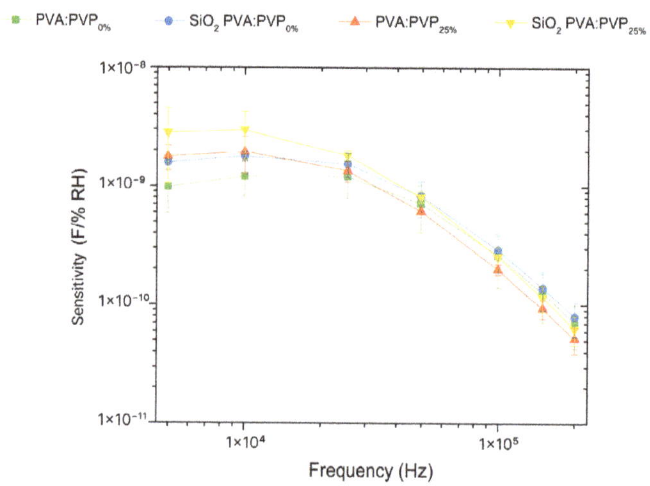

Figure 5. Comparison of the capacitance sensitivities exhibited by PVA:PVP$_{0\%}$-ILGPE-, SiO_2 PVA:PVP$_{0\%}$-CILGPE-, PVA:PVP$_{25\%}$-ILGPE-, and SiO_2 PVA:PVP$_{25\%}$-CILGPE-based capacitors at the evaluated frequencies.

In addition to the analysis of the impact of PVP and SiO_2 nanoparticles on the sensitivity, the hysteresis parameter has also been analysed. Hysteresis refers to the highest deviation between the obtained ILGPE and SiO_2 CILGPE capacitor responses during desorption and adsorption processes. Factors such as porous structures or surface morphology of the sensing layer mainly dominate the hysteresis parameter [42]. Therefore, hysteresis is a significant parameter in evaluating the sensing performance and understanding the absorption principle of the humidity sensor. Initially, to obtain the adsorption responses, the capacitors were exposed to a controlled environment in which the RH was gradually increased from 20 %RH to 90 %RH over a specific time. Afterwards, to measure the desorption responses, the relative humidity was progressively decreased from 90 %RH to 20 %RH. The impedance spectra of the ILGPE and SiO_2 CILGPE capacitors were obtained at each relative humidity variation. After calculating the capacitance values at each specific %RH during the adsorption and desorption processes, the hysteresis value was calculated using the following Equation (4) [1,13]:

$$H = \frac{max\left(\gamma_{x\%_D} - \gamma_{x\%_A}\right)}{S} \quad (4)$$

where $\gamma_{x\%_D}$ and $\gamma_{x\%_A}$ are the obtained capacitance values during desorption and adsorption at the steady humidity level, respectively, and S is the sensitivity. Figure 6 compares the hysteresis characteristics exhibited by the fabricated devices at the evaluated frequencies.

As is noticeable from Figure 6, SiO_2 PVA:PVP$_{0\%}$ and SiO_2 PVA:PVP$_{25\%}$ capacitors exhibit lower hysteresis values than their electrolyte counterparts without embedded SiO_2 nanoparticles in the electrolyte. This finding indicates that SiO_2 CILGPE capacitors are more susceptible to humidity changes than ILGPE-based capacitors, exposing the dominant role of silica nanoparticles in determining the hysteresis. SiO_2 nanofillers could generate pathways that can facilitate the diffusion of water molecules since they indicate a connection between the different levels of the sensing layer, thereby promoting water molecule diffusion. Furthermore, it is essential to highlight the constructive contribution of PVP in SiO_2 CILGPE capacitors. The simultaneous existence of SiO_2 nanobeads and PVP in the PVA-based ILGPE capacitor exhibits the lowest hysteresis values in this study, revealing the favourable combination of these materials. The reduction in hysteresis may be primarily attributed to the amorphous nature of PVP. The inherent amorphous structure of PVP disrupts the crystallinity of the PVA matrix, increasing the amorphousness of the SiO_2 CILGPE and leading to a reduction in zones with a crystalline phase that can hinder the diffusion of water molecules, thus facilitating their transport. Moreover, the hydrophilicity of PVP stimulates the formation of additional hydrogen bonds with water, affecting the retention and release of water molecules and increasing the susceptibility of SiO_2 CILGPE to %RH variations. These findings are consistent with those of [16,42], which indicate that high surface-to-volume-ratio materials result in excellent humidity-sensing performance, although their intrinsic chemical and physical characteristics are more susceptible to changes, which directly influence the hysteresis.

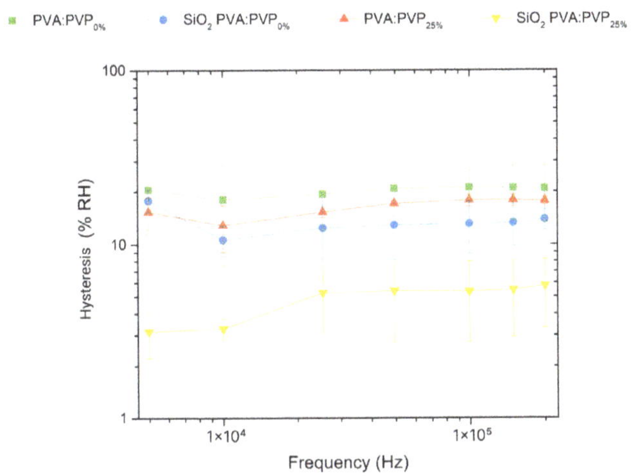

Figure 6. Comparison of the capacitance hysteresis values exhibited by $PVA:PVP_{0\%}$-ILGPE-, SiO_2 $PVA:PVP_{0\%}$-CILGPE-, $PVA:PVP_{25\%}$-ILGPE-, and SiO_2 $PVA:PVP_{25\%}$-CILGPE-based capacitors at different possible working frequencies.

As a summary, Figure 7 depicts the capacitance responses of the $PVA:PVP_{0\%}$, SiO_2 $PVA:PVP_{0\%}$, $PVA:PVP_{25\%}$, and SiO_2 $PVA:PVP_{25\%}$ capacitors during the adsorption and desorption processes at the frequencies that exhibit higher sensitivities and lower hysteresis values. Consequently, for $PVA:PVP_{0\%}$ capacitors, the optimal working frequency is 50 kHz, while for SiO_2 $PVA:PVP_{0\%}$, $PVA:PVP_{25\%}$, and SiO_2 $PVA:PVP_{25\%}$ capacitors, it is 10 kHz. Notably, the $PVA:PVP_{0\%}$ ILGPE capacitor shows a capacitance response with a sensitivity of about 739 pF/%RH. As explained above, the incorporation of silicon dioxide nanoparticles stimulates increased sensibility and decreased hysteresis. As a result, the SiO_2 $PVA:PVP_{0\%}$ CILGPE capacitor exhibits capacitance sensitivities of about 1780 pF/%RH with a hysteresis value of 10.64 %RH.

Furthermore, adding PVP also enhances the sensing performance, leading to capacitance sensitivities of 1950 pF/%RH with maximum hysteresis of about 12.89 %RH. Finally, as is noticeable, the SiO_2 $PVA:PVP_{25\%}$-ILGPE-based capacitor shows the highest capacitance sensitivity with the lowest hysteresis values, exhibiting capacitance sensitivities of 2730 pF/%RH with the lowest maximum hysteresis about 3.28 %RH, respectively, thereby revealing the positive effect of combining PVP and SiO_2 fillers. The lowest hysteresis value can be mainly ascribed to the presence of SiO_2 nanoparticles. As explained earlier, SiO_2 nanoparticles may facilitate the diffusion of water molecules by connecting different levels of the sensing layer (top and bottom). In addition to this effect, the presence of PVP, which can increase the amorphousness, could reduce the barriers to water molecule movement. Irregular arrangements may provide more pathways for water molecule diffusion, which can reduce the hysteresis.

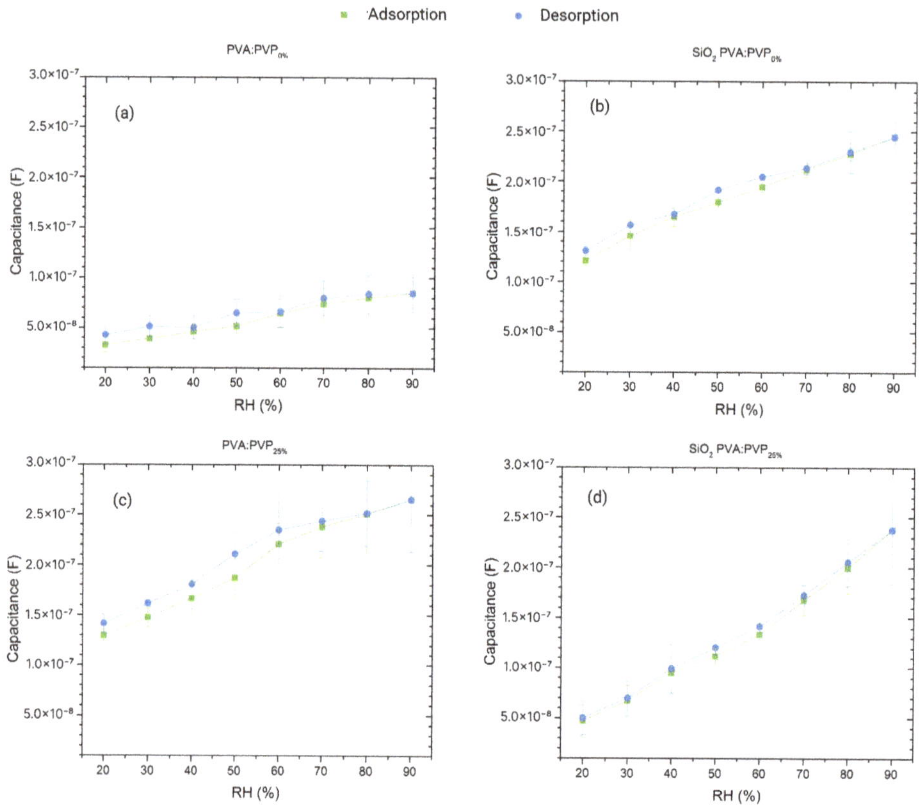

Figure 7. Comparison of the capacitance responses during the adsorption and desorption processes exhibited by (**a**) PVA:PVP0%-ILGPE-, (**b**) SiO_2 PVA:PVP0%-CILGPE-, (**c**) PVA:PVP25%-ILGPE-, and (**d**) SiO_2 PVA:PVP25%-CILGPE-based capacitors at their potential operating frequencies.

On the other hand, regarding the higher sensitivity, the coexistence of PVP and SiO_2 nanoparticles increases the amorphousness of the sensing layer. Therefore, their presence can increase the surface area where water molecules can be adsorbed. Additionally, as both materials are hydrophilic, their coexistence further increases the active sites available for the interaction between water molecules and the sensing layer, increasing the susceptibility to humidity changes.

Additionally, Table 1 provides an overview of the linearity performance exhibited by $PVA:PVP_{0\%}$, SiO_2 $PVA:PVP_{0\%}$, $PVA:PVP_{25\%}$, and SiO_2 $PVA:PVP_{25\%}$ capacitors at 10 kHz within the suggested detection range. The regression analysis conducted on ILGPE- and SiO_2 CILGPE-based capacitors demonstrates the existence of a linear relationship between the relative humidity and capacitance responses. The resulting high R-squared values show a reliable and accurate linear fit, further emphasising the linear performance within the proposed sensing range. Furthermore, regarding the detection limit, the criteria for selecting sensing limits is based on discarding the relative humidity levels that reveal differences lower than 3% compared to their previous %RH levels. Consequently, the $PVA:PVP_{0\%}$-based capacitor is suggested to operate within the 20–70 %RH sensing range. From Table 1, it can be concluded that the addition of SiO_2 and PVP could expand the humidity-sensing span, revealing the positive impact of these materials on the sensing range. Additionally, as the sensitivity can also be defined as the slope of the calibrated fitting response of a sensor [1], this table also shows the sensitivity values of the calibrated responses.

Table 1. Comparison of the regression models, R-squared, and calibrated sensitivities obtained during the fitting of the capacitance responses of the $PVA{:}PVP_{0\%}$-ILGPE-, SiO_2 $PVA{:}PVP_{0\%}$-CILGPE-, $PVA{:}PVP_{25\%}$-ILGPE-, and SiO_2 $PVA{:}PVP_{25\%}$-CILGPE-based capacitors and their preferred sensing range at 10 kHz.

Identifier	Sensing Range (%RH)	Capacitance Response		
		Fitting Function	**R^2**	**Calibrated Sensitivity (nF/%RH)**
$PVA{:}PVP_{0\%}$	20–70	$0.79 \times 10^{-9}x + 0.16 \times 10^{-7}$	0.989	0.79
SiO_2 $PVA{:}PVP_{0\%}$	20–90	$1.71 \times 10^{-9}x + 9.26 \times 10^{-8}$	0.994	1.71
$PVA{:}PVP_{25\%}$	20–90	$2.04 \times 10^{-9}x + 8.87 \times 10^{-8}$	0.988	2.04
SiO_2 $PVA{:}PVP_{25\%}$	20–90	$2.66 \times 10^{-9}x - 1.38 \times 10^{-8}$	0.985	2.66

As a result of the findings discussed above, it can be concluded that the most suitable electrolyte for acting as a humidity-sensing layer is the SiO_2 $PVA{:}PVP_{25\%}$ CILGPE in terms of sensitivity and hysteresis. Nevertheless, it is important to investigate the reliability of a sensor in the context of humidity sensing. To evaluate the reliability of the optimised SiO_2 CILGPE, the long-term stability, the effect of the temperature on the capacitance response, and the impact of the cyclability on the hysteresis parameter have been investigated.

The cyclability of the optimised sensor has been examined since it is a crucial parameter in determining the reliability and durability of a sensor. Cyclability refers to the ability exhibited by a sensor to maintain its sensing performance over many cycles of usage. In particular, the cyclability has been analysed by measuring the adsorption and desorption responses for ten cycles to determine their impact on the hysteresis parameter. The study of the hysteresis parameter across several cycles provides a first insight into how consistent and reliable the capacitive humidity response can be over repeated cycles. Additionally, understanding how the cyclability affects the hysteresis can give information about the sensor's durability and stability. Figure 8 shows the obtained hysteresis over ten consecutive cycles, providing insights into the stability and repeatability of the optimised sensor. In particular, Figure 8 illustrates that the capacitance response of a SiO_2 $PVA{:}PVP_{25\%}$ capacitor exhibits similar hysteresis during ten cycles, showing hysteresis values around 3.6 %RH. This finding may indicate that, in terms of hysteresis, the SiO_2 $PVA{:}PVP_{25\%}$ capacitor could exhibit a consistent and reliable humidity-sensing performance over ten cycles.

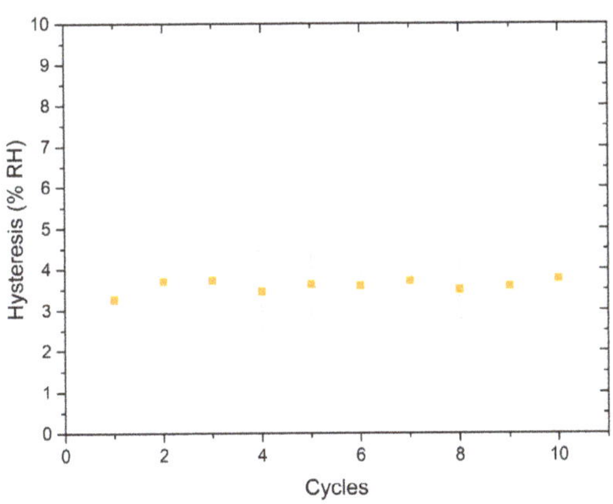

Figure 8. Exhibited evolution of the hysteresis parameter over ten cycles.

Furthermore, to analyse the long-term stability, a SiO_2 $PVA{:}PVP_{25\%}$-based capacitor was exposed to natural environmental conditions for five months after its fabrication. A comparison between the responses of the SiO_2 $PVA{:}PVP_{25\%}$ five months after the fabrication process and in its initial state is displayed in Figure 9a. This analysis provides initial insights into the ability of the SiO_2 PVA:PVP$_{25\%}$ CILGPE to maintain its sensitivity and hysteresis characteristics.

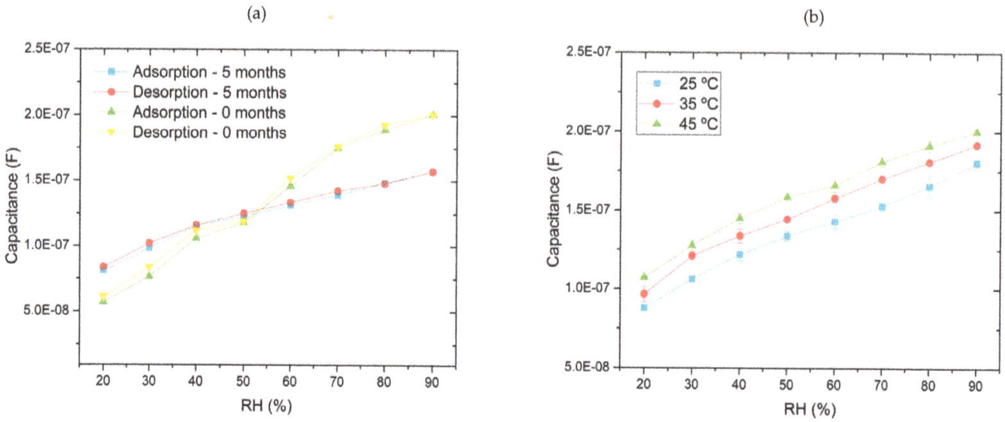

Figure 9. Comparison of (**a**) a capacitance response after five months and (**b**) capacitance responses of the SiO_2 PVA:PVP$_{25\%}$ capacitor at 25 °C, 35 °C, and 45 °C.

Figure 9a compares the initial and the five-month capacitance responses, exhibiting a humidity-sensing deterioration. After five months, the capacitance response shifts to higher values and displays a reduced sensitivity. Notably, the initial capacitance responses exhibit a calibrated sensitivity of 2140 pF/%RH, while five months later, the same device presents a reduction of about 50% in its sensitivity, thereby revealing a sensitivity of 1024 pF/%RH. However, despite this observed sensitivity degradation, the SiO_2 PVA:PVP$_{25\%}$ CILGPE-based capacitors still display a linear behaviour, thereby indicating the possibility of compensation for this degradation. This alteration in the humidity capacitance response indicates some concerns that merit further investigation. Therefore, it is advisable to delve deeper through additional research aimed at studying and mitigating this degradation.

Considering that temperature can impact humidity sensors' long-term stability and durability, a five-month-old device has also been employed in temperature-dependence analysis. This approach provides insights into SiO_2 CILGPE-based capacitors' performance and endurance under fluctuating thermal conditions, revealing more information about their reliability over time. As is widely known, most hygroscopic materials reveal variations in their characteristics under different temperatures, exhibiting a dependency on temperature. This effect is particularly noticeable in impedimetric and capacitance sensors. Therefore, the evaluation of how temperature affects humidity-sensing responses may result in the mitigation of its impact. Thus, the reliability of humidity sensors can be improved by compensating for the potential erroneous humidity readings. To determine the effect of the temperature variation on the humidity-sensing performance of the optimised sensing layer, IS measurements of RH at three different temperatures (25 °C, 35 °C, and 45 °C) have been performed at the selected working frequency. Figure 9b illustrates the resulting capacitive humidity responses at the proposed working frequency of 10 kHz under different temperatures. Notably, Figure 9b shows that the obtained capacitance responses of the optimised sensing layer are shifted to higher values with the increase in ambient temperature. This behaviour has been previously reported and indicates that an increase in temperature can provide additional energy to charge carriers, leading to a stronger thermal movement of the water molecules [13,43].

As a consequence, there is an increase in the charge carriers' mobility, which can also increase the capacitance values. Furthermore, as can be noticed in Figure 9b, the obtained capacitance responses at different temperatures depict similar linear trends without exhibiting significant sensitivity variations. This finding indicates a suitable temperature dependence to function as a humidity sensor since the deviation change promoted by the temperature increase can be easily compensated.

Finally, Figure 10 and Table 2 compare the SiO_2 PVA:PVP$_{25\%}$ capacitors' sensing properties with those previously exhibited in the SOA. Figure 10 shows that, when compared with some recently reported capacitive humidity sensors based on pyromellitic dianhydride (PMDA)-oxydianiline (ODA)/TiO_2 [13], polyaniline (PANI)/1% Cu–ZnS [33], poly(ionic liquid) [44], PPy/graphene oxide (GO) [9], graphene oxide quantum dots (GOQD)/Nafion [11], PVA/cellulose acetate butyrate (CAB) [6], keratine/1% carbon fibres (CF) [10], graphene–carbon nanotube (CNT)–silicone adhesive (SA) [34], PVDF-$BaTiO_3$ [15], GO-doped poly(vinylidene fluoride-trifluoroethylene) (P(VDF-TrFE))/lithium chloride (LiCl) [41], or PVP+CAB [8], the SiO_2 PVA:PVP$_{25\%}$-CILGPE capacitor has relevant detection parameters. These prior studies combine nanofillers or salts with polymers or use ILs to enhance sensitivity and hysteresis in humidity sensors. For instance, incorporating carbon fibres into a keratin matrix resulted in a capacitance sensitivity of 633.12 pF/%RH across a wide sensing range. Similarly, the PPy and graphene oxide composite yielded a low-hysteresis humidity sensor with a sensitivity of 1670.3 pF/%RH. Furthermore, the polymerisation of an ionic liquid or adding salt into a PVA matrix to create microstructures revealed capacitance sensitivities of 1.8 nF/%RH with low hysteresis and 1.41 pF/%RH, respectively.

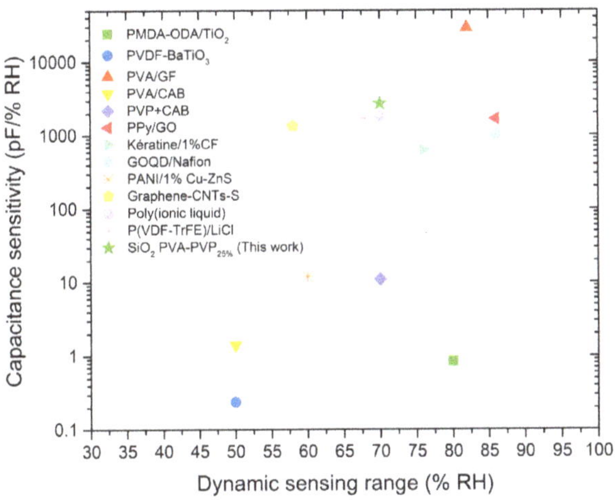

Figure 10. Comparison of the sensing parameters exhibited by some reported humidity sensors in SOA and the SiO_2 PVA:PVP$_{25\%}$ capacitor.

Table 2. Comparison of sensitivity, hysteresis, sensing ranges, and fabrication methods of capacitive-type humidity sensors in the literature and the SiO_2 PVA:PVP$_{25\%}$-CILGPE-based capacitor. TW (this work).

Sensing Material	Fabrication Method	Sensing Range (% RH)	Sensitivity (pF/% RH)	Hysteresis (% RH)	Ref.
PMDA-ODA/TiO_2	Spin Coating	10–90	0.85	0.95	[13]
PVDF-$BaTiO_3$	Spin Coating	40–90	0.2416	2.5	[15]
Mesoporous silica-PEDOT	-	11–95	~120	6	[16]
PVA/GF	Spin + Spray Coating	40–90	29,000	low	[5]
PVA/CAB	-	10–90	1.41	-	[6]
PVP+CAB	Inkjet Printing + Doctor Blade	20–80	11.4	-	[8]
PPy/GO	Drop Casting	11–97	1670.3	1.12	[9]
Keratine/1% CF	Drop Casting	16–92	633.12	1.33	[10]

Table 2. *Cont.*

Sensing Material	Fabrication Method	Sensing Range (% RH)	Sensitivity (pF/% RH)	Hysteresis (% RH)	Ref.
Nafion/GOQD	Drop Casting	11.3–97.3	1025.97	4	[11]
PANI/ 1% Cu–ZnS	Spin Coating	30–90	12	1.56	[33]
Graphene–CNT–SA	Drop Casting + Doctor Blade	36–94	1336.7 −84.5 Ω/%RH	- -	[34]
Poly(ionic liquid)	Bar Printing	10–80	1800	Low	[44]
GO-doped P(VDF-TrFE)/LiCl	Drop Casting	25–93	1708.8	-	[41]
SiO_2 PVA:$PVP_{25\%}$	Drop Casting	20–90	2660	3.28	TW

In contrast, this study explores the impact of silica nanoparticles, PVP, and their combination on an ILGPE, revealing the beneficial effects of the simultaneous presence of these materials and exhibiting a higher calibrated capacitance sensitivity with a value of 2660 pF/%RH and low hysteresis. Therefore, this work introduces a novel composite that blends polymers, nanofillers, and an ionic liquid. The resulting composite is based on eco-friendly materials and uses an easy and fast fabrication process. This approach exhibits an enhancement of capacitance sensitivity and hysteresis. Finally, as could be concluded from Table 2 and Figure 10, the SiO_2 PVA:$PVP_{25\%}$-CILGPE-based device outperforms some reported capacitive humidity sensors in terms of sensitivity. Approximately 64% of the studies summarised in the table reported capacitance sensitivities below 1500 pF/%RH. The good sensitivity and hysteresis of the SiO_2 PVA:$PVP_{25\%}$-CILGPE-based capacitor, combined with its simple and cost-effective fabrication process and possible easy compensation of its temperature dependence, position the optimised device as a viable option for a humidity sensor.

2.4. Humidity-Sensing Mechanism

Previous research has established a link between the electrical and physical behaviours of humidity sensors by fitting their complex impedance spectroscopy data using equivalent circuits, allowing for the investigation of the sensing mechanism [9,16,43]. Consequently, to investigate the working mechanism of the PVA:$PVP_{0\%}$, SiO_2 PVA:$PVP_{0\%}$, and PVA:$PVP_{25\%}$ capacitors and the impact of the addition of SiO_2 nanoparticles, PVP, or both on the sensing mechanism, the variation in the values of the elements that compose the equivalent circuit have been analysed as a function of %RH. As previously explained, the obtained impedance spectra can be modelled using an equivalent circuit. This electrical circuit consists of a resistance (R_s) in series with the parallel combination of a CPE and a leakage resistance (R_p). The R_s parameter corresponds to the bulk resistance of the electrolyte and the electrodes, while the CPE is attributed to the double-layer capacitances. Alterations in humidity mainly affect the CPE, which is noticeable in the Q_{CPE} parameter and the value of R_s. Consequently, the impact of moisture on both parameters has been analysed, as any changes in the electrical characteristics of the electrolyte are reflected in these parameters.

On the one hand, Figure 11 shows the evident tendency followed by the values of R_s and Q_{CPE} with changes in humidity. Notably, Figure 11a illustrates the obtained R_s values for PVA:$PVP_{0\%}$, SiO_2 PVA:$PVP_{0\%}$, PVA:$PVP_{25\%}$, and SiO_2 PVA:$PVP_{25\%}$ capacitors across different %RH levels, exhibiting their respective trends. As displayed in Figure 11a, there is an evident reduction in R_s values as humidity increases. This finding aligns with the impedance modulus trend and indicates an enhancement in the charge conduction that reduces the electrolyte resistance. This improved charge conduction could be the result of an increase in the charge content and enhanced ionic mobility. The higher ion density can be attributed to the ability of PVA, PVP, and SiO_2 nanoparticles to absorb and interact with water molecules from the environment. This effect is also noticeable in the behaviour of the Q_{CPE} parameter relative to relative humidity levels. Figure 11b shows that Q_{CPE} values increase with rising humidity. This behaviour is widely exhibited by capacitive humidity sensors [16,33,34]. The higher Q_{CPE} values due to the %RH increase also indicate an enhancement in charge density, thereby promoting the formation and enhancement of electrical double-layer capacitance.

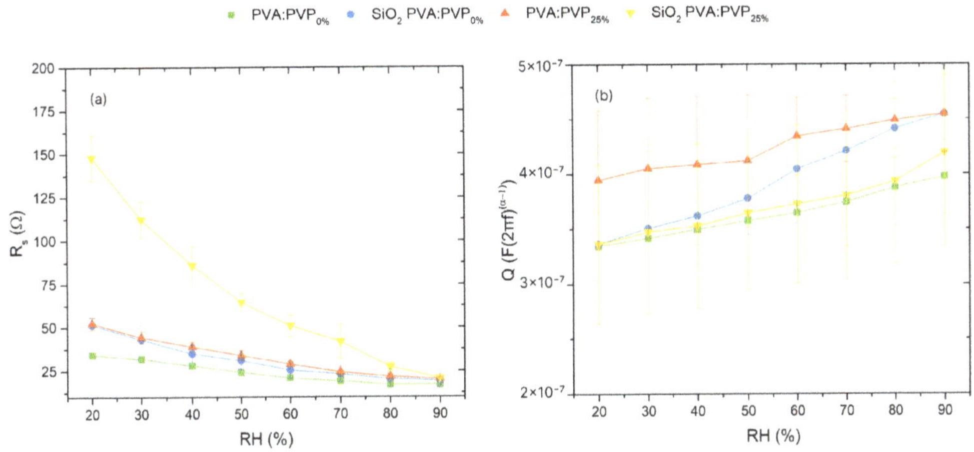

Figure 11. Trends in the (**a**) R_S (series resistance) and (**b**) Q_{CPE} (ideal admittance of the CPE) values at different relative humidity levels.

Therefore, the humidity-sensing mechanism works as depicted in Figure 12. Initially, at low %RH levels, an initial interaction transpires between the surface of the sensing layer and the water molecules present in the environment, leading to some water molecules being trapped at the outermost levels of the sensing layer (Figure 12a,b). This first interaction forms a chemisorbed layer formed by hydrogen bonds between water molecules and the hydroxyl groups in the sensing material, as explained in [1,20,45]. This interaction is mainly attributed to the inherent hydrophilicity of the sensing layer, which has a minor impact on the charge conduction since chemisorbed molecules are not free to move [1]. As the humidity rises, the sensing layer transitions from chemisorption to physisorption interactions. At this point, the physically adsorbed water molecule layers add to the chemically adsorbed layer, forming a continuous water molecule layer. These additional water molecules start to contribute to the charge conduction of the sensing layer due to the conduction of some extra ions [20,45]. This process leads to a decrease in the R_s parameter and an increase in the Q_{CPE} values. With further increases in the humidity, a fraction of the confined water molecules on the most superficial levels start to penetrate the polymeric layer, revealing the existence of an internal–external density balance of water molecules until a state of equilibrium is reached (Figure 12c). At this point, the electrical behaviour is mainly controlled by charge conduction, resulting in even lower R_s and higher Q_{CPE} values. This finding is attributed to improved charge conduction due to (i) enhanced charge mobility, as the presence of continuous water molecules may allow higher mobility of the ions in the sensing layer, and (ii) a higher effective charge density, owing to the extra ions (H_3O^+, H^-) contributing to the conduction. Consequently, higher relative humidity levels imply an increase in charge density inside the sensing layer, reducing the impedance of the polymer electrolyte and boosting capacitance values (Figure 12d,e).

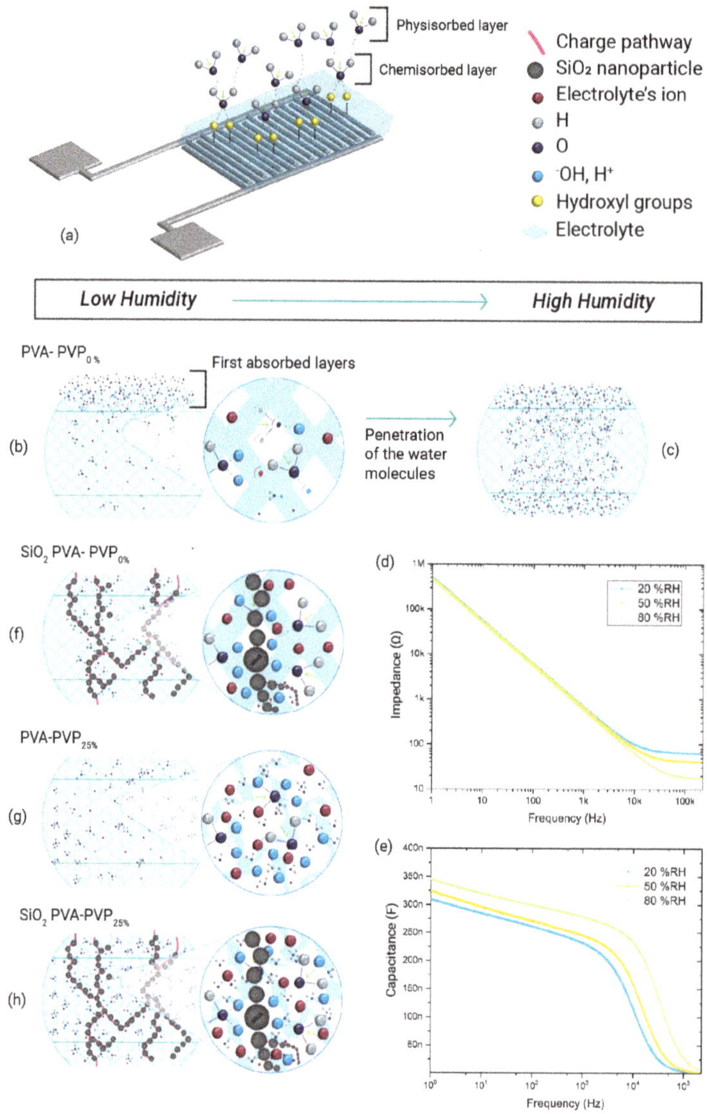

Figure 12. Proposed operation principle of the fabricated ILGPE and SiO_2 CILGPE sensors. (**a**) Graphical representation of a fabricated ILGPE sensor and the first chemisorbed and physisorbed water layers. (**b**) longitudinal section of PVA:PVP0% ILGPEs at low humidity levels. (**c**) longitudinal section of PVA:PVP0% ILGPEs at high humidity levels, illustrating the penetration of water molecules with the increase in humidity. (**d**) comparison of the impedance vs. frequency curves at different frequencies. Higher humidity levels imply lower impedance values. (**e**) comparison of the capacitance vs. fre-quency curves at different frequencies. Higher humidity levels result in higher capacitance values. (**f**) longitudinal section of SiO_2 PVA:PVP0% CILGPEs. The grey spheres, connecting the top and bottom of the sensing layer, represent the SiO_2 nanoparticle pathways that can facilitate the diffu-sion of water molecules. (**g**) longitudinal section of PVA:PVP25% ILGPEs. The addition of PVP dis-rupts the crystallinity of PVA, increasing the amorphousness and, hence, the available active sites to interact with water molecules. (**h**) longitudinal section of SiO_2 PVA:PVP25% CILGPEs. This type of CILGPE combines the benefits of incorporating silica nanoparticles and PVP.

In contrast, during a decrease in %RH, this internal–external density balance is disrupted, and the trapped water molecules inside the sensing layer try to escape adsorption and initiate diffusion. This diffusion process occurs gradually, commencing with freeing the trapped water molecules at the outermost levels. Subsequently, the trapped molecules at the deeper layers start to ascend and diffuse, attaining equilibrium, decreasing the number of charges precedent from the water molecules' interactions, increasing the impedance, decreasing the capacitance, and, hence, increasing the R_s and decreasing the Q_{CPE} values.

This proposed sensing mechanism is consistent with those discussed in [10,16,20,33,40,42,45]. These previous studies explain that the humidity-sensing mechanism of polymeric sensors is governed by the interaction between the humidity-sensitive layer and water molecules, which is mainly explained by chemisorption and physisorption processes, as well as Grotthus chain reactions. The hydrophilic polymer possesses active sites that facilitate the interaction with water molecules, promoting an increase in the number of charge carriers within the host polymer.

On the other hand, Figure 11a,b also illustrate the impact of the addition of SiO_2, PVP, and both materials to the PVA-based ILGPE on the sensing mechanism. Firstly, Figure 11a displays how the composition of the sensing layer affects the R_s behaviour at different %RH levels. In particular, as %RH increases from 20 to 90%, the R_s values for PVA:$PVP_{0\%}$, SiO_2 PVA:$PVP_{0\%}$, PVA:$PVP_{25\%}$, and SiO_2 PVA:$PVP_{25\%}$ capacitors ranged from 34 Ω, 51 Ω, 52 Ω, and 148 Ω to 17 Ω, 19 Ω, 20 Ω, and 21 Ω, respectively. Notably, incorporating PVP and SiO_2 fillers leads to higher R_s values. Higher R_s values could suggest a hindrance to charge transport at lower humidity levels, indicating that charge carriers require more energy to be transported within the polymer electrolyte [12]. This effect is most pronounced at low humidity levels, where PVP and SiO_2 nanoparticles may attempt to retain the charge due to their hydrophilic properties, impeding the mobility of some charge carriers.

Secondly, Figure 11b shows the evolution of the CPE element with the humidity, depending on the composition of the electrolyte, through the Q_{CPE} parameter. At 20 %RH and 90 %RH, the PVA:$PVP_{0\%}$, SiO_2 PVA:$PVP_{0\%}$, PVA:$PVP_{25\%}$, and SiO_2 PVA:$PVP_{25\%}$ capacitors exhibit Q_{CPE} values range from 0.32 µF, 0.34 µF, 0.39 µF, and 0.34 µF to 0.39 µF, 0.45 µF, 0.45 µF, and 0.42 µF, respectively. This trend aligns with the abovementioned capacitance responses. The higher Q_{CPE} values may be ascribed to the amorphousness caused by adding PVP and SiO_2 nanoparticles. Additionally, the higher variations in Q_{CPE} values with increasing humidity can also be attributed to the expanded air-contact area on the surface of the sensing layer, enhancing the sensitivity of the devices. This hypothesis is consistent with the results from the surface topography analysis and the observed capacitance responses.

In summary, the results exhibit that adding PVP to a PVA-based gel polymer electrolyte leads to higher capacitance values in response to humidity variation (Figure 11b). This effect could be ascribed to PVP's amorphous nature and hydrophilic properties. Firstly, the excellent miscibility between PVP and PVA due to hydrogen bonding formation, coupled with the amorphousness of PVP, decreases the degree of crystallinity of the PVA matrix [37–39]. The reduction in the crystalline phases and increase in the amorphous phase enhance the ionic conductivity of the ILGPE [36], which can stimulate a capacitance Increase. Secondly, the amorphousness induced by PVP increases the number of available active sites per volume in the sensing layer. This, combined with the hydrophilic nature of PVP, may enhance the interaction between the sensing layer and the water molecules, leading to a higher charge density within the electrolyte layer and, hence, higher variations in the capacitance (Figure 12g,h).

Moreover, SiO_2 CILGPE capacitors show more significant changes in Q_{CPE} values than ILGPE-based capacitors. This effect can be mainly attributed to the active sites for ion exchange provided by SiO_2, which modifies the interaction of charge carriers with the sensing layer. As revealed in [20], embedding SiO_2 nanoparticles into a polymer matrix offers several beneficial effects, including an increased charge contribution owing to their hydrophilic nature and the creation of new effective charge transport pathways, among others. These effects provided by the presence of SiO_2 nanoparticles can contribute to the overall sensing mechanism. Incorporating nanoparticles in the hydrophilic polymer matrix enhances the available surface area for water interaction, increasing the amount of water molecules adhered to the sensing layer. This is achieved by introducing active sites that stimulate water adsorption [1,16], leading to improved binding of water molecules. Consequently, the probability of capturing water molecules increases, resulting in an augmented charge density within the humidity-sensitive layer.

Furthermore, as depicted in Figure 12f,h, the addition of silicon dioxide nanoparticles can introduce nanostructures that establish connections between different levels of the sensing layer. These silica structures can act as pathways for charge carriers and water molecules, facilitating their movement through the sensing layer. Consequently, these nanostructures can play a crucial role during desorption, providing a path for water molecules to traverse between the bottom and surface of the humidity-detection layer, assisting in the diffusion of water molecules into the environment and enhancing the water molecules' equilibrium. This structural effect can improve the responsiveness of the humidity sensing, leading to a reduction in the hysteresis parameter. This hypothesis is consistent with the results shown in the analysis of the hysteresis parameter.

3. Conclusions

In this work, different ionic liquid gel polymer electrolytes (ILGPEs) consisting of CH3COONH4, BmImBr, SiO_2 nanoparticles, and different concentrations of PVA and PVP have been fabricated and drop-casted over interdigitated aluminium electrodes to fabricate ILGPE-based and SiO_2 CILGPE-based capacitors. The capacitors have been characterised via the IS method in the humidity range of 20–90 %RH and modelled using a Cole–Cole model as an equivalent circuit. The results show that using PVP and SiO_2 nanoparticles enhances the sensitivity. Notably, SiO_2 CILGPE capacitors exhibit higher sensitivity and lower hysteresis values than their counterparts due to (i) an increased air-contact sensing area, (ii) an increase in the number of active sites per volume available in the sensing layer stimulated by the disruption of the crystalline structure of PVA and the hydroxyl groups in the nanoparticles, and (iii) additional charge transportation pathways, which boost the ion conduction and facilitating the water molecule diffusion, therefore revealing the dominant role of silica in the hysteresis. Furthermore, the coexistence of PVP and silica nanoparticles further enhances the hysteresis and sensitivity by combining their hydrophilicity and ability to increase the amorphousness of the PVA:$PVP_{0\%}$ ILGPE. This amalgamation provides additional active sites for water interaction and pathways, further increasing the interaction with water molecules and facilitating their diffusion.

Consequently, the results suggest that the SiO_2 PVA:$PVP_{25\%}$ CILGPE is the optimal electrolyte, exhibiting calibrated capacitance sensitivities around 2660 pF/%RH. These findings, coupled with its easy and fast fabrication process and environmentally friendly materials, outperform some reported capacitive humidity sensors in terms of sensitivity without compromising the hysteresis. Nevertheless, future research could investigate the optimisation of nanoparticle concentrations or the use of hydrophilic porous fillers, such as HNTs, to further improve the sensing properties of the optimal humidity-sensitive electrolyte.

4. Materials and Methods

4.1. Fabrication Process Steps

Figure 13 illustrates the fabrication process employed to produce the ILGPE and SiO_2 CILGPE capacitors. The fabrication process initiates with the creation of interdigitated aluminium electrodes with a total electrode area of 0.1962 cm^2 (step 1). The fabricated electrolytes are deposited onto the electrodes by a drop-casting method. The first ionic liquid gel polymer electrolyte is based on the composition shown in [32,46]. Its formulation contains polyvinyl alcohol (PVA), ammonium acetate (CH_3COONH_4), and 1-butyl-3-methylimidazolium bromide (BmImBr). The second ILGPE is also composed of polyvinylpyrrolidone (PVP), CH_3COONH_4, and BmImBr. Furthermore, monodisperse SiO_2 nanoparticles in a colloidal suspension are employed to create composite ILGPEs. The PVA and PVP serve as host polymers, ammonium acetate functions as salt and plasticiser, 1-butyl-3-methylimidazolium bromide acts as the ionic liquid, and silica nanobeads act as doping fillers.

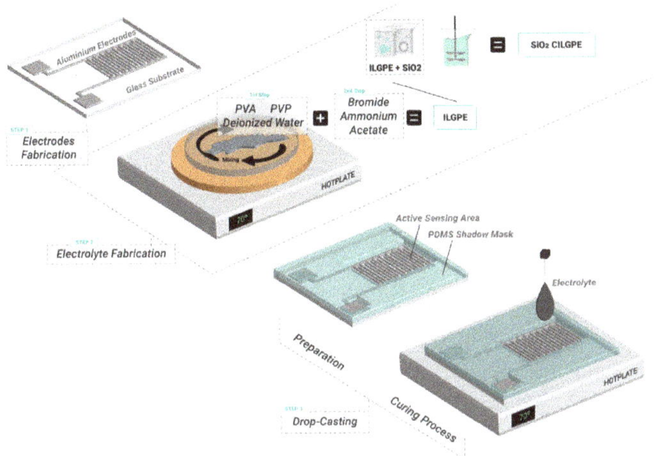

Figure 13. Fabrication method to manufacture ILGPE and CILGPE capacitors. Step 1: fabrication of aluminium interdigitated electrodes. Step 2: electrolyte fabrication. Step 3: electrolyte deposition by drop casting and curing at 70 °C.

The preparation of the PVA:PVP$_{0\%}$ and SiO_2 PVA:PVP$_{0\%}$ electrolytes follows the fabrication processes exhibited in [32,46], respectively. As shown in step 2 of Figure 10, the preparation method of ILGPEs and SiO_2 CILGPEs starts by dissolving the host polymer (Hp) in 10 mL of deionised water, followed by stirring the mixture for 10 min at 70 °C. Immediately, a blend of 420 mg of CH_3COONH_4 in a weight ratio of 70:30 for Hp:CH_3COONH_4 is added, and 4.24 g of 1-butyl-3-methylimidazolium bromide is incorporated into the Hp–CH_3COONH_4 mixture. The final blend is manually stirred and heated at 70 °C for 15 min. Afterwards, for the SiO_2 CILGPEs, a volume ratio of 50:50 is used for the Hp–CH_3COONH_4–BmImBr–H_2O:SiO_2 nanobeads blend. In particular, 1 mL of SiO_2 nanoparticles is mixed with the Hp–CH_3COONH_4–BmImBr aqueous mixture using an ultrasound bath for 10 min. Finally, as illustrated in the third step of Figure 10, the resulting gel polymer electrolytes are deposited onto the active electrode area using drop casting and subsequently cured at 70 °C for 35 min and 15 min, respectively. The resultant gel polymer electrolyte layers present a thickness between 300 and 600 μm and a length and width of 0.4 mm and 1 mm, respectively. Table 3 summarises the codification of the developed electrolyte-based capacitors and the composition of the electrolytes that compose them.

Table 3. Weight and volume ratios of the incorporated materials in the fabricated gel polymer electrolytes. The host polymer matrix is mixed with CH_3COONH_4 in a weight ratio of 70:30. BmImBr is added to the resulting mixture using a weight ratio of 75:25.

Identifier	Weight Ratio of Host Polymer (Hp) PVA:PVP	Volume Ratio of Hp–CH_3COONH_4–BmImBr–H_2O:SiO_2 Nanoparticles
PVA:PVP$_{0\%}$	100:0	100:0
SiO_2 PVA:PVP$_{0\%}$		50:50
PVA:PVP$_{25\%}$	75:25	100:0
SiO_2 PVA:PVP$_{25\%}$		50:50

4.2. Methodology

The electrical performance of the developed ILGPE- and SiO_2 CILGPE-based capacitors was measured using a HIOKI IM3590 (HIOKI, Dallas, TX, USA) impedance analyser. The IS measurements have been performed in the frequency range from 0.1 Hz to 200 kHz, applying a sinusoidal signal with an amplitude of 50 mV and without bias voltage.

In particular, the IS responses of three samples from different batches of PVA:$PVP_{0\%}$-ILGPE-, SiO_2 PVA:$PVP_{0\%}$-CILGPE-, PVA:$PVP_{25\%}$-ILGPE-, and SiO_2 PVA:$PVP_{25\%}$-CILGPE-based capacitors have been fabricated by the same abovementioned method and measured. The measurement setup and process follow similar configurations to those exhibited in [13]. First, the resultant capacitors were connected to a measurement board and placed inside a commercial environmental chamber at 25 °C. Coaxial cables connected the impedance analyser and the measurement board. The chamber's humidity was increased from 20 %RH to 90 %RH to obtain the adsorption responses. On the contrary, to obtain the desorption responses, the relative humidity was decreased to 20 %RH in steps of 10 %RH.

The IS measurements obtained from PVA:$PVP_{0\%}$, SiO_2 PVA:$PVP_{0\%}$, PVA:$PVP_{25\%}$, and SiO_2 PVA:$PVP_{25\%}$ capacitors at different relative humidity levels have been fitted to show how the humidity affects the impedance characteristics of the devices and their potential as humidity sensors.

Author Contributions: Conceptualisation, S.B. and M.C.M.; Methodology, S.B. and M.C.M.; Formal analysis, S.B., M.C.M. and M.D.-P.; Investigation, M.C.M.; Data curation, M.C.M.; Writing—original draft preparation, M.C.M., S.B., A.O. and M.D.-P.; Writing—review and editing, M.C.M., S.B., A.O. and M.D.-P.; Visualisation, M.C.M.; Supervision, S.B.; Project administration, M.C.M. and S.B.; Funding acquisition, S.B. and M.D.-P. All authors have read and agreed to the published version of the manuscript.

Funding: This work has been supported by the Secretariat for Universities and Research of the Ministry of Research and Universities, the European Social Fund, and the projects TED2021-131552B-C22 and PID2021-126719OB-C42.

Institutional Review Board Statement: Not applicable.

Informed Consent Statement: Not applicable.

Data Availability Statement: All data and materials are available on request from the corresponding author. The data are not publicly available due to ongoing research using a part of the data.

Acknowledgments: The conceptual figures have been partly designed and produced by Octavi Prats Reig.

Conflicts of Interest: The authors declare no conflicts of interest.

References

1. Arman Kuzubasoglu, B. Recent Studies on the Humidity Sensor: A Mini Review. *ACS Appl. Electron. Mater.* **2022**, *4*, 4797–4807. [CrossRef]
2. Qin, J.; Yin, L.J.; Hao, Y.N.; Zhong, S.L.; Zhang, D.L.; Bi, K.; Zhang, Y.X.; Zhao, Y.; Dang, Z.M. Flexible and Stretchable Capacitive Sensors with Different Microstructures. *Adv. Mater.* **2021**, *33*, 2008267. [CrossRef] [PubMed]
3. Blank, T.A.; Eksperiandova, L.P.; Belikov, K.N. Recent Trends of Ceramic Humidity Sensors Development: A Review. *Sens. Actuators B Chem.* **2016**, *228*, 416–442. [CrossRef]
4. Ku, C.-A.; Chung, C.-K.; Ku, C.-A.; Chung, C.-K. Advances in Humidity Nanosensors and Their Application: Review. *Sensors* **2023**, *23*, 2328. [CrossRef] [PubMed]
5. Rahman, S.A.; Khan, S.A.; Rehman, M.M.; Kim, W.Y. Highly Sensitive and Stable Humidity Sensor Based on the Bi-Layered PVA/Graphene Flower Composite Film. *Nanomaterials* **2022**, *12*, 1026. [CrossRef] [PubMed]
6. Yao, S.; Zhang, Q.; Ding, W.; Zhang, Y.; Zhang, J.; Wang, C. Capacitive Humidity Sensors Based on PVA/CAB Compound Film with Potassium Acetate-Enhanced Sensitivity. In Proceedings of the 2022 15th International Congress on Image and Signal Processing, BioMedical Engineering and Informatics, CISP-BMEI 2022, Beijing, China, 5–7 November 2022; Institute of Electrical and Electronics Engineers Inc.: Piscataway, NJ, USA.
7. Jianping, M.; Weiyuan, G.; Ziyan, H.; Xiaozong, D.; Hongyan, Z. High-Performance ZnO Humidity Sensor Synthesized by Coprecipitation with PVP as Surfactant for Human Respiration Detection. *Optoelectron. Lett.* **2023**, *19*, 1–7. [CrossRef]
8. Zhang, Z.; Chen, M.; Alem, S.; Tao, Y.; Chu, T.Y.; Xiao, G.; Ramful, C.; Griffin, R. Printed Flexible Capacitive Humidity Sensors for Field Application. *Sens. Actuators B Chem.* **2022**, *359*, 131620. [CrossRef]

9. Wang, M.; Zhang, D.; Yang, A.; Wang, D.; Zong, X. Fabrication of Polypyrrole/Graphene Oxide Hybrid Nanocomposite for Ultrasensitive Humidity Sensing with Unprecedented Sensitivity. *J. Mater. Sci. Mater. Electron.* **2019**, *30*, 4967–4976. [CrossRef]
10. Hammouche, H.; Achour, H.; Makhlouf, S.; Chaouchi, A.; Laghrouche, M. A Comparative Study of Capacitive Humidity Sensor Based on Keratin Film, Keratin/Graphene Oxide, and Keratin/Carbon Fibers. *Sens. Actuators A Phys.* **2021**, *329*, 112805. [CrossRef]
11. Li, X.; Ni, L.; Chen, N.; Liu, J.; Li, W.; Xian, Y. Enhanced Sensitivity of Humidity Sensor Using Nafion/Graphene Oxide Quantum Dot Nanocomposite. *Measurement* **2021**, *181*, 109566. [CrossRef]
12. Si, R.; Xie, X.; Li, T.; Zheng, J.; Cheng, C.; Huang, S.; Wang, C. TiO_2/(K,Na)NbO_3 Nanocomposite for Boosting Humidity-Sensing Performances. *ACS Sens.* **2020**, *5*, 1345–1353. [CrossRef]
13. Liu, M.-Q.; Wang, C.; Kim, N.-Y.; Corres, J.M.; Arregui, F.J. High-Sensitivity and Low-Hysteresis Porous MIM-Type Capacitive Humidity Sensor Using Functional Polymer Mixed with TiO2 Microparticles. *Sensors* **2017**, *17*, 284. [CrossRef]
14. Cheng, B.; Tian, B.; Xie, C.; Xiao, Y.; Lei, S. Highly Sensitive Humidity Sensor Based on Amorphous Al_2O_3 Nanotubes. *J. Mater. Chem.* **2011**, *21*, 1907–1912. [CrossRef]
15. Mallick, S.; Ahmad, Z.; Karwan, W.Q.; Rehaman, A.; Shakoor, R.A.; Touati, F.; Al-Muhtaseb, S.A. Effect of $BaTiO_3$ on the Sensing Properties of PVDF Composite-Based Capacitive Humidity Sensors. *Ceram. Int.* **2020**, *46*, 2949–2953. [CrossRef]
16. Qi, R.; Zhang, T.; Guan, X.; Dai, J.; Liu, S.; Zhao, H.; Fei, T. Capacitive Humidity Sensors Based on Mesoporous Silica and Poly(3,4-Ethylenedioxythiophene) Composites. *J. Colloid. Interface Sci.* **2020**, *565*, 592–600. [CrossRef]
17. Duan, Z.; Zhao, Q.; Wang, S.; Yuan, Z.; Zhang, Y.; Li, X.; Wu, Y.; Jiang, Y.; Tai, H. Novel Application of Attapulgite on High Performance and Low-Cost Humidity Sensors. *Sens. Actuators B Chem.* **2020**, *305*, 127534. [CrossRef]
18. Duan, Z.; Jiang, Y.; Zhao, Q.; Wang, S.; Yuan, Z.; Zhang, Y.; Liu, B.; Tai, H. Facile and Low-Cost Fabrication of a Humidity Sensor Using Naturally Available Sepiolite Nanofibers. *Nanotechnology* **2020**, *31*, 355501. [CrossRef] [PubMed]
19. Afify, A.S.; Dadkhah, M.; Tulliani, J.-M. Sol-Gel Synthesis of Iron-Doped Sepiolite as a Novel Humidity-Sensing Material. *Ceramics* **2022**, *5*, 575–592. [CrossRef]
20. Duan, Z.; Zhao, Q.; Wang, S.; Huang, Q.; Yuan, Z.; Zhang, Y.; Jiang, Y.; Tai, H. Halloysite Nanotubes: Natural, Environmental-Friendly and Low-Cost Nanomaterials for High-Performance Humidity Sensor. *Sens. Actuators B Chem.* **2020**, *317*, 128204. [CrossRef]
21. Yang, Y.; Chen, Y.; Leng, F.; Huang, L.; Wang, Z.; Tian, W. Recent Advances on Surface Modification of Halloysite Nanotubes for Multifunctional Applications. *Appl. Sci.* **2017**, *7*, 1215. [CrossRef]
22. Wu, Y.; Zhang, Y.; Ju, J.; Yan, H.; Huang, X.; Tan, Y. Polymers Advances in Halloysite Nanotubes-Polysaccharide Nanocomposite Preparation and Applications. *Polymers* **2019**, *11*, 987. [CrossRef] [PubMed]
23. Saleem, A.; Muhammad, N.; Ullah, Z.; Khan, A.S.; Rahim, A. Applications of Ionic Liquids in Sensors and Biosensors. In *Industrial Applications of Green Solvents*; Materials Research Forum LLC Location: Millersville, PA, USA, 2019; Volume 2, pp. 29–50.
24. Pei, Y.; Zhang, Y.; Ma, J.; Fan, M.; Zhang, S.; Wang, J. *Materials Today Nano*; Elsevier Ltd.: Amsterdam, The Netherlands, 2022; p. 100159.
25. Fernandes, L.C.; Correia, D.M.; Pereira, N.; Tubio, C.R.; Lanceros-Méndezméndez, S. Highly Sensitive Humidity Sensor Based on Ionic Liquid–Polymer Composites. *ACS Appl. Polym. Mater.* **2019**, *1*, 2723–2730. [CrossRef]
26. Esteves, C.; Palma, S.I.C.J.; Costa, H.M.A.; Alves, C.; Santos, G.M.C.; Ramou, E.; Carvalho, A.L.; Alves, V.; Roque, A.C.A. Tackling Humidity with Designer Ionic Liquid-Based Gas Sensing Soft Materials. *Adv. Mat.* **2022**, *34*, 2107205. [CrossRef]
27. Mishra, K.; Devi, N.; Siwal, S.S.; Zhang, Q.; Alsanie, W.F.; Scarpa, F.; Thakur, V.K. Ionic Liquid-Based Polymer Nanocomposites for Sensors, Energy, Biomedicine, and Environmental Applications: Roadmap to the Future. *Adv. Sci.* **2022**, *9*, 2202187. [CrossRef] [PubMed]
28. Wang, C.; He, J.; Zou, L.; Wang, H.; Wang, Y.; Li, Y.V. Plasticizing Effect of 1-Butyl-3-Methylimidazolium Bromide Content on Poly(Vinyl Alcohol). *Polym. Int.* **2022**, *72*, 500–507. [CrossRef]
29. Poonam, D.; Mahapure, S.W.; Gosavi, R.C. Aiyer Studies on PVP, PVA and Their NAg Composites Based Humidity Sensors. In Proceedings of the International Symposium on Physics and Technology of Sensors, Pune, India, 7–10 March 2015.
30. Yi, Y.; Chaio, M.; Mahmoud, K.A.; Wu, L.; Wang, B. Preparation and Characterization of PVA:PVP Conductive Hydrogels Formed by Freeze–Thaw Processes as a Promising Material for Sensor Applications. *J. Mater. Sci.* **2022**, *57*, 8029–8038. [CrossRef]
31. Véliz Noboa, B.M. *Novel Metal Insulator Metal Capacitors Based on Electrosprayed Colloidal Nanoparticles*; Universitat Politècnica de Catalunya: Barcelona, Spain, 2020.
32. Cedeño Mata, M.; Orpella, A.; Domínguez-Pumar, M.; Bermejo, S. Space-Charge Limited Ionic Conductivity Enhancement in Gel Polymer Electrolyte Capacitors by Embedding Nanoparticles. *Electrochim. Acta* **2021**, *393*, 138952. [CrossRef]
33. Hemalatha, P.; Jolly, B.; Zubair, A.; Shoaib, M.; Farid, T.; Noora, A.-T. Capacitive Type Humidity Sensor Based on PANI Decorated Cu–ZnS Porous Microspheres. *Talanta* **2020**, *219*, 121361. [CrossRef]
34. Chani, M.T.S.; Karimov, K.S.; Asiri, A.M. Impedimetric Humidity and Temperature Sensing Properties of the Graphene–Carbon Nanotubes–Silicone Adhesive Nanocomposite. *J. Mater. Sci: Mater. Electron.* **2019**, *30*, 6419–6429. [CrossRef]
35. Barsoukov, E.; Macdonald, J.R. *Impedance Spectroscopy: Theory, Experiment, and Applications*; Wiley: Hoboken, NJ, USA, 2018; ISBN 9781119074083.
36. Winie, T.; Arof, A.K.; Thomas, S. *Polymer Electrolytes: Characterization Techniques and Energy Applications*; Wiley: Hoboken, NJ, USA, 2019; ISBN 9783527342006.

37. Oikonomou, P.; Sanopoulou, M.; Papadokostaki, K.G. Blends of Poly(Vinyl Alcohol) and Poly(Vinyl Pyrrolidone): Interrelation between the Degree of Hydration and Thermal and Mechanical Properties. *Ind. Eng. Chem. Res.* **2021**, *60*, 14203–14212. [CrossRef]
38. Mondal, D.; Masud, M.; Mollick, R.; Bhowmick, B.; Maity, D.; Bain, M.K.; Rana, D.; Mukhopadhyay, A.; Dana, K.; Chattopadhyay, D. Effect of Poly(Vinyl Pyrrolidone) on the Morphology and Physical Properties of Poly(Vinyl Alcohol)/Sodium Montmorillonite Nanocomposite Films. *Prog. Nat. Sci. Mater. Int.* **2013**, *23*, 579–587. [CrossRef]
39. Phetdaphat, B. Super-Tough Biodegradable Poly/Vinyl Alcohol)/Poly(Vinyl Pyrrolidone) Blends Plasticized by Glycerol and Sorbitol. *J. Appl. Polym. Sci.* **2018**, *135*, 46406. [CrossRef]
40. Mir, M.A.; Shah, M.A.; Ganai, P.A. Nanoporous Anodic Alumina (NAA) Prepared in Different Electrolytes with Different Pore Sizes for Humidity Sensing. *J. Solid. State Electrochem.* **2020**, *24*, 1679–1686. [CrossRef]
41. Ganbold, E.; Sharma, P.K.; Kim, E.S.; Lee, D.N.; Kim, N.Y. Capacitive Humidity Sensor with a Rapid Response Time on a GO-Doped P(VDF-TrFE)/LiCl Composite for Noncontact Sensing Applications. *Chemosensors* **2023**, *11*, 122. [CrossRef]
42. Xia, J.; Wang, X.; Wang, X.; Majer-Baranyi, K.; Zhang, X. Hysteresis Dynamic Modeling and Analysis of Flexible Nano Silver-Polyvinyl Alcohol Humidity Sensor Based on the Microscopic Process and Langmuir-Fick Theory. *ACS Omega* **2022**, *7*, 14994–15004. [CrossRef] [PubMed]
43. Mondal, P.P.; Mahapatra, P.L.; Das, S.; Saha, D. Study on the Novel Capacitive Moisture Sensing Behaviour of Nickel Chromite Nanoparticle Based Thick Film. *Measurement* **2020**, *163*, 107992. [CrossRef]
44. Park, S.J.; Jeon, J.Y.; Ha, T.J. Wearable Humidity Sensors Based on Bar-Printed Poly(Ionic Liquid) for Real-Time Humidity Monitoring Systems. *Sens. Actuators B Chem.* **2022**, *354*, 131248. [CrossRef]
45. Sekulić, D.L.; Ivetić, T.B. Characterization of an Impedance-Type Humidity Sensor Based on Porous SnO2/TiO2 Composite Ceramics Modified with Molybdenum and Zinc. *Sensors* **2023**, *23*, 8261. [CrossRef]
46. Liew, C.W.; Ramesh, S.; Arof, A.K. Characterization of Ionic Liquid Added Poly(Vinyl Alcohol)-Based Proton Conducting Polymer Electrolytes and Electrochemical Studies on the Supercapacitors. *Int. J. Hydrogen Energy* **2015**, *40*, 852–862. [CrossRef]

gels

MDPI

Article

Fabrication and Performance Study of 3D-Printed Ceramic-in-Gel Polymer Electrolytes

Xiubing Yao [1,†], Wendong Qin [1,†], Qiankun Hun [1], Naiyao Mao [1], Junming Li [1,*], Xinghua Liang [1,*], Ying Long [2] and Yifeng Guo [3]

1 Guangxi Key Laboratory of Automobile Components and Vehicle Technology, Guangxi University of Science & Technology, Liuzhou 545006, China; 18300084899@163.com (X.Y.); qinwendong@163.com (W.Q.); 13383371921@163.com (Q.H.); maonaiyao@qmail.com (N.M.)
2 Liuzhou Institute of Technology, Liuzhou 545616, China; longying@lzhit.edu.cn
3 School of Mechanical Engineering, Chengdu University, Chengdu 610106, China; guobujia2000@163.com
* Correspondence: jmli@gxust.edu.cn (J.L.); lxh304@126.com (X.L.)
† These authors contributed equally to this work.

Abstract

Solid-state electrolytes (SSEs) have emerged as a promising solution for next-generation lithium-ion batteries due to their excellent safety and high energy density. However, their practical application is still hindered by critical challenges such as their low ionic conductivity and high interfacial resistance at room temperature. The innovative application of 3D printing in the field of electrochemistry, particularly in solid-state electrolytes, endows energy storage devices with attractive characteristics. In this study, ceramic-in-gel polymer electrolytes (GPEs) based on PVDF-HFP/PAN@LLZTO were fabricated using a direct ink writing (DIW) 3D printing technique. Under the optimal printing conditions (printing speed of 40 mm/s and fill density of 70%), the printed electrolyte exhibited a uniform and dense sponge-like porous structure, achieving a high ionic conductivity of 5.77×10^{-4} S·cm^{-1}, which effectively facilitated lithium-ion transport. A structural analysis indicated that the LLZTO fillers were uniformly dispersed within the polymer matrix, significantly enhancing the electrochemical stability of the electrolyte. When applied in a $LiFePO_4$ | GPEs | Li cell configuration, the electrolyte delivered excellent electrochemical performance, with high initial discharge capacities of 168 mAh·g^{-1} at 0.1 C and 166 mAh·g^{-1} at 0.2 C, and retained 92.8% of its capacity after 100 cycles at 0.2 C. This work demonstrates the great potential of 3D printing technology in fabricating high-performance GPEs. It provides a novel strategy for the structural design and industrial scalability of lithium-ion batteries.

Keywords: 3D printing; porous structure; lithium-ion batteries

1. Introduction

With the rapid development of electric vehicles, portable electronic devices, and renewable energy sources, the demand for energy storage devices with high energy density, strong safety, and long life cycles is increasing. Solid-state lithium-ion batteries (SSLIBs), which utilize solid-state electrolytes instead of traditional liquid electrolytes, offer improved thermal stability and safety and have attracted increasing attention as next-generation energy storage devices [1,2]. However, current all-solid-state batteries still face multiple challenges regarding their ionic conductivity, interfacial stability, and manufacturing processes, making large-scale commercialization difficult [3]. Gel polymer electrolytes (GPEs), combining the structural integrity of solid-state systems with the ionic conductivity of liquid-like phases, provide a promising intermediate solution. GPEs typically consist of a polymer matrix, lithium salt, and a small amount of plasticizer or solvent retained within the polymer network, forming a quasi-solid-state phase that enhances ion transport and electrode interfacial compatibility [4]. This class of materials maintains mechanical flexibility while exhibiting high ionic conductivity, making them attractive for safe and efficient lithium-ion batteries.

There are various preparation methods for gel electrolytes, such as solution casting [5], the Phase Inversion Method [6], and electrospinning technology [7]. Although these traditional processes are mature, they still have limitations in terms of uniformity, structure controllability, and interface design. With its flexible processing and precise structural control capabilities, 3D printing technology has gradually been applied to the fabrication of multifunctional and shape-customized gel polymer electrolytes [8].

Poly(vinylidene fluoride)-hexafluoropropylene (PVDF-HFP) is a copolymer based on crystalline -VDF and amorphous -HFP segments [9]. PVDF-HFP is widely studied due to its low crystallinity and glass transition temperature (−62 °C). Its repeating unit, -CH_2-CF_2-, has a strong electron-withdrawing group and exhibits high polarity [10]. PVDF-HFP also possesses excellent chemical stability, excellent mechanical properties, and a high dielectric constant [11]; its amorphous regions help improve ionic conductivity [12]. However, a single polymer system can no longer meet the growing demands for modern dielectric materials [13]. Polyacrylonitrile (PAN), due to its strong polar electron-withdrawing nitrile group, C≡N, has strong molecular interactions with lithium salts and thus exhibits high ionic conductivity [14]. It also performs well in terms of its thermal stability and interfacial oxidation [15]. Notably, PVDF-HFP has excellent ionic conductivity, high elongation at break, and good miscibility with other polymers [16]. A composite system of PVDF-HFP and PAN can thus effectively balance ionic transport and mechanical robustness. To further improve the performance of GPEs, ceramic fillers such as garnet-type $Li_{6.4}La_3Zr_{1.4}Ta_{0.6}O_{12}$ (LLZTO) are commonly introduced. LLZTO provides high intrinsic ionic conductivity and enhances the structural stability of the composite electrolyte [17]. The resulting composite structure, combining organic polymers with inorganic ceramic fillers, exhibits both gel-like ionic conduction pathways and solid-state mechanical strength.

In the early stage of applying 3D printing to polymer electrolytes, most studies focused on electrochemical performance. Aaron J. Blake [18] established a method for 3D printing porous electrolytes using a dry-phase conversion technique. These electrolytes demonstrated similar high-rate performance (e.g., 5 C) to commercial polyolefin separators but had better wettability and thermal stability. Meng Cheng [19] proposed a new approach to fabricate hybrid solid-state electrolytes using high-temperature direct ink writing (DIW) without solvent evaporation. This method improved both the capacity and rate performance. Later, Zixian Liu [20] developed slurries based on lithium aluminum titanium phosphate (LATP), poly(ethylene oxide) (PEO), and lithium bis(trifluoromethanesulfonyl)imide (LiTFSI), exhibiting favorable rheological properties for direct ink writing (DIW). The printed solid-state electrolytes, applied directly onto $LiFePO_4$ cathodes, achieved a high discharge capacity of 150 mAh g^{-1} at 0.5 C. Despite the promising electrochemical performance of LATP-based ceramics, their long-term cycling stability remains a challenge. Bo Chen [21] developed three LLZTO slurry formulations that enabled the printing of complex 3D battery structures with 150 μm resolution. These slurries had a high modulus, high yield stress, and clear shear-thinning behavior. Changgang Li [22] prepared PEO-LATP polymer electrolytes using 3D printing and introduced various lithium salts to stabilize the electrode/electrolyte interfaces, suppressing $LiNi_{0.8}Co_{0.1}Mn_{0.1}O_2$ (NCM811) degradation and dendrite formation. The printed solid-state batteries could cycle 4000 times at 2 C.

Currently, 3D printing research on solid-state electrolytes remains focused on semi-solid systems (e.g., PEO-based), with limited studies on the systematic control of process parameters or in situ solidification strategies in non-PEO systems. These gaps hinder the practical development of high-safety, all-climate solid-state batteries.

In this work, we used DIW to design and fabricate 3D-printed gel polymer electrolytes. PVDF-HFP/PAN@LLZTO inks were prepared, their rheology was measured, and the printing parameters were optimized. Physical and electrochemical properties were compared across different thicknesses. The porous structure provided continuous channels for lithium-ion transport. The 90 μm thick electrolyte showed the highest ionic conductivity and excellent electrochemical performance. This research provides a new method for enhancing GPEs' performance and exploring novel paths for all-solid-state battery development.

2. Results and Discussion

In specific applications, batteries often need to be customized to different geometries under different conditions. Three-dimensional printing provides an effective solution to this technical challenge by enabling the fabrication of gel polymer electrolytes with customized geometries through simple adjustments to the printing program. As shown in Figure 1a–c, under consistent process parameters, we printed circular, maze-shaped, convex, and square GPEs. Through 3D printing technology, efficient material utilization, customizable structural design, and functional integration can be achieved, providing a new perspective for improving gel polymer electrolyte performance.

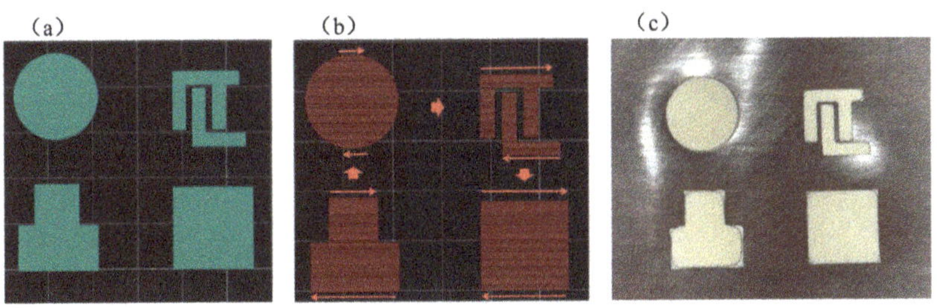

Figure 1. GPEs with customized shapes: (**a**) schematic; (**b**) printing routine; (**c**) macroscopic image.

Ink with good rheological properties is the key to being able to perform continuous 3D printing, and observing the flow state of the paste ink during extrusion is one of the indicators for evaluating the rheological properties [20]. The prepared ink was loaded into a syringe and the continuous extrusion ability of the ink was observed, as shown in Figure 2a. Under a pressure of 0.06 MPa, continuous ink extrusion could be observed within 6 s. The viscosity of the paste has a great influence on the rheology within the coating gap, as well as the quality of the wet film, and with an increase in the shear rate, the electrolyte ink exhibits shear-thinning characteristics. During the printing process, the shear-thinning property favors the continuous extrusion of the slurry when the coating gap is subjected to high shear, while the near-zero shear viscosity, which is elevated when the shear is very low, ensures good stability of the shape of the wet film aggregate during the drying process [23]. Figure 2b shows the rheological properties of the slurry, which can be seen to exhibit shear thinning, which is essential for successful printing. To determine the optimal process parameters, nine sets of experiments were conducted with the printing speed and infill density as variables. The printing speed was set to 20, 40, or 60 mm/s, and the infill density was selected as 50%, 70%, or 90%. The 3D printing process was then optimized based on the ionic conductivity results, as summarized in Table 1. As shown in Figure 2c, the bulk resistance obtained from Nyquist plots indicates that the optimal printing parameters are a speed of 40 mm/s and a filling density of 70%, corresponding to the highest ionic conductivity (Sample No. 5, 5.77×10^{-4} S·cm^{-1}) calculated using Equation (1).

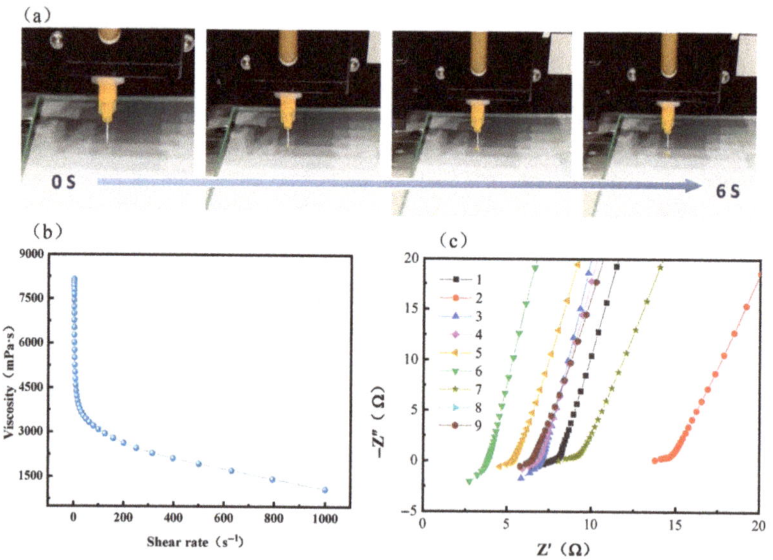

Figure 2. (**a**) Extrusion of slurry from syringe; (**b**) apparent viscosity of slurry as function of shear rate value; (**c**) Nyquist plots of different process parameters.

Table 1. Experiments with different parameter settings.

Number	Printing Speed (mm/s)	Filling Density (%)	Ionic Conductivity (S·cm^{-1})
1	20	50	2.9×10^{-4}
2	20	70	1.98×10^{-4}
3	20	90	3.18×10^{-4}
4	40	50	2.8×10^{-4}
5	40	70	5.77×10^{-4}
6	40	90	2.93×10^{-4}
7	60	50	1.09×10^{-4}
8	60	70	9.41×10^{-5}
9	60	90	1.47×10^{-4}

In order to investigate the effect of different thicknesses on the electrolyte, three different thicknesses of electrolyte films, 50 (±10) μm, 90 (±10) μm, and 130 (±10) μm, were prepared by controlling the air pressure during printing. As can be seen from Figure 3a–c, the prepared polymer electrolyte films were white in color with a flat and smooth surface; could recover their original state when arbitrarily bent, coiled, or folded; and had good plasticity and flexibility. To evaluate the flame-retardant properties of the GPEs, vertical combustion tests were conducted. As shown in Figure 3b,c, GPE_50, with its loosely porous internal structure, readily ignited upon exposure to a flame. Once the flame was removed, it underwent significant shrinkage and deformation. In contrast, as depicted in Figure 3e,f,h,i, GPE_90 and GPE_130, due to their relatively denser porous structures and greater thicknesses, exhibited reduced flammability when exposed to a flame. Moreover, they retained more intact morphologies after the flame was removed, demonstrating improved flame-retardant properties.

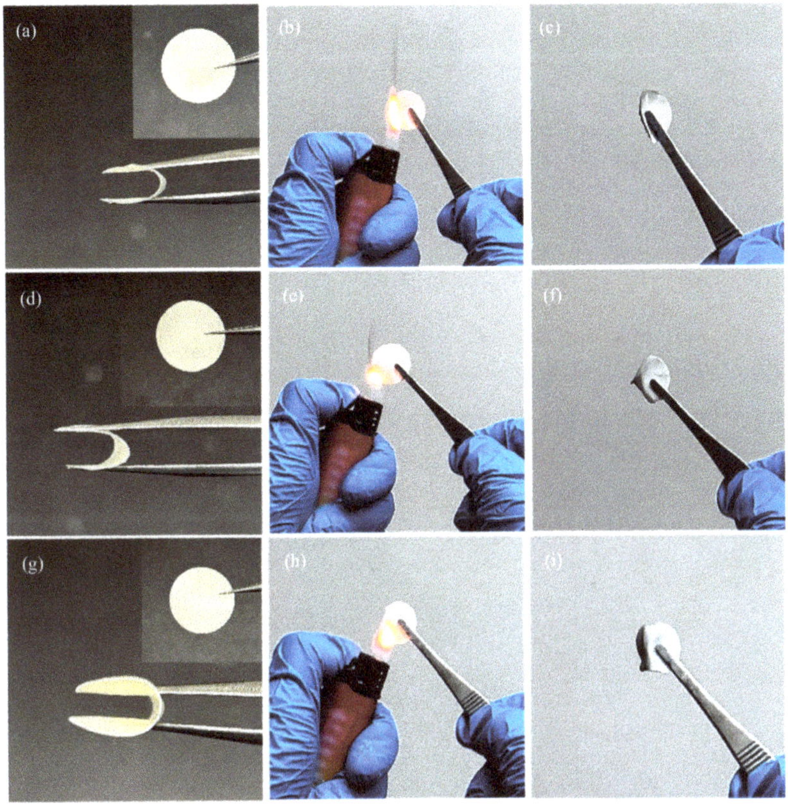

Figure 3. (**a–c**) Front and folded views, ignition moment, and morphology after flame removal of GPE_50; (**d–f**) front and folded views, ignition moment, and morphology after flame removal of GPE_90; (**g–i**) front and folded views, ignition moment, and morphology after flame removal of GPE_130.

The structural characteristics of electrolyte membranes with different thicknesses were investigated, as shown in Figure 4. All samples exhibited uniform sponge-like porous structures, which were attributed to the breath figure method. During the natural drying process of the freshly printed wet membrane in ambient air, solvent evaporation leads to a localized temperature drop, causing water droplets to condense on the surface. Polymer chains accumulate around these droplets, forming polymer shells. As drying progresses, the water droplets undergo orderly rearrangement under the influence of surface capillary forces, eventually organizing into regular hexagonal arrays, leaving behind a porous structure after evaporation [24]. This porous morphology is advantageous for constructing continuous lithium-ion transport channels. In particular, the presence of larger pores enhances the interfacial contact area between the electrolyte and electrode, facilitating fast ion diffusion and efficient charge transfer kinetics [25]. As shown in Figure 4a,b, GPE_50 exhibited a fluffy and uniform sponge-like porous structure. The Energy Dispersive X-ray Spectroscopy (EDS) mapping in Figure 4c confirmed the homogeneous distribution of F, La, Ta, and Zr elements without visible aggregation, suggesting good dispersion of LLZTO fillers within the polymer matrix. Figure 4d,e reveal that the GPE_90 membrane presented a denser sponge-like structure, while its corresponding EDS mapping (Figure 4f) further confirmed uniform elemental distribution, again suggesting effective filler dispersion. Figure 4g,h demonstrate that GPE_130 exhibited a dense sponge-like structure; however, Figure 4i reveals partial agglomeration of La elements, which may have resulted from excessively high printing pressure disrupting the ink's homogeneity and causing filler aggregation. Overall, GPE_90, fabricated under optimal printing pressure, demonstrated balanced porosity and high compactness, contributing to efficient and uniform lithium-ion transport while maintaining strong mechanical integrity. In addition, the small and uniform particle size of LLZTO fillers facilitates their homogeneous dispersion in the polymer matrix, which is beneficial for forming continuous Li^+ transport pathways and improving electrochemical performance [26]. The uniform distribution of fillers is critical for enhancing both the electrochemical performance and thermal stability of the electrolyte [27]. Notably, Figure 4f shows that the F element was more evenly distributed and more broadly covered in GPE_90, suggesting that PVDF-HFP was better dispersed in this sample, leading to more effective integration between the polymer and ceramic fillers. This further highlights the morphological and structural advantages of the GPE_90 membrane.

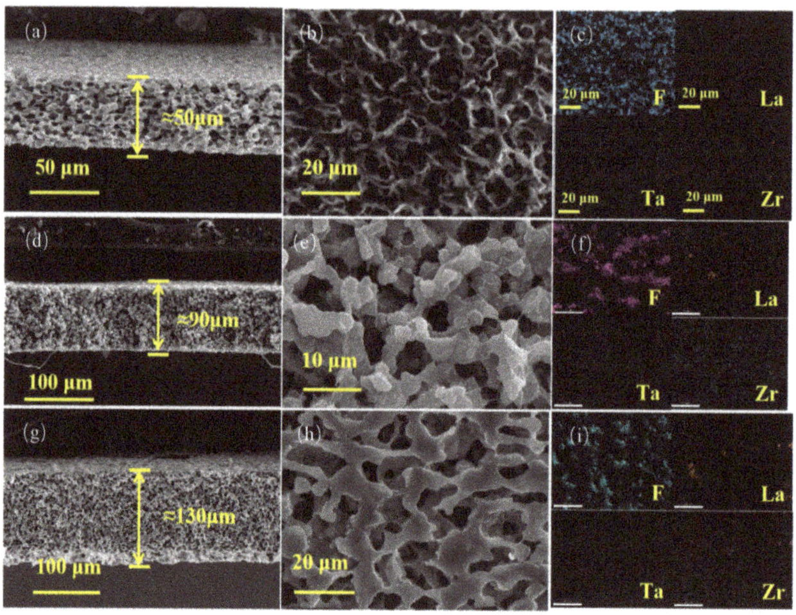

Figure 4. (**a–c**) Cross-sectional SEM, surface SEM, and EDS maps of GPE_50; (**d–f**) cross-sectional SEM, surface SEM, and EDS maps of GPE_90; (**g–i**) cross-sectional SEM, surface SEM, and EDS maps of GPE_130.

To further illustrate the amorphous nature of the GPEs, an X-ray diffraction (XRD) analysis was performed. Figure 5a shows the XRD patterns of LLZTO, PAN, PVDF-HFP, GPE_50, GPE_90, and GPE_130. The results show that the characteristic peaks of LLZTO remained clearly observable in the electrolyte, and the main diffraction peaks of pure PVDF-HFP were observed at $2\theta = 18.2°$, $20.0°$, and $26.6°$, which correspond to the planes of semicrystalline polymers (100), (020), and (110), respectively. After blending, the characteristic peaks of PVDF-HFP were weakened, indicating that the PVDF-HFP had reduced crystallinity, which may have resulted from molecular interactions with PAN and the lithium salt. A slight shift of the (020) diffraction peak of PVDF-HFP to lower 2θ angles was observed after blending, which can be attributed to the interaction between PVDF-HFP and PAN chain segments, as well as the Lewis acid–base interaction between LLZTO and the polymer matrix [28]. The diffraction peaks of LLZTO stabilized in the electrolyte, and the crystal structure of LLZTO did not change, which indicates that LLZTO retained its structural integrity and electrochemical compatibility within the electrolyte film.

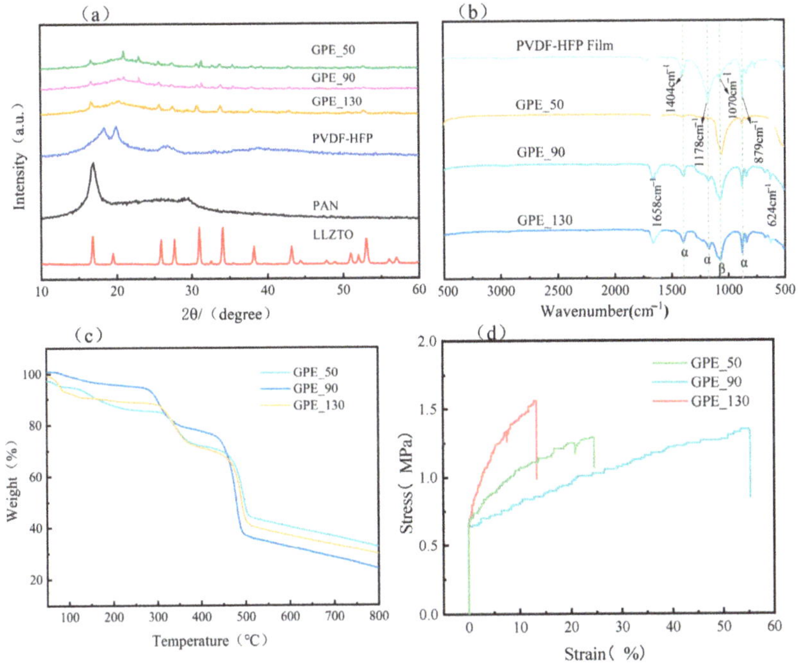

Figure 5. (**a**) XRD patterns of GPEs with different thicknesses; (**b**) FTIR spectra of the pure PVDF-HFP membrane and GPEs with different thicknesses; (**c**) thermogravimetric analysis of GPE membranes; (**d**) stress–strain curves of GPEs with different thicknesses.

FTIR spectroscopy was employed to investigate the molecular interactions among the polymer matrix, $LiClO_4$, and LLZTO in the GPEs, as shown in Figure 5b. The peak at 1658 cm^{-1} corresponds to C=O stretching in PAN, and the peak at 624 cm^{-1} corresponds to the presence of ClO_4^- anions in the complexes, indicating coordination between the polymer and lithium salt [29]. In comparison with pure PVDF-HFP membranes, the α-phase peaks at 1404 cm^{-1} (C-H bending in CH_2), 1178 cm^{-1} (-CF_2 symmetric stretching) [30], and 879 cm^{-1} (C-C backbone vibration) were reduced in intensity in each GPE, while the β-phase (1070 cm^{-1}, C-C stretching) [31] was enhanced. This could be attributed to the LLZTO filler in the GPEs promoting the enhancement of the β-phase, which facilitates the dissociation and migration of the lithium salt ions and improves the ionic conductivity. The decrease in the peak intensities at 1178 cm^{-1} and 1404 cm^{-1} suggests the dehydrochlorination of the PVDF-HFP-HFP strands, which could be attributed to the fact that the LLZTO filler plays the role of a Lewis base, inducing partial chemical dehydrofluorination, thus enhancing the interactions among the polymer matrix, lithium salt, and LLZTO particles [32].

The TG curves of the GPEs (Figure 5c) show three main weight loss stages. Below 300 °C is the decomposition of the residual solvent in the electrolyte and water absorbed from the air; from 300 °C is the second stage of thermal decomposition of PAN; and from 400 °C is the third stage, the decomposition of PVDF-HFP. It was previously reported that pure PVDF-HFP can remain stable without decomposition at 450 °C, while the GPEs slightly decomposed before reaching this temperature. This may be due to the addition of $LiClO_4$ and LLZTO reducing their crystallinity [25], which is the same as the XRD results. As the temperature increased, at 800 °C, GPE_50 had the lowest residue (24%), while GPE_90 had the highest residue (32%). This may be attributed to the dense porous structure within GPE_90, which enhances the thermal stability of the electrolyte [33]. In a sense, the thermodynamic properties of the prepared gel polymer electrolyte membranes can meet the requirements of most solid-state lithium-metal battery applications.

The mechanical stability of the GPEs is a pivotal factor in maintaining the structural stability of a lithium-ion battery during cycling. As depicted in Figure 5d, GPE_130 exhibited the highest tensile strength of 1.5 MPa with an elongation at break of 15%, while GPE_90 demonstrated a moderate tensile strength of 1.2 MPa along with a significantly higher elongation of 55%. This mechanical behavior plays a crucial role in suppressing lithium dendrite growth during cycling.

Ionic conductivity is an important indicator of electrochemical performance, and the ionic conductivity associated with different printing parameters is shown in Figure 6a. The bulk resistances of GPE_50, GPE_90, and GPE_130 were 6.877, 5.988, and 10.37 Ω, respectively, and according to Equation (1), the ionic conductivities were calculated to be 3.62×10^{-4}, 7.48×10^{-4}, and 6.24×10^{-4} S cm^{-1} (Figure 6b). These values demonstrate that the obtained GPEs, particularly GPE_90, exhibited relatively high ionic conductivity. For comparison, previous studies have reported ionic conductivity values of 4.24×10^{-4} S cm^{-1} and 7.4×10^{-4} S cm^{-1} for PVDF-HFP-based gel electrolytes under similar conditions [34,35], confirming the validity of our results. The enhanced performance of GPE_90 can be attributed to the homogeneous distribution of LLZTO fillers within the polymer matrix, which facilitates the formation of continuous lithium-ion conductive pathways [36]. In addition, the moderate printing pressure used in preparing GPE_90 results in a well-developed porous structure, which further supports efficient ion transport. Therefore, GPE_90 achieved the highest ionic conductivity among the tested samples. The equivalent circuit model used for fitting the impedance spectra included R1 representing the electrolyte resistance, R2 and R3 as the electrolyte–electrode interface resistances, and CE1 and CE2 as constant phase elements accounting for the non-ideal capacitive behavior. The simulated impedance response was directly calculated using ZView 3.1 software, and the good agreement between the experimental and simulated curves supports the validity of the fitted model. The Bode plots (Figure 6b–d) further confirmed that GPE_90 had low impedance and the highest phase angle across the frequency range, suggesting superior ion transport and interfacial behavior.

Electrochemical stability is the basis for matching high-voltage electrodes. Linear scanning voltammetric curves were used to evaluate the oxidative stability of the electrolyte films. The ionic conductivities associated with different printing parameters are shown in Figure 7c. GPE_90 had a higher decomposition voltage (4.6 V), and GPE_50 and GPE_130 had similar electrochemical windows (4.5 V), which shows that there was no significant difference among the 3D-printed electrolytes. Further, there is a stabilization window below 4.5 V vs. Li/Li^+, which suggests that they are able to match Li-FePhosphate batteries, and oxidative decomposition does not occur.

The multiplicity performance of $LiFePO_4$ | GPE | Li batteries at room temperature (25 °C) is shown in Figure 7c. It can be seen that the prepared GPEs all achieved a high specific capacity of discharge (above 160 mA·g^{-1}) at low multiplicity, demonstrating good multiplicity performance. GPE_90 exhibited a higher specific capacity of discharge due to its higher ionic conductivity (the initial specific capacity of discharge can be up to 168 mAh·g^{-1}), and GPE_130, due to its excessive thickness, led to a cell performance decrease, discharging the least specific capacity at low multiplicity. However, with an increase in current density, the growth of lithium dendrites was promoted, the internal side reaction of GPE_90 increased, and the battery exhibited unstable charging and discharging states with fast decay. Meanwhile, the GPE_50 discharge specific capacity decreased relatively slowly, which may be due to the lesser thickness shortening the ion transport path and reducing the ion transport resistance. Table 2 shows a performance comparison between the GPE films in this study and the existing solid-state electrolytes. It can be seen that the electrochemical performance of 3D printing-based solid-state electrolytes still needs to be further improved.

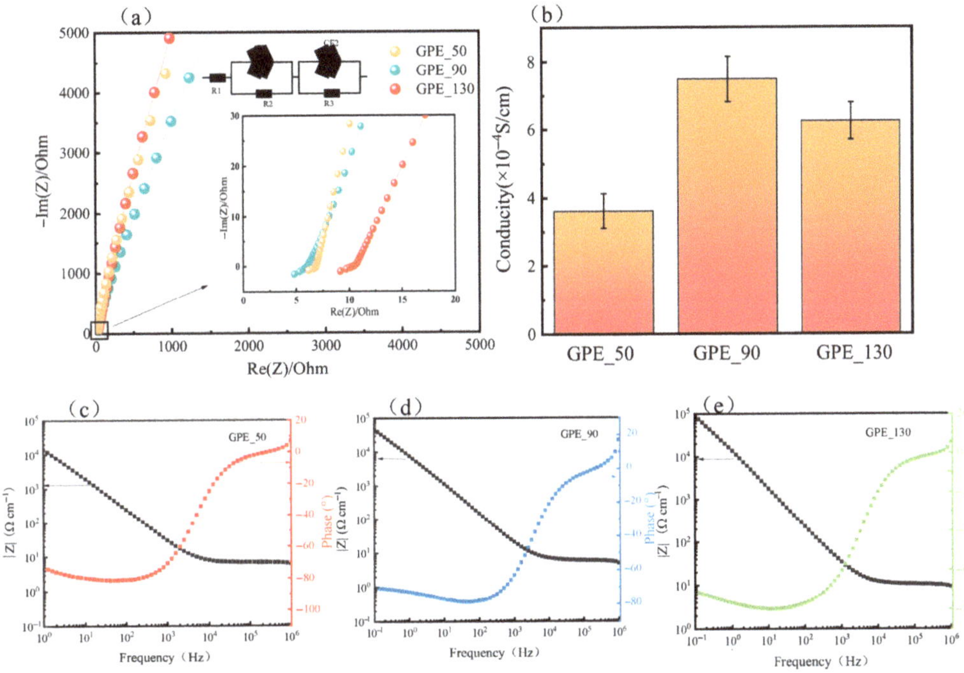

Figure 6. Electrochemical impedance spectroscopy (EIS) results for different GPE samples: (**a**) Nyquist plots for various GPEs; (**b**) ionic conductivity of GPE_90, with error bars indicating standard deviations; (**c**) Bode plot for GPE_50; (**d**) Bode plot for GPE_130; (**e**) Bode plot for GPE_90.

The GPEs with different thicknesses were cycled for 100 cycles at 0.2 C magnification, and the results are shown in Figure 7c,e. The results show that GPE_90 was chemically stable at low magnification, and its capacity was maintained at 152.8 mAh g^{-1} after 100 cycles, with a capacity retention rate of 92.8%. The capacity of GPE_50 and GPE_130 decayed faster, with capacity retention rates of 56.7% and 61.1%, respectively, after 100 cycles. The side reaction between residual DMF and lithium metal during the cycling process resulted in a faster capacity decay of the battery. The thicker GPE_130 showed significant capacity decay after 30 cycles, which could be attributed to the fact that the thicker electrolyte reduced the risk of lithium dendrite puncture, but due to its internal interlayer structural defects, it led to the capacity decay of the battery. On the other hand, GPE_90 exhibited higher discharge capacity and lower capacity decay with stronger cycling stability and dendrite resistance due to the moderate printing air pressure, which formed a good internal structure and balance between continuous Li^+ channels and dendrite resistance. The charging/discharging curves (Figure 7d) under different cycles corresponding to the performance of the GPE_90 cell at 0.2 C at room temperature (25 °C) show that the voltage plateaus in the first 50 cycles tended to overlap (the polarization voltage was 0.1 V) with small differences, indicating the reversibility and stability of GPE_90 in long-time operation [37]. The polarization voltage of charge/discharge reached 0.2 V at the end of the 100-lap cycle, which may be due to the deterioration of the electrode–electrolyte interface contact due to the volume change and interfacial side reaction during the cycling process, which increases the interfacial resistance and leads to an increase in the polarization voltage. In addition, lithium dendrite formation is a critical challenge impacting the long-term cycling stability and safety of lithium-metal batteries. The enhanced cycling performance and dendrite suppression observed in GPE_90 can be attributed to its optimized internal microstructure achieved through the 3D printing process. This structure promotes uniform Li^+ ion flux, effectively inhibiting the nucleation and growth of lithium dendrites, thereby reducing the risk of short circuits and capacity fading under moderate current densities. However, at elevated current densities, the accelerated growth of lithium dendrites and intensified side reactions contribute to the observed faster capacity decay [38,39]. Figure 8 demonstrates a comparison of the discharge performance of different processes and material systems at 0.1 C multiplication.

Table 2. Performance comparison of existing solid-state electrolytes.

SSEs	Method	0.1 C (mAh g^{-1})	0.2 C (mAh g^{-1})	0.5 C (mAh g^{-1})	After Cycles (mAh g^{-1})	Ref.
PPLP10%	Solution casting	171.20 (25 °C)	\	167.20 (25 °C)	157.30, 97.8% (1 C, 100)	[40]
PAP/PEP	Solution casting	168.00 (25 °C)	150.00 (25 °C)	120.40 (25 °C)	140.00, 93.3% (0.2 C, 100)	[29]
PHLP	Solution casting	148.52 (25 °C)	141.74 (25 °C)	118.68 (25 °C)	123.16, 88.29% (0.33 C, 200)	[41]
LLZO/LIC LISE	In situ sintering	167.20 (60 °C)	165.50 (60 °C)	\	144.20, 89.0% (0.5 C, 200)	[42]
PVDF-HFP/SBA-15 (10%)	3D printing	139.00 (25 °C)	\	\	\	[43]
LiPF6 Gel10	3D printing	\	145.00 (25 °C)	144 (25 °C)	117.00, 96% (1 C, 150)	[44]
SPE-SD-40 °C	3D printing	100.00 (40 °C)	40.00 (40 °C)	\	\	[45]
GPE-CQDs	3D printing	79.00 (25 °C)	78.00 (25 °C)	74.00 (25 °C)	74.00, 92% (0.2 C, 200)	[46]
H-ionogel-LLZO	3D printing	160.00 (50 °C)	153.00 (50 °C)	148.00 (50 °C)	125.00, 88.1% (0.5 C, 200)	[47]
This work (GPE_90)	3D printing	168.10 (25 °C)	166.90 (25 °C)	147.00 (25 °C)	152.80, 92.8% (0.2 C, 100)	This Work

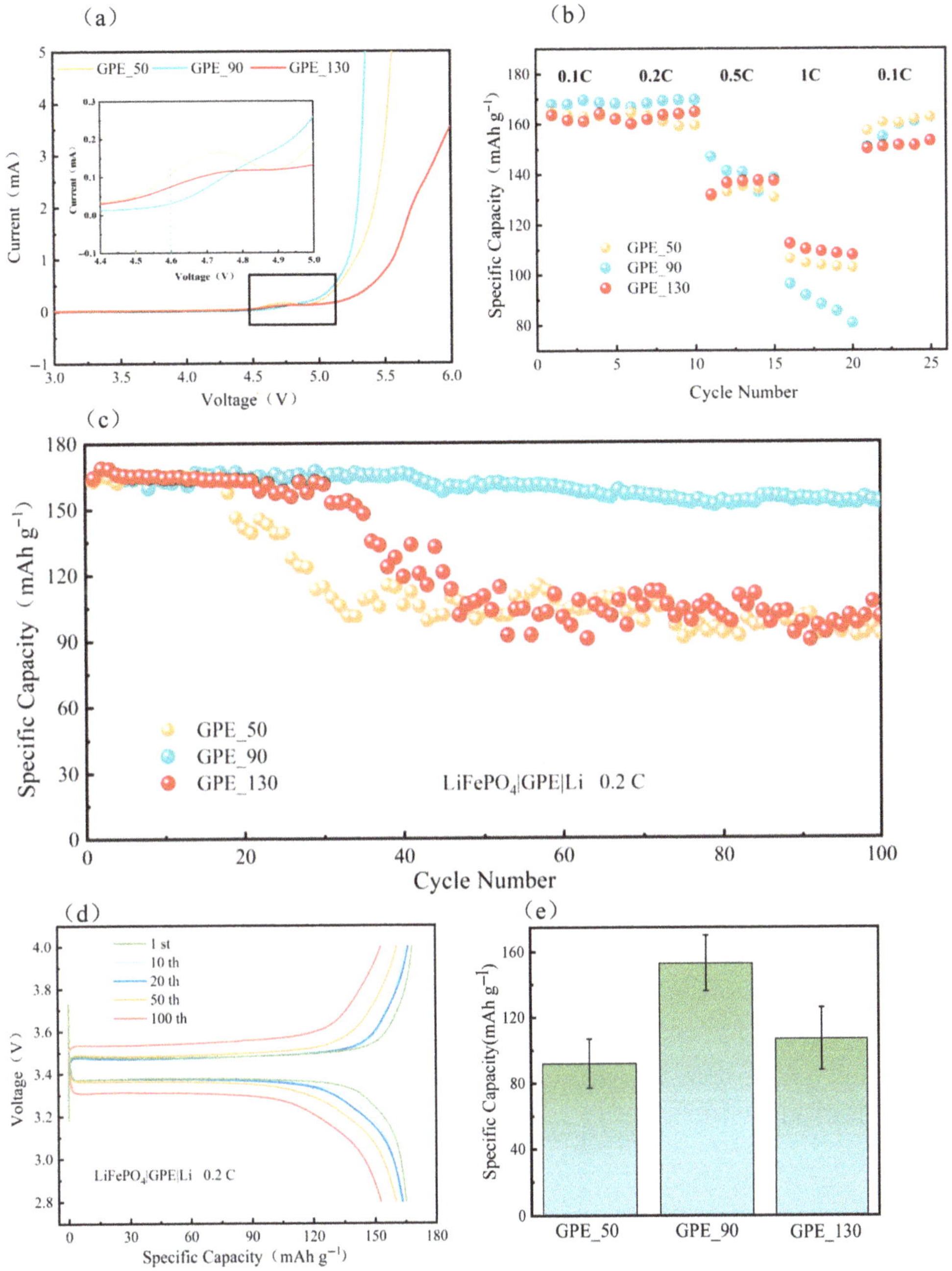

Figure 7. (**a**) Voltammetric linear scan curves of solid electrolytes with different thicknesses; (**b**) GPEs' multiplicity performance for different thicknesses; (**c**) cycling plots of GPEs with different thicknesses; (**d**) charge/discharge curves of GPE_90 with different numbers of cycling turns; (**e**) residual capacity after cycling, with error bars indicating standard deviations.

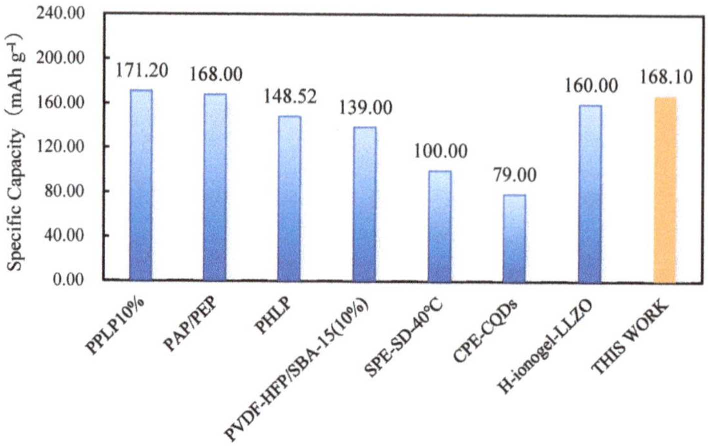

Figure 8. Comparison of discharge performance of existing solid-state electrolytes at 0.1 C multiplication rate.

3. Conclusions

In this study, PVDF-HFP/PAN@LLZTO gel polymer electrolytes were successfully prepared by 3D printing technology, and their rheological properties, physical properties, and electrochemical properties were systematically explored. It was shown that the optimized paste formulation and rheological properties ensured the controllability and stability of the 3D printing process, which enabled continuous extrusion and high-precision molding. The shear-thinning property enabled the electrolyte slurry to have good rheological behavior at high shear rates, while the appropriate near-zero shear viscosity ensured the stability of the wet film morphology during the drying process. The optimized 3D printing process parameters (printing speed of 40 mm/s, filling density of 70%) significantly enhanced the ionic conductivity of the solid electrolytes, among which the GPE_90 sample exhibited the highest ionic conductivity (7.48×10^{-4} S cm^{-1}) and lower interfacial impedance. The SEM results showed that the electrolytes with different thicknesses exhibited a uniform sponge-like porous structure, and the LLZTO fillers were uniformly distributed in the polymer matrix, improving the lithium-ion transport efficiency. The thermogravimetric (TG) analysis results further verified the thermal stability of the GPEs, which can meet the application requirements of lithium-metal batteries. Additionally, the printed membranes exhibited adequate mechanical strength to meet the basic handling and assembly requirements of quasi-solid-state batteries. The GPEs had a wide electrochemical window (4.6 V), good multiplicity performance (0.1 C@168 mAh·g^{-1},0.2 C@166 mAh·g^{-1}), and good cycling performance (92.8% capacity retention after 100 cycles). Furthermore, considering its industrial application potential, the DIW 3D printing process used here offers distinct advantages in fabricating customized electrolyte membranes with precise control over their structure and composition. Scaling this process for industrial manufacturing will require further optimization of the printing speed, slurry formulation, and post-processing steps to ensure reproducibility and throughput, which will be a focus of subsequent research efforts. In summary, this study demonstrated the feasibility and advantages of 3D printing technology in the preparation of customized GPEs, which provides new ideas and technical support for the development of high-performance quasi-solid-state lithium-metal batteries.

4. Materials and Methods

4.1. Materials

PVDF, PVDF-HFP (Mn = 600,000, Arkema, Colombes, France), poly(acrylonitrile) (PAN, Mw = 85,000, Shanghai, China), N,N-Dimethylformamide (DMF, ≥99.9%), lithium perchlorate ($LiClO_4$) (99.99%, Aladdin Chemical Co., Shanghai, China), N-methyl pyrrolidone (NMP, ≥99.9%), Super-P (≥99.5%), and $LiFePO_4$ (LFP, ≥99.5%) were all obtained from Krud in Shanghai, China. Lithium lanthanum zirconium tantalum oxide (LLZTO, average particle size ~500 nm) was also obtained from Krud.

4.2. Preparation of Gel Polymer Electrolytes (GPEs)

Preparation of ink: A certain amount of PVDF-HFP was weighed and dissolved in DMF on a magnetic stirrer at 45 °C. In order to obtain well-dispersed printable ink, PAN powder was mixed with the PVDF-HFP at a weight ratio of 7:3, and LLZTO (10 wt% relative to the total polymer mass) was thoroughly ground with the polymer blend using an agate mortar and pestle. After the PVDF-HFP was dissolved, the ground PAN and LLZTO powders were added. The polymer matrix and filler were stirred well, then $LiClO_4$ (30% by mass to polymer) was added, stirred well for 4 h, and ultrasonicated for 30 min for defoaming.

Preparation of GPEs: The preparation process of the composite polymer electrolyte is shown in Figure 9. Once the ink was prepared, it was loaded into a syringe equipped with an extrusion nozzle tip (inner diameter of 0.34 mm), and a 3D printer with pneumatically driven extrusion was used to perform the DIW process on quartz glass. In order to obtain the best printing parameters, nine sets of experiments were conducted to optimize the printing parameters, including the speed and filling density. The printed composite polymer electrolytes were placed in a drying oven and dried for 3 h, after which they were cut into 19 mm discs and placed in a glove box (H_2O < 0.01 ppm, O_2 < 0.01 ppm) to rest for 24 h before use.

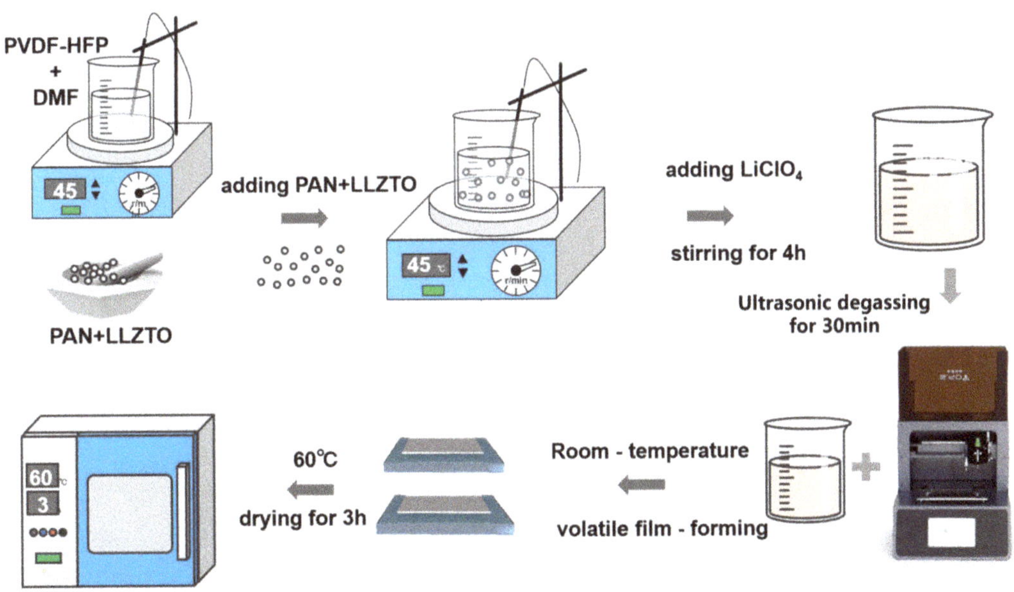

Figure 9. Preparation of gel polymer electrolyte membrane.

Gel polymer electrolyte films with controlled nominal thicknesses of 50 ± 10 μm, 90 ± 10 μm, and 130 ± 10 μm (systematically designated GPE_50, GPE_90, and GPE_130) were fabricated through precise adjustment of the printing pressure while maintaining all other processing parameters consistent with the reference methodology.

4.3. Assembling the Battery

$LiFePO_4$ (LFP) was used as the cathode and lithium metal as the anode for cell assembly. All components, including LFP electrodes, lithium foil, stainless steel spacers, and GPE membranes, were dried at 60 °C for 12 h prior to use. During assembly, approximately 10 μL of liquid electrolyte (1 M $LiPF_6$ in EC/DMC/EMC = 1:1:1 Vol% with 10% FEC 2% VC) was drop-cast onto the surface of the GPEs to promote polymer swelling and improve interfacial contact. The assembled CR2025 coin cells were rested for 12 h before subsequent electrochemical characterization.

4.4. Material Characterization

Rheological properties critical to the printing performance of the pastes were measured using a rotational rheometer (MCR302, Anton Paar, Graz, Austria) equipped with a 27 mm diameter cylinder. The viscosity of the slurries was tested within a shear rate range of 0.1 to 1000 s^{-1} at 25 °C to evaluate their shear-thinning behavior and printability. The crystalline structures of the dried GPE membranes and raw materials (PVDF-HFP, PAN, and LLZTO) were analyzed by X-ray diffraction (XRD; DX-2700, Dandong, China) with Cu Kα radiation at 40 kV and 30 mA, using a scan rate of 0.2°/s across a 2θ range of 10–90°. The surface and cross-sectional morphologies of GPE_50, GPE_90, and GPE_130 were observed and analyzed using scanning electron microscopy (SEM; Phenom Spectrum G2, Shanghai, China) after the samples were fractured in liquid nitrogen and sputter-coated with gold. Elemental mapping was performed using energy-dispersive X-ray spectroscopy (EDS; Phenom Spectrum G2, Shanghai, China) to assess the dispersion of ceramic particles. Fourier transform infrared (FTIR; Spectrum 100, PerkinElmer, Shelton, CT, USA) spectroscopy was used to determine the chemical bonding of GPE_50, GPE_90, and GPE_130, in the range of 500–3500 cm^{-1} using the ATR mode. The thermal stability of the GPEs was investigated using thermogravimetric analysis (TGA; METTLER TOLEDO DSC 3+) in a nitrogen atmosphere with ceramic crucibles, over a temperature range of 50 °C to 800 °C at a heating rate of 10 °C min^{-1}.

4.5. Electrochemical Properties

The GPEs were assembled into a symmetric cell structure of steel sheet (SS)/gel polymer electrolyte/steel sheet (SS) and tested for their electrochemical impedance spectra by performing AC impedance experiments at room temperature. The spectra were measured according to the equation

$$\sigma = \frac{L}{R \cdot S} \tag{1}$$

where σ is the ionic conductivity ($S \cdot cm^{-1}$); L is the thickness of the electrolyte membrane (cm); R is the bulk impedance of the material (Ω); and S is the area of the gel polymer electrolyte (cm^2). The electrolyte membrane was assembled into a lithium/gel polymer electrolyte/steel sheet battery structure for linear scanning (LSV) experiments with an initial potential of 2 V, a termination potential of 6 V, and a scanning rate of 5 mV/s.

Author Contributions: Conceptualization, X.Y. and W.Q.; methodology, Q.H.; software, X.Y. and W.Q.; validation, X.Y., Y.L. and X.L.; formal analysis, N.M.; investigation, Q.H. and N.M.; resources, Y.L. and X.L.; data curation, Q.H.; writing—original draft preparation, X.Y.; writing—review and editing, W.Q. and J.L.; visualization, X.L. and Y.L.; supervision, Q.H. and Y.G.; project administration,

Y.L. and J.L.; funding acquisition, X.L., J.L. and Y.G. All authors have read and agreed to the published version of the manuscript.

Funding: This research was funded by the Regional Collaboration R&D Program of Sichuan Province under Grant (No. 2024YFHZ0209), Fund Project of the Guangxi Key R&D Programme Projects (No. AB24010346), National Natural Science Foundation of China (No. 22462003, 52161033, 22262005), Guangxi key laboratory of Automobile Components and Vehicle technology (2023GK-LACVTZZ02), Guangxi Key Laboratory of Automobile Components and Vehicle Technology (No. 2022GKLACVTZZ04, 2023GKLACVTZZ10), Innovation Project of Guangxi Graduate education (YCSW2025576), and Ph.D. Programs Foundation of Guangxi University of Science and Technology (Grant No. 23Z04).

Institutional Review Board Statement: Not applicable.

Informed Consent Statement: Not applicable.

Data Availability Statement: The original contributions presented in this study are included in the article. Further inquiries can be directed to the corresponding author.

Conflicts of Interest: The authors declare no conflicts of interest.

References

1. Wu, D.; Luo, M.; Yang, R.; Hu, X.; Lu, C. Achieve high dielectric and energy-storage density properties by employing cyanoethyl cellulose as fillers in PVDF-based polymer composites. *Materials* **2023**, *16*, 4201. [CrossRef] [PubMed]
2. Kim, H.; Lee, S.-H.; Kim, J.-M.; Yoon, C.S.; Sun, Y.-K. High-energy-density, long-life li-metal batteries via application of external pressure. *ACS Energy Lett.* **2023**, *8*, 2970–2978. [CrossRef]
3. Tan, D.H.; Banerjee, A.; Chen, Z.; Meng, Y.S. From nanoscale interface characterization to sustainable energy storage using all-solid-state batteries. *Nat. Nanotechnol.* **2020**, *15*, 170–180. [CrossRef] [PubMed]
4. Zhou, Q.; He, M.; Gao, S.; Hou, W.; Ma, Y.; Huo, H.; Du, C.; Yin, G.; Zuo, P. Molecular Regulation and Intermolecular Chemistry in Gel Polymer Electrolytes for High-Voltage Lithium Batteries. *Adv. Sci.* **2025**, *12*, 2417169. [CrossRef]
5. Cheng, Y.; Cai, Z.; Xu, J.; Sun, Z.; Wu, X.; Han, J.; Wang, Y.H.; Wang, M.S. Zwitterionic cellulose-based polymer electrolyte enabled by aqueous solution casting for high-performance solid-state batteries. *Angew. Chem.* **2024**, *136*, e202400477. [CrossRef]
6. Kataoka, H.; Saito, Y.; Sakai, T.; Quartarone, E.; Mustarelli, P. Conduction Mechanisms of PVDF-Type Gel Polymer Electrolytes of Lithium Prepared by a Phase Inversion Process. *J. Phys. Chem. B* **2000**, *104*, 11460–11464. [CrossRef]
7. Li, W.; Wu, Y.; Wang, J.; Huang, D.; Chen, L.; Yang, G. Hybrid gel polymer electrolyte fabricated by electrospinning technology for polymer lithium-ion battery. *Eur. Polym. J.* **2015**, *67*, 365–372. [CrossRef]
8. Yan, J.; Huang, S.; Von Lim, Y.; Xu, T.; Kong, D.; Li, X.; Yang, H.Y.; Wang, Y. Direct-ink writing 3D printed energy storage devices: From material selectivity, design and optimization strategies to diverse applications. *Mater. Today* **2022**, *54*, 110–152. [CrossRef]
9. Nguyen, H.B.T.; Ding, L.; Pohle, B.; Schmeida, T.; Nguyen, H.B.A.; Mikhailova, D. Ternary PEO/PVDF-HFP-Based Polymer Electrolytes for Li-Ion Batteries. *Batteries* **2025**, *11*, 45. [CrossRef]
10. Halder, B.; Mohamed, M.G.; Kuo, S.-W.; Elumalai, P. Review on composite polymer electrolyte using PVDF-HFP for solid-state lithium-ion battery. *Mater. Today Chem.* **2024**, *36*, 101926. [CrossRef]
11. Wu, Y.; Li, Y.; Wang, Y.; Liu, Q.; Chen, Q.; Chen, M. Advances and prospects of PVDF based polymer electrolytes. *J. Energy Chem.* **2022**, *64*, 62–84. [CrossRef]
12. He, Z.; Rault, F.; Lewandowski, M.; Mohsenzadeh, E.; Salaün, F. Electrospun PVDF nanofibers for piezoelectric applications: A review of the influence of electrospinning parameters on the β phase and crystallinity enhancement. *Polymers* **2021**, *13*, 174. [CrossRef] [PubMed]
13. Yang, Y.; Qiao, J.; Sun, H.; Yang, W.; Wei, L.; Zhao, X. Induced Electron Traps via the PCBM in P (VDF-HFP) Composites to Enhance Dielectric and Energy Storage Performance. *Polymers* **2024**, *16*, 3030. [CrossRef] [PubMed]
14. Sivaraj, P.; Abhilash, K.; Nalini, B.; Perumal, P.; Selvin, P.C. Free-standing, high Li-ion conducting hybrid PAN/PVdF/$LiClO_4$/$Li_{0.5}La_{0.5}TiO_3$ nanocomposite solid polymer electrolytes for all-solid-state batteries. *J. Solid State Electrochem.* **2021**, *25*, 905–917. [CrossRef]
15. Hu, L.; Gao, X.; Wang, H.; Song, Y.; Zhu, Y.; Tao, Z.; Yuan, B.; Hu, R. Progress of Polymer Electrolytes Worked in Solid-State Lithium Batteries for Wide-Temperature Application. *Small* **2024**, *20*, 2312251. [CrossRef]
16. Farooqui, U.; Ahmad, A.; Hamid, N. Effect of polyaniline (PANI) on Poly (vinylidene fluoride-co-hexaflouro propylene)(PVDF-co-HFP) polymer electrolyte membrane prepared by breath figure method. *Polym. Test.* **2017**, *60*, 124–131. [CrossRef]

17. Wang, L.; Wu, J.; Bao, C.; You, Z.; Lu, Y.; Wen, Z. Interfacial engineering for high-performance garnet-based solid-state lithium batteries. *SusMat* **2024**, *4*, 72–105. [CrossRef]
18. Blake, A.J.; Kohlmeyer, R.R.; Hardin, J.O.; Carmona, E.A.; Maruyama, B.; Berrigan, J.D.; Huang, H.; Durstock, M.F. 3D printable ceramic–polymer electrolytes for flexible high-performance Li-ion batteries with enhanced thermal stability. *Adv. Energy Mater.* **2017**, *7*, 1602920. [CrossRef]
19. Cheng, M.; Jiang, Y.; Yao, W.; Yuan, Y.; Deivanayagam, R.; Foroozan, T.; Huang, Z.; Song, B.; Rojaee, R.; Shokuhfar, T. Elevated-temperature 3D printing of hybrid solid-state electrolyte for Li-ion batteries. *Adv. Mater.* **2018**, *30*, 1800615. [CrossRef]
20. Liu, Z.; Tian, X.; Liu, M.; Duan, S.; Ren, Y.; Ma, H.; Tang, K.; Shi, J.; Hou, S.; Jin, H. Direct ink writing of Li1. 3Al0. 3Ti1. 7 (PO4) 3-based solid-state electrolytes with customized shapes and remarkable electrochemical behaviors. *Small* **2021**, *17*, 2002866. [CrossRef]
21. Chen, B.; Willenbacher, N. High-precision direct ink writing of $Li_{6.4}La_3Zr_{1.4}Ta_{0.6}O_{12}$. *J. Eur. Ceram. Soc.* **2022**, *42*, 7491–7500. [CrossRef]
22. Li, C.; Deng, S.; Feng, W.; Cao, Y.; Bai, J.; Dou, Q.; Tian, X.; Dong, Y.; Xia, F. Lithium salt regulation enabling 3D printed solid-state electrolytes to achieve ultra-long cycle performance of $LiNi_{0.8}Mn_{0.1}Co_{0.1}O_2$. *Chem. Eng. J.* **2024**, *494*, 152770. [CrossRef]
23. Singer, C.; Schmalzbauer, S.; Daub, R. Influence of the slurry composition on thin-film components for the wet coating process of sulfide-based all-solid-state batteries. *J. Energy Storage* **2023**, *68*, 107703. [CrossRef]
24. Zhang, J.; Sun, B.; Huang, X.; Chen, S.; Wang, G. Honeycomb-like porous gel polymer electrolyte membrane for lithium ion batteries with enhanced safety. *Sci. Rep.* **2014**, *4*, 6007. [CrossRef]
25. Wang, D.; Cai, D.; Zhong, Y.; Jiang, Z.; Zhang, S.; Xia, X.; Wang, X.; Tu, J. A three-dimensional electrospun Li6. 4La3Zr1. 4Ta0. 6O12–poly (vinylidene fluoride-hexafluoropropylene) gel polymer electrolyte for rechargeable solid-state lithium ion batteries. *Front. Chem.* **2021**, *9*, 751476.
26. Park, J.-H.; Kim, M.; Kim, M.-Y.; Jeong, J.; Jung, H.-G.; Yoon, W.Y.; Chung, K.Y. Correlation between the particle size of Li1.3Al0.3Ti1.7(PO4)3 solid electrolyte and lithium-ion transport in composite cathodes for all-solid-state lithium-ion batteries. *Chem. Eng. J.* **2024**, *481*, 148436. [CrossRef]
27. Zhu, C.; Liu, T.; Qian, F.; Han, T.Y.-J.; Duoss, E.B.; Kuntz, J.D.; Spadaccini, C.M.; Worsley, M.A.; Li, Y. Supercapacitors based on three-dimensional hierarchical graphene aerogels with periodic macropores. *Nano Lett.* **2016**, *16*, 3448–3456. [CrossRef]
28. Beshahwured, S.L.; Wu, Y.-S.; Wu, S.-h.; Chien, W.-C.; Jose, R.; Lue, S.J.; Yang, C.-C. Flexible hybrid solid electrolyte incorporating ligament-shaped Li6. 25Al0. 25La3Zr2O12 filler for all-solid-state lithium-metal batteries. *Electrochim. Acta* **2021**, *366*, 137348. [CrossRef]
29. Hun, Q.; Lan, L.; Lu, X.; Hu, Q.; Liang, X.; Guo, Y.; Wang, Y. Bilayer Heterostructure Electrolytes Were Prepared by a UV-Curing Process for High Temperature Lithium-Ion Batteries. *Polymers* **2024**, *16*, 2972. [CrossRef]
30. Satapathy, S.; Pawar, S.; Gupta, P.; Varma, K. Effect of annealing on phase transition in poly (vinylidene fluoride) films prepared using polar solvent. *Bull. Mater. Sci.* **2011**, *34*, 727–733. [CrossRef]
31. Janakiraman, S.; Surendran, A.; Ghosh, S.; Anandhan, S.; Venimadhav, A. Electroactive poly (vinylidene fluoride) fluoride separator for sodium ion battery with high coulombic efficiency. *Solid State Ion.* **2016**, *292*, 130–135. [CrossRef]
32. Xu, Y.; Wang, K.; An, Y.; Liu, W.; Li, C.; Zheng, S.; Zhang, X.; Wang, L.; Sun, X.; Ma, Y. Rapid ion transport induced by the enhanced interaction in composite polymer electrolyte for all-solid-state lithium-metal batteries. *J. Phys. Chem. Lett.* **2021**, *12*, 10603–10609. [CrossRef] [PubMed]
33. Wu, J.; Wu, X.; Wang, W.; Wang, Q.; Zhou, X.; Liu, Y.; Guo, B. Dense PVDF-type polymer-in-ceramic electrolytes for solid state lithium batteries. *RSC Adv.* **2020**, *10*, 22417–22421. [CrossRef] [PubMed]
34. Mouraliraman, D.; Shaji, N.; Praveen, S.; Nanthagopal, M.; Ho, C.W.; Varun Karthik, M.; Kim, T.; Lee, C.W. Thermally Stable PVDF-HFP-Based Gel Polymer Electrolytes for High-Performance Lithium-Ion Batteries. *Nanomaterials* **2022**, *12*, 1056. [CrossRef]
35. Yi, L.; Chen, X.; Zou, C.; Liu, J.; Cao, X.; Tao, X.; Zang, Z.; Zheng, L.; Liu, L.; Wang, X. Construction and performance of PVDF-HFP-matrix comb-like gel copolymer electrolyte for high-performance quasi-solid-state lithium-metal batteries. *J. Energy Storage* **2023**, *68*, 107810. [CrossRef]
36. Duan, T.; Cheng, H.; Sun, Q.; Liu, Y.; Nie, W.; Chu, Y.; Xu, Q.; Lu, X. Reinforcing interfacial compatibility of LLZTO/PVDF-HFP composite electrolytes by chemical interaction for solid-state lithium metal batteries. *J. Power Sources* **2024**, *589*, 233789. [CrossRef]
37. Nguyen, M.C.; Nguyen, H.L.; Duong, T.P.M.; Kim, S.H.; Kim, J.Y.; Bae, J.H.; Kim, H.K.; Lim, S.N.; Ahn, W. Highly Safe, Ultra-Thin MOF-Based Solid Polymer Electrolytes for Superior All-Solid-State Lithium-Metal Battery Performance. *Adv. Funct. Mater.* **2024**, *34*, 2406987. [CrossRef]
38. Lu, Q.; He, Y.-B.; Yu, Q.; Li, B.; Kaneti, Y.V.; Yao, Y.; Kang, F.; Yang, Q.-H. Dendrite-Free, High-Rate, Long-Life Lithium Metal Batteries with a 3D Cross-Linked Network Polymer Electrolyte. *Adv. Mater.* **2017**, *29*, 1604460. [CrossRef]
39. Fan, W.; Li, N.W.; Zhang, X.; Zhao, S.; Cao, R.; Yin, Y.; Xing, Y.; Wang, J.; Guo, Y.G.; Li, C. A dual-salt gel polymer electrolyte with 3D cross-linked polymer network for dendrite-free lithium metal batteries. *Adv. Sci.* **2018**, *5*, 1800559. [CrossRef]

40. Xu, K.; Xu, C.; Jiang, Y.; Cai, J.; Ni, J.; Lai, C. Sandwich structured PVDF-HFP-based composite solid electrolytes for solid-state lithium metal batteries. *Ionics* **2022**, *28*, 3243–3253. [CrossRef]
41. Lee, H.; Song, Y.-W.; Kim, M.-Y.; Lee, J.; Ryu, J.; Noh, Y.; Kim, S.-J.; Kim, J.; Lim, J. Composite solid electrolyte with improved ionic conductivity and high lithium transference number through reduced PVDF-HFP crystallinity. *Solid State Ion.* **2024**, *411*, 116571. [CrossRef]
42. Kou, W.; Guo, Z.; Li, W.; Liu, S.; Zhang, J.; Zhang, X.; Wu, W.; Wang, J. Highly conductive thin lamellar Li7La3Zr2O12/Li3InCl6 composite inorganic solid electrolyte for high-performance all-solid-state lithium battery. *J. Membr. Sci.* **2023**, *687*, 122080. [CrossRef]
43. Robert, V.; Loganathan, S.; Sivashanmugam, A.; Thomas, S.; Valapa, R.B. 3D Printed P (VdF-HFP)/SBA-15 composite-based gel polymer electrolyte versus commercial celgard: A comparative investigation on thermal stability and electrochemical performance for lithium-ion battery application. *J. Am. Ceram. Soc.* **2024**, *107*, 5903–5922. [CrossRef]
44. Bae, J.; Oh, S.; Lee, B.; Lee, C.H.; Chung, J.; Kim, J.; Jo, S.; Seo, S.; Lim, J.; Chung, S. High-performance, printable quasi-solid-state electrolytes toward all 3D direct ink writing of shape-versatile Li-ion batteries. *Energy Storage Mater.* **2023**, *57*, 277–288. [CrossRef]
45. Bourseau, F.; Grugeon, S.; Lafont, U.; Dupont, L. 3D Printing of solvent-free PEO-Polyolefin solid polymer electrolyte by Fused Filament Fabrication. *Virtual Phys. Prototyp.* **2024**, *19*, e2409975. [CrossRef]
46. Zhang, Y.; Zheng, H.; Ding, H.; Jabbar, K.A.; Gao, L.; Zhao, G. Ceria Quantum Dot Filler-Modified Polymer Electrolytes for Three-Dimensional-Printed Sodium Solid-State Batteries. *Polymers* **2024**, *16*, 1707. [CrossRef]
47. Wei, L.; Feng, Y.; Ge, S.; Liu, S.; Ma, Y.; Yan, J. Three-Dimensionally Printed Ionogel-Coated Ceramic Electrolytes for Solid-State Lithium Batteries. *ACS Nano* **2025**, *19*, 5789–5800. [CrossRef]

gels

MDPI

Article

Interpretable Prediction and Analysis of PVA Hydrogel Mechanical Behavior Using Machine Learning

Liying Xu [1,*], Siqi Liu [2], Anqi Lin [2], Zichuan Su [3] and Daxin Liang [2,*]

[1] School of Food Engineering, Harbin University, Harbin 150086, China
[2] Key Laboratory of Bio-Based Material Science and Technology (Ministry of Education), Northeast Forestry University, Harbin 150040, China
[3] College of Chemistry and Chemical Engineering, China University of Petroleum (East China), Qingdao 266580, China
* Correspondence: lyxutk@163.com (L.X.); daxin.liang@nefu.edu.cn (D.L.); Tel.: +86-181-82806182 (L.X.)

Abstract

Polyvinyl alcohol (PVA) hydrogels have emerged as versatile materials due to their exceptional biocompatibility and tunable mechanical properties, showing great promise for flexible sensors, smart wound dressings, and tissue engineering applications. However, rational design remains challenging due to complex structure–property relationships involving multiple formulation parameters. This study presents an interpretable machine learning framework for predicting PVA hydrogel tensile strain properties with emphasis on mechanistic understanding, based on a comprehensive dataset of 350 data points collected from a systematic literature review. XGBoost demonstrated superior performance after Optuna-based optimization, achieving R^2 values of 0.964 for training and 0.801 for testing. SHAP analysis provided unprecedented mechanistic insights, revealing that PVA molecular weight dominates mechanical performance (SHAP importance: 84.94) through chain entanglement and crystallization mechanisms, followed by degree of hydrolysis (72.46) and cross-linking parameters. The interpretability analysis identified optimal parameter ranges and critical feature interactions, elucidating complex non-linear relationships and reinforcement mechanisms. By addressing the "black box" limitation of machine learning, this approach enables rational design strategies and mechanistic understanding for next-generation multifunctional hydrogels.

Keywords: PVA hydrogel; interpretable machine learning; SHAP analysis; feature importance; hyperparameter optimization

1. Introduction

Hydrogels, three-dimensional cross-linked polymer networks capable of absorbing substantial water while maintaining structural integrity, have emerged as versatile materials across biomedical, electronic, and environmental domains [1,2]. Among various systems, PVA hydrogels have attracted considerable attention due to their exceptional biocompatibility, tunable mechanical properties, and chemical stability [3]. The unique combination of hydrogen bonding networks and crystalline regions enables physically cross-linked formation without toxic chemical cross-linkers [4–6], allowing mechanical properties ranging from soft, tissue-like materials to robust, load-bearing structures [7]. This tunability positions PVA hydrogels as promising candidates for flexible sensors, smart wound dressings, drug delivery systems, and tissue engineering scaffolds [8–10].

Recent advances have demonstrated the remarkable versatility of PVA hydrogels across diverse technological applications. PVA-based conductive hydrogels have been successfully employed in human–machine interaction systems for intelligent rehabilitation training, leveraging their excellent sensing capabilities and biocompatibility [11]. Self-healing ionic conductive PVA hydrogels have shown exceptional performance in wearable sensor applications, offering multi-sensing capabilities while maintaining transparency and mechanical robustness [12]. Additionally, PVA hydrogels have been integrated into sophisticated sensing platforms, such as SERS sensors for antitumor drug detection and robust conductive systems based on hydrotalcite nanocomposites [13]. These applications highlight the potential of PVA hydrogels to address complex technological challenges across healthcare, environmental monitoring, and biomedical diagnostics. However, the full realization of PVA hydrogels' potential in these applications depends critically on precise control and understanding of their mechanical properties and modulus characteristics. A comprehensive investigation of the structure–property relationships governing PVA hydrogel mechanics would significantly contribute to advancing these emerging fields and enabling rational design of next-generation hydrogel-based devices [14–16].

Machine learning (ML) approaches have demonstrated remarkable success in materials science for predicting complex material properties and understanding structure–property relationships [17]. SHAP analysis has emerged as a powerful interpretability technique to address the "black box" problem in ML models, enabling researchers to gain mechanistic insights into feature contributions and underlying material behavior. However, comprehensive ML studies focusing specifically on PVA hydrogel mechanical property prediction remain limited [18].

To address these limitations, this study presents a machine learning framework for predicting PVA hydrogel tensile strain properties based on key formulation and processing parameters including PVA molecular weight, degree of hydrolysis, concentration, freeze–thaw cycles, cross-linking methods, and conductive content. Through systematic literature screening of 46 research papers and collection of 350 data points, we developed an Extreme Gradient Boosting (XGBoost) model and employed Shapley Additive Explanations (SHAP) analysis to elucidate mechanistic contributions of individual features. The analysis revealed that PVA molecular weight dominates mechanical performance, followed by degree of hydrolysis and cross-linking parameters, while feature interactions significantly influence optimal parameter ranges. This work establishes an interpretable framework for rational design of these materials.

2. Results and Discussion

2.1. Analysis of Input Features Prior to Model Development

In this study, a comprehensive dataset was assembled encompassing key formulation and processing parameters that influence PVA hydrogel mechanical properties. The continuous variables include PVA molecular weight (PVA_MW), degree of hydrolysis (PVA_DH), PVA concentration (PVA_Conc), freeze–thaw cycles (FC), cross-linking method (CLM), and conductive content (CC). Additionally, categorical variables were incorporated for chemical cross-linkers (Borax, $CaCl_2$, EG, GL, MBA, PBA) and conductive types (AgNPs, CNF, CNT, Ionic_only, LDH, MXene, PANI, PEDOT:PSS, TA, rGO, PPy), which were subsequently encoded as binary variables for machine learning analysis. The target variable is tensile strain (Tensile_Strain).

The distribution analysis of continuous features reveals important characteristics of the assembled dataset (Figure 1). PVA_MW shows a broad distribution from 70,000 to 160,000 Da, with a concentration around 100,000–120,000 Da (Figure 1a), while PVA_DH exhibits a narrow distribution centered at 98–99% (Figure 1b), indicating high hydrolysis degrees across samples. PVA_Conc demonstrates a multimodal distribution ranging from 2% to 18% w/w, with distinct peaks at lower and higher concentrations (Figure 1c). FC displays relatively uniform distribution across 0–5 cycles (Figure 1d), CLM shows bimodal distribution indicating different cross-linking strategies (Figure 1e), and CC exhibits a highly skewed distribution with most samples containing low conductive content (0–5%) and fewer samples with higher loadings up to 20% (Figure 1f). The diverse distribution patterns observed across all features demonstrate the comprehensive nature of the dataset, encompassing a wide range of formulation parameters that are representative of practical PVA hydrogel synthesis conditions. These distribution characteristics provide a solid foundation for robust machine learning model training and ensure that the developed predictive framework can capture the complex relationships between processing parameters and mechanical properties across the entire experimental space [18–20].

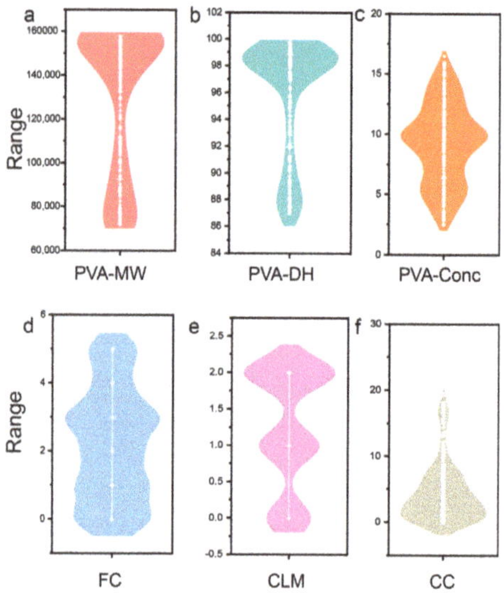

Figure 1. Distribution analysis of continuous input features. (**a**) PVA_MW, (**b**) PVA_DH, (**c**) PVA_Conc, (**d**) FC, (**e**) CLM, (**f**) CC.

The Pearson correlation matrix reveals important relationships between input features that inform feature selection and model interpretation (Figure 2). Among the continuous variables, most features exhibit relatively weak correlations with each other, indicating good feature independence. PVA_MW shows minimal correlation with other continuous parameters, while PVA_DH demonstrates slight positive correlations with several chemical cross-linkers. PVA_Conc exhibits weak correlations with most features, suggesting its independent contribution to hydrogel properties [21,22]. FC shows moderate correlations with certain conductive materials, particularly with some nanomaterials, indicating potential interactions between freeze–thaw processing and filler incorporation [23–25].

The categorical features representing chemical cross-linkers and conductive types generally show weak to moderate correlations among themselves, which is expected given their binary encoding nature [4,7,26]. Notable correlations exist between certain conductive materials, such as moderate positive correlations among some carbon-based materials (CNF, CNT, rGO) and conducting polymers (PANI, PEDOT:PSS). Most correlation coefficients remain below 0.5, indicating an absence of severe multicollinearity issues that could compromise model performance [27–30]. The relatively low inter-feature correlations suggest that each parameter contributes unique information to the prediction model, justifying the inclusion of all features in the machine learning framework [13,31].

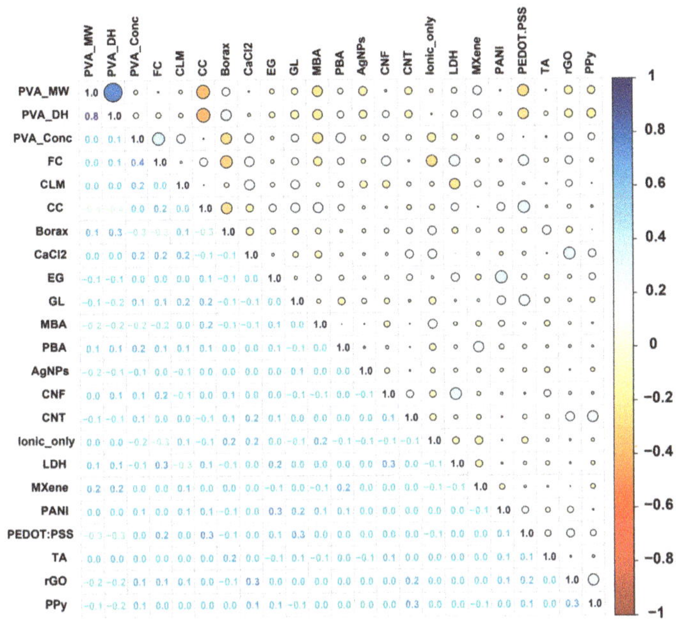

Figure 2. Pearson correlation matrix of input features.

2.2. Model Development

To establish an effective predictive framework for PVA hydrogel tensile strain properties, six different machine learning algorithms were evaluated and compared: AdaBoost, Gradient Boosting Regression (GBR), K-Nearest Neighbors (KNN), Support Vector Machine (SVM), Extreme Gradient Boosting (XGBoost), and Multi-Layer Perceptron (MLP). Figure 3 presents the predicted versus true value plots for all models, providing comprehensive performance evaluation across both training and testing datasets.

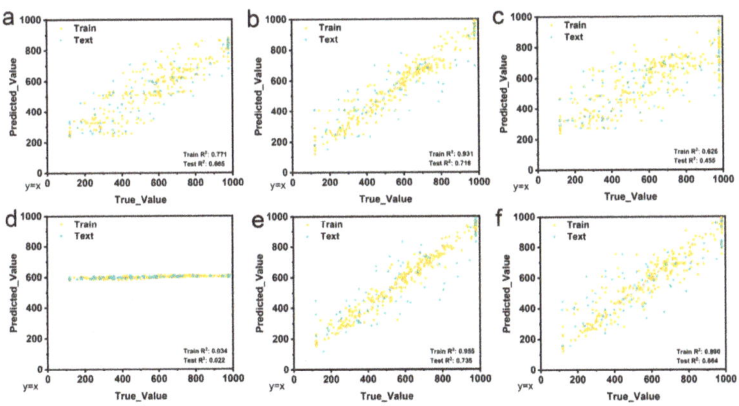

Figure 3. Machine learning model performance comparison for tensile strain prediction. (**a**) AdaBoost, (**b**) GBR, (**c**) KNN, (**d**) SVM, (**e**) XGBoost, (**f**) MLP.

The performance analysis reveals significant variations among different algorithms. AdaBoost demonstrates moderate predictive capability with a training R^2 of 0.771 and test R^2 of 0.665 (Figure 3a), indicating reasonable model performance but with notable generalization gap. GBR exhibits exceptional training performance with R^2 of 0.931 but shows substantial overfitting with test R^2 dropping to 0.718 (Figure 3b), suggesting limited generalizability to unseen data. KNN displays poor overall performance with extremely low training R^2 of 0.034 and test R^2 of 0.022 (Figure 3c), with predictions clustered around a constant value near 600, indicating the algorithm's inability to capture the underlying data patterns effectively [32,33].

SVM shows promising results with a high training R^2 of 0.955 and reasonable test performance R^2 of 0.735 (Figure 3d), demonstrating good predictive capability but with some overfitting tendency. MLP achieves excellent training performance with an R^2 of 0.890 and maintains good generalization with a test R^2 of 0.664 (Figure 3e), showing balanced performance across both datasets. Most notably, XGBoost delivers outstanding performance, with the highest training R^2 of 0.955 and a superior test R^2 of 0.735 (Figure 3f), demonstrating both excellent predictive accuracy and robust generalization capability [34,35].

The comparative analysis clearly indicates that XGBoost outperforms other algorithms in terms of both accuracy and generalization ability. Unlike GBR which shows severe overfitting, XGBoost maintains consistent performance between training and testing phases, suggesting better model stability. The superior performance of XGBoost can be attributed to its advanced ensemble learning mechanism, effective handling of feature interactions, and built-in regularization techniques that prevent overfitting. Additionally, XGBoost's inherent feature importance calculation and compatibility with SHAP analysis make it an ideal choice for subsequent mechanistic interpretation studies [36]. Based on these comprehensive performance evaluations, XGBoost was selected as the optimal model for further optimization and detailed analysis of PVA hydrogel property prediction.

2.3. Optimized XGBoost Model Performance and Feature Importance Analysis

Following the initial model comparison, the XGBoost algorithm was subjected to systematic hyperparameter optimization using the Optuna framework to maximize predictive performance. The optimized XGBoost model demonstrates exceptional performance, with significantly improved metrics compared to the preliminary version (Figure 4a). The training R^2 increased to 0.964 while maintaining excellent generalization capability with a test R^2 of 0.801, representing a substantial improvement over the initial XGBoost performance. The model achieves low prediction errors with MAE of 30.534 and RMSE of 46.864, indicating high accuracy in tensile strain prediction across the entire range of experimental values. The strong alignment between predicted and true values for both training and testing datasets confirms the model's robustness and reliability for practical applications [37,38].

The feature importance analysis reveals critical insights into the relative contributions of different parameters to PVA hydrogel tensile strain properties (Figure 4b). PVA_MW emerges as the most influential feature with an importance score of approximately 0.19, highlighting the fundamental role of molecular weight in determining mechanical properties through polymer chain entanglement and crystallization behavior [13,22,39]. rGO follows as the second most important feature with a score of ~0.11, indicating the significant impact of reduced graphene oxide incorporation on hydrogel mechanics [40–42]. $CaCl_2$ shows notable importance (~0.07), reflecting the crucial role of ionic cross-linking in modulating mechanical properties. PVA_Conc demonstrates moderate importance (~0.06), confirming the expected influence of polymer concentration on network density and mechanical performance [26,29,43]. Interestingly, several conductive materials including Borax, MBA, and various nanomaterials show relatively lower but still meaningful contributions, suggesting their secondary but important roles in fine-tuning mechanical properties. The residual analysis demonstrates overall model reliability with generally random scatter around zero for most data points (Figure 4c). Training data exhibit uniform residual distribution within ±100 units. The residual analysis reveals some data points that deviate from the ideal zero-residual line, which can be attributed to several factors. Experimental variations across different research groups, including differences in testing protocols and sample preparation methods, contribute to prediction deviations. Additionally, variations in measurement techniques and equipment precision across literature sources, as well as inherent material property differences from various PVA suppliers, influence the residual distribution. These factors highlight the challenges of developing predictive models from literature-based datasets and suggest areas for future model refinement through more standardized experimental protocols.

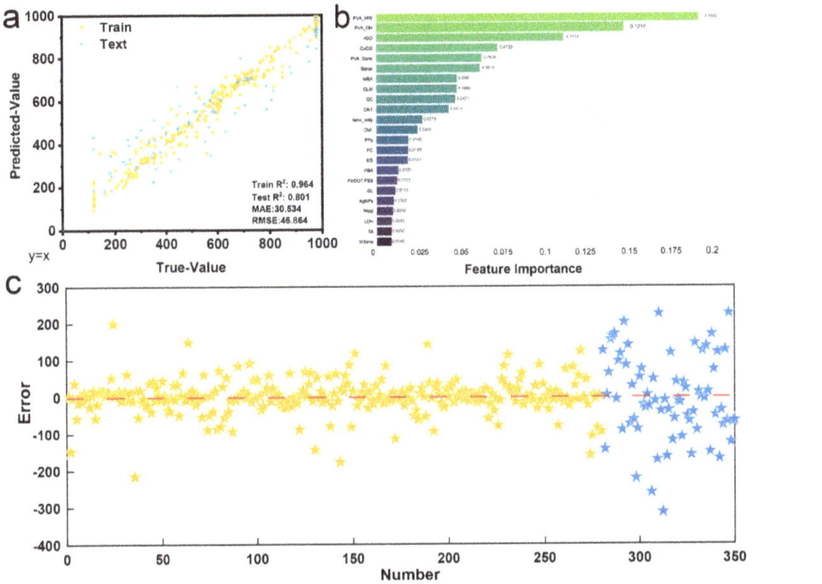

Figure 4. (**a**) Predicted vs. true values with performance metrics, (**b**) feature importance ranking, (**c**) residual analysis for training and test datasets.

2.4. Feature Importance and Interpretability Analysis

To gain a deeper mechanistic understanding of the XGBoost model predictions and elucidate the underlying structure–property relationships, SHAP analysis was employed to quantify individual feature contributions to tensile strain predictions. The SHAP analysis provides both global feature importance and local explanations for individual predictions [17,19], offering unprecedented insights into PVA hydrogel mechanical behavior.

The global SHAP importance analysis reveals the hierarchical contributions of features to tensile strain properties, providing critical insights for rational hydrogel design (Figure 5a). SHAP values represent quantitative measures of feature importance, where higher values indicate greater contribution to model predictions. PVA_MW emerges as the dominant factor with a SHAP value of 84.94, indicating that molecular weight accounts for the largest proportion of predictive influence among all features, fundamentally governing mechanical performance through polymer chain entanglement dynamics and crystalline domain formation [12,31,44]. This finding suggests that molecular weight selection should be the primary consideration in hydrogel formulation, with higher molecular weights promoting enhanced chain entanglement density and load transfer efficiency. PVA_DH demonstrates substantial importance (72.46), directly influencing hydrogen bonding capacity and crystallization tendency higher DH values facilitate stronger intermolecular interactions and more organized crystalline structures, leading to improved mechanical integrity [45–47]. The significant contributions of CLM (33.75) and $CaCl_2$ (32.78) highlight the critical role of cross-linking strategies, where ionic cross-linking through Ca^{2+} ions creates additional physical junction points that enhance network connectivity and stress distribution [24,48,49]. Notably, rGO shows considerable importance (30.25), indicating that graphene incorporation provides effective reinforcement through strong interfacial interactions and load transfer mechanisms [22,31,39].

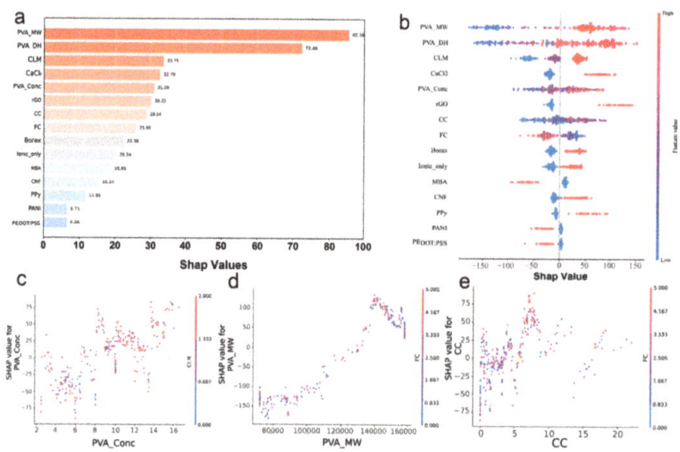

Figure 5. (**a**) Global SHAP feature importance values, (**b**) SHAP summary plot showing individual feature contributions across all samples, (**c**–**e**) SHAP dependence plots for PVA_Conc, PVA_MW, and CC, respectively, with color coding indicating interaction effects.

The SHAP summary plot provides mechanistic insights into how each feature influences tensile strain across different value ranges (Figure 5b). PVA_MW exhibits predominantly positive SHAP values for higher molecular weights (red regions), confirming that increased chain length enhances mechanical performance through greater entanglement density and improved stress transfer pathways [12,31,44]. The distribution pattern suggests an optimal molecular weight threshold above which mechanical benefits become pronounced, guiding formulation strategies toward higher MW PVA selection. PVA_DH shows a more complex pattern with both positive and negative contributions, indicating that while high hydrolysis degrees generally promote hydrogen bonding, excessive hydrolysis may lead to increased crystallinity that could compromise flexibility [45–47]. This suggests an optimal DH range that balances hydrogen bonding strength with network flexibility. The binary cross-linking features (CLM, $CaCl_2$) display clear positive contributions when present, emphasizing the necessity of appropriate cross-linking for mechanical enhancement [24,48,49]. Importantly, the rGO distribution shows strong positive SHAP values when present, but also reveals that not all graphene incorporations yield equivalent benefits, suggesting that processing conditions and dispersion quality critically influence reinforcement effectiveness [22,31,39]. For conductive additives, the varied SHAP distributions indicate that material selection and loading optimization are crucial for achieving desired mechanical properties while maintaining conductivity. Following the primary features, the secondary parameters reveal important fine-tuning mechanisms for PVA hydrogel optimization. Borax and MBA demonstrate consistent positive SHAP contributions when present, confirming their effectiveness as cross-linking agents that enhance network connectivity through borate complexation and covalent bonding, respectively [24,50]. The freeze–thaw cycles (FC) show moderate positive effects, indicating that controlled crystallization during freeze–thaw processing creates beneficial physical cross-links [25,51]. Among the conductive materials, CNT and CNF exhibit predominantly positive SHAP values, suggesting that these carbon-based nanofillers provide dual benefits of electrical conductivity and mechanical reinforcement. However, the conducting polymers (PANI, PEDOT:PSS, PPy) display more variable SHAP distributions [30,42,52], indicating that their mechanical contributions are highly dependent on processing conditions and dispersion quality, emphasizing the need for careful optimization when incorporating these materials for multifunctional applications.

The SHAP dependence plots reveal complex non-linear relationships and feature interactions for key parameters (Figure 5c–e). The PVA_Conc dependence plot (Figure 5c) demonstrates a complex relationship where moderate concentrations (6–10%) show predominantly positive SHAP values, while both very low and very high concentrations tend to have neutral or negative effects. This reflects optimal chain entanglement density at moderate concentrations, where sufficient polymer chains enable effective cross-linking without chain crowding effects that hinder molecular mobility. The color coding by CLM reveals important interaction effects, suggesting that the optimal concentration range depends significantly on the cross-linking method employed [25,49]. The PVA_MW dependence plot (Figure 5d) shows a generally positive correlation between molecular weight and SHAP values, with the strongest positive effects occurring at molecular weights above 120,000 Da. This threshold corresponds to the critical molecular weight for chain entanglement in PVA, above which enhanced mechanical properties result from increased entanglement density and crystallization potential. Interestingly, the interaction with FC reveals that the beneficial effect of high molecular weight is enhanced when combined with appropriate freeze–thaw cycling [5,41]. The CC dependence plot (Figure 5e) exhibits a non-monotonic relationship where low to moderate conductive content (2–8%) provides positive contributions to tensile strain, while higher loadings show diminishing or negative effects, likely due to filler agglomeration and network disruption [25,53].

These SHAP analyses collectively provide mechanistic insights into PVA hydrogel design principles. For enhanced mechanical performance, we recommend prioritizing high molecular weight PVA above 120,000 Da combined with optimized freeze–thaw cycles of 2–4 cycles to maximize chain entanglement and crystallization benefits, maintaining moderate PVA concentrations between 6 and 10% with appropriate ionic cross-linking agents such as $CaCl_2$ for optimal network connectivity, and incorporating low-to-moderate conductive filler loadings of 2–8% to achieve reinforcement without network disruption. The dominant influence of PVA_MW and PVA_DH confirms the fundamental importance of polymer characteristics, while the significant contributions of cross-linking parameters and conductive additives highlight the critical role of formulation optimization. The identified parameter interactions suggest that synergistic combinations, particularly high MW PVA with controlled freeze–thaw processing and moderate filler loading with ionic cross-linking, can achieve superior mechanical properties compared to individual parameter optimization. The revealed interaction effects between features emphasize the need for holistic design approaches rather than individual parameter optimization, providing valuable guidance for rational hydrogel design and property tailoring, future experimental validation of these design principles would benefit from advanced characterization techniques such as MALLS (multi-angle laser light scattering) for precise molecular weight determination and conformational analysis, as demonstrated by Macarie and Pekar for vinyl-containing polymers, which could provide deeper insights into the structure–property relationships governing these synergistic effects [54].

3. Conclusions

This study successfully developed an interpretable machine learning framework for predicting PVA hydrogel tensile strain properties. XGBoost emerged as the optimal model after comprehensive evaluation, achieving exceptional performance with R^2 values of 0.964 for training and 0.801 for testing datasets following Optuna-based hyperparameter optimization. The low prediction errors (MAE: 30.534, RMSE: 46.864) demonstrate reliable predictive capability for practical applications.

SHAP analysis revealed critical design principles governing PVA hydrogel mechanical behavior. SHAP importance values quantify individual feature contributions to model predictions, with higher values indicating greater influence on the target property. PVA molecular weight emerged as the dominant factor (SHAP importance: 84.94), followed by the degree of hydrolysis (72.46) and cross-linking parameters, emphasizing the fundamental roles of polymer chain characteristics and structural connectivity. The significant impact of rGO incorporation demonstrates effective graphene-based reinforcement strategies [40,48,55]. The identified optimal parameter ranges and revealed feature interactions provide valuable guidance for rational hydrogel design, emphasizing holistic optimization approaches [36,37].

The interpretable machine learning framework established here advances both fundamental understanding and practical applications of PVA hydrogels. This work represents the first comprehensive application of SHAP analysis to elucidate structure–property relationships in PVA hydrogels, providing quantitative mechanistic insights that were previously inaccessible through conventional analytical methods. By bridging complex formulation parameters with mechanical properties, this approach enables accelerated materials discovery and rational design strategies that surpass conventional design paradigms by providing predictive capability combined with mechanistic understanding. The novel integration of feature importance analysis with polymer physics principles establishes a new paradigm for interpretable materials design, addressing the critical gap between machine learning predictions and mechanistic understanding in soft materials research. Future advancement of predictive models would benefit from more comprehensive molecular characterization, including polydispersity indices, conformational analysis through GPC-MALLS, and detailed molecular weight distribution data, enabling higher-quality datasets and more accurate data-driven experimental design strategies. The methodology can be extended to other hydrogel systems and property predictions, potentially transforming advanced soft materials development for flexible electronics, biomedical devices, and smart material applications.

4. Materials and Methods

4.1. Data Collection and Preprocessing

This study utilized Web of Science as the primary data source, employing 'PVA hydrogel', 'polyvinyl alcohol hydrogel mechanical properties', and 'PVA hydrogel tensile strain' as search keywords to systematically collect experimental data from peer-reviewed literature published between 2017 and 2024. After comprehensive screening and quality assessment, 46 research papers were selected that provided complete formulation parameters and mechanical testing data for PVA hydrogels [3–13,21–31,39–53,55–63].

The collected data encompassed six continuous variables: PVA molecular weight (PVA_MW), degree of hydrolysis (PVA_DH), PVA concentration (PVA_Conc), freeze–thaw cycles (FC), cross-linking method (CLM), and conductive content (CC). Chemical cross-linkers were categorized into six types: Borax, $CaCl_2$, ethylene glycol (EG), glutaraldehyde (GL), N,N′-methylenebisacrylamide (MBA), and polyboronic acid (PBA). Conductive materials were classified into eleven categories: silver nanoparticles (AgNPs), carbon nanofibers (CNF), carbon nanotubes (CNT), ionic conductors only (Ionic_only), layered double hydroxide (LDH), MXene, polyaniline (PANI), poly(3,4-ethylenedioxythiophene):poly(styrenesulfonate) (PEDOT:PSS), tannic acid (TA), reduced graphene oxide (rGO), and polypyrrole (PPy). One-hot encoding was applied to all categorical features to create binary numerical representations for machine learning analysis. Missing values were systematically handled using domain-specific approaches. For PVA_MW and PVA_DH parameters, when papers reported only one value, the missing parameter was estimated using established the molecular weight-degree of hydrolysis relationships from manufacturer specifications or the same research group's previous work. For PVA_Conc, when not explicitly reported, concentrations were calculated from solid content percentages, polymer-to-solvent ratios, or preparation protocols described in experimental sections. The final dataset incorporated 23 input features.

4.2. Model Development and Evaluation Methodology

The dataset was split into training and testing sets using an 80:20 ratio to evaluate model generalization performance. Six machine learning algorithms were systematically evaluated: AdaBoost (ADB), Gradient Boosting Regression (GBR), K-Nearest Neighbors (KNN), Support Vector Machine (SVM), XGBoost, and Multi-Layer Perceptron (MLP). Tree-based ensemble models were emphasized for their effectiveness in handling mixed-type features and capturing complex non-linear relationships in polymer systems.

The optimal XGBoost model was selected and subjected to hyperparameter optimization using the Optuna framework to maximize predictive accuracy. Model performance was quantified using root mean squared error (RMSE), mean absolute error (MAE), and coefficient of determination (R^2). RMSE measures prediction error magnitude, MAE represents average absolute deviation, and R^2 quantifies the proportion of variance explained by the model. The RMSE, MAE, and R^2 were calculated using Equations (1)–(3):

$$RMSE = \sqrt{\frac{\sum_{i=1}^{n}\left(y_t - y_p\right)^2}{n}} \quad (1)$$

$$MAE = \frac{\sum_{i=1}^{n}\left|y_t - y_p\right|}{n} \quad (2)$$

$$R^2 = 1 - \frac{\sum_{i=1}^{n}\left(y_p - y_t\right)^2}{\sum_{i=1}^{n}\left(y_t - y_m\right)^2} \quad (3)$$

where y_p represents the predicted output value, y_t denotes the reported true output value, y^m represents the mean of observed output values, n indicates the number of samples in the training or testing datasets. SHAP (Shapley Additive Explanations) analysis was implemented to provide interpretable insights into feature contributions and mechanistic relationships. SHAP values quantify individual feature impacts on predictions, enabling both global feature importance ranking and local explanations of formulation effects on tensile strain properties, thereby addressing the "black box" limitation of machine learning models.

Author Contributions: Conceptualization, L.X. and S.L.; methodology, L.X. and Z.S.; software, S.L. and D.L.; validation, L.X., S.L. and D.L.; formal analysis, S.L. and A.L.; investigation, D.L. and A.L.; resources, L.X.; data curation, S.L. and D.L.; writing—original draft preparation, S.L.; writing—review and editing, L.X., D.L., and Z.S.; visualization, S.L. and A.L.; supervision, L.X. and Z.S.; project administration, L.X. funding acquisition, L.X. All authors have read and agreed to the published version of the manuscript.

Funding: This work was supported by the National Natural Science Foundation of China (22378089), the Natural Science Foundation of Heilongjiang Province (LH2023B007).

Institutional Review Board Statement: Not applicable.

Informed Consent Statement: Not applicable.

Data Availability Statement: The data presented in this study are available on request from the corresponding authors.

Conflicts of Interest: The authors declare no conflicts of interest.

References

1. Yin, S.; Yao, D.R.; Song, Y.; Heng, W.; Ma, X.; Han, H.; Gao, W. Wearable and Implantable Soft Robots. *Chem. Rev.* **2024**, *124*, 11585–11636. [CrossRef]
2. Li, H.; Tan, P.; Rao, Y.; Bhattacharya, S.; Wang, Z.; Kim, S.; Gangopadhyay, S.; Shi, H.; Jankovic, M.; Huh, H.; et al. E-Tattoos: Toward Functional but Imperceptible Interfacing with Human Skin. *Chem. Rev.* **2024**, *124*, 3220–3283. [CrossRef]
3. Pan, X.; Pan, J.; Li, X.; Wang, Z.; Ni, Y.; Wang, Q. Tough Supramolecular Hydrogels Crafted via Lignin-Induced Self-Assembly. *Adv. Mater.* **2024**, *36*, 2406671. [CrossRef] [PubMed]
4. Li, Z.; Yin, F.; He, W.; Hang, T.; Li, Z.; Zheng, J.; Li, X.; Jiang, S.; Chen, Y. Anti-Freezing, Recoverable and Transparent Conductive Hydrogels Co-Reinforced by Ethylene Glycol as Flexible Sensors for Human Motion Monitoring. *Int. J. Biol. Macromol.* **2023**, *230*, 123117. [CrossRef] [PubMed]
5. Wang, Y.; Zhang, L.; Lu, A. Highly Stretchable, Transparent Cellulose/PVA Composite Hydrogel for Multiple Sensing and Triboelectric Nanogenerators. *J. Mater. Chem. A* **2020**, *8*, 13935–13941. [CrossRef]
6. Wang, H.; Lu, J.; Huang, H.; Fang, S.; Zubair, M.; Peng, Z. A Highly Elastic, Room-Temperature Repairable and Recyclable Conductive Hydrogel for Stretchable Electronics. *J. Colloid Interface Sci.* **2021**, *588*, 295–304. [CrossRef]
7. Zheng, B.; Zhou, H.; Wang, Z.; Gao, Y.; Zhao, G.; Zhang, H.; Jin, X.; Liu, H.; Qin, Z.; Chen, W.; et al. Fishing Net-Inspired Mutiscale Ionic Organohydrogels with Outstanding Mechanical Robustness for Flexible Electronic Devices. *Adv. Funct. Mater.* **2023**, *33*, 2213501. [CrossRef]
8. Yu, W.; Chen, J.; Gao, Q.; Guo, Y.; Zhang, S.; Pan, Y.; Nie, B.; Zhang, X.; Jiang, L.; Qiu, J.; et al. Multifunctional PVA/SA-Based Hydrogels Integrating High Stretchability, Conductivity, and Antibacterial Activity for Human-Machine Interactive Flexible Sensors. *Chem. Eng. J.* **2025**, *519*, 164695. [CrossRef]
9. Liu, X.; Wu, Z.; Alomar, T.S.; AlMasoud, N.; Liu, X.; Han, X.; Guo, N.; Weng, L.; Gao, J.; Algadi, H.; et al. Poly (Vinyl Alcohol)/Carboxylated Cellulose Nanofibers Composite Hydrogel Flexible Strain Sensors. *Int. J. Biol. Macromol.* **2025**, *309*, 142902. [CrossRef]
10. Hasan, M.Z.; Xu, C.; Motaleb, K.Z.M.A.; Ahmed, M.F.; Zhuang, J.; Tan, S.; Janutėnienė, J.; Bashar, M.M.; Tu, H.; Luo, L.; et al. Extraction and Incorporation of Cellulose Microfibers from Textile Wastes into MXene-Enhanced PVA-Borax Hydrogel for Multifunctional Wearable Sensors. *Int. J. Biol. Macromol.* **2025**, *295*, 139640. [CrossRef]
11. Zhao, Y.; Zhang, X.; Hao, Y.; Zhao, Y.; Ding, P.; Zhai, W.; Dai, K.; Zheng, G.; Liu, C.; Shen, C. Multifunctional PVA/PNIPAM Conductive Hydrogel Sensors Enabled Human-Machine Interaction Intelligent Rehabilitation Training. *Adv. Compos. Hybrid Mater.* **2024**, *7*, 245. [CrossRef]
12. Qing, X.; Liu, Z.; Vananroye, A.; Franceschini, F.; Bouropoulos, N.; Katsaounis, A.; Taurino, I.; Fardim, P. Self-Healing and Transparent Ionic Conductive PVA/Pullulan/Borax Hydrogels with Multi-Sensing Capabilities for Wearable Sensors. *Int. J. Biol. Macromol.* **2025**, *284*, 137841. [CrossRef] [PubMed]
13. Zhang, B.; Yang, J.; Jiang, Q.; Zheng, R.; Qiu, L.; Chen, J.; You, R.; Yan, Z.; Lu, Y. AuNPs/C-CNF/PVA Hydrogel SERS Sensor for Comprehensive Detection of Antitumor Drug Separation and Enrichment. *Chem. Eng. J.* **2025**, *516*, 164282. [CrossRef]
14. Huang, H.; Han, L.; Li, J.; Fu, X.; Wang, Y.; Yang, Z.; Xu, X.; Pan, L.; Xu, M. Super-Stretchable, Elastic and Recoverable Ionic Conductive Hydrogel for Wireless Wearable, Stretchable Sensor. *J. Mater. Chem. A* **2020**, *8*, 10291–10300. [CrossRef]
15. Chen, S.; Dong, Y.; Ma, S.; Ren, J.; Yang, X.; Wang, Y.; Lü, S. Superstretching MXene Composite Hydrogel as a Bidirectional Stress Response Thixotropic Sensor. *ACS Appl. Mater. Interfaces* **2021**, *13*, 13629–13636. [CrossRef]
16. Zhao, L.; Luo, B.; Gao, S.; Liu, Y.; Lai, C.; Zhang, D.; Guan, W.; Wang, C.; Chu, F. Stretchable, Self-Adhesive, and Conductive Hemicellulose-Based Hydrogels as Wearable Strain Sensors. *Int. J. Biol. Macromol.* **2024**, *282*, 137313. [CrossRef]
17. Zhu, C.; Bamidele, E.A.; Shen, X.; Zhu, G.; Li, B. Machine Learning Aided Design and Optimization of Thermal Metamaterials. *Chem. Rev.* **2024**, *124*, 4258–4331. [CrossRef]

18. Zheng, J.-J.; Li, Q.-Z.; Wang, Z.; Wang, X.; Zhao, Y.; Gao, X. Computer-Aided Nanodrug Discovery: Recent Progress and Future Prospects. *Chem. Soc. Rev.* **2024**, *53*, 9059–9132. [CrossRef]
19. Xu, B.; Yang, G. Interpretability Research of Deep Learning: A Literature Survey. *Inf. Fusion* **2025**, *115*, 102721. [CrossRef]
20. Yan, B.; Shi, Y.; Liu, S.; Wang, W.; Ba, Z.; Li, H.; Wang, J.; Xiao, Z.; Peng, T.; Liang, D.; et al. Development and Optimization of Lithium-Ion Sieves through Machine Learning in Complex Brine Systems. *Sep. Purif. Technol.* **2025**, *374*, 133726. [CrossRef]
21. Xu, B.; Zhang, Y.; Li, J.; Wang, B.; Li, R.; Cheng, D.; Chang, G. Self-Healing PVA/Chitosan/MXene Triple Network Hydrogel for Strain and Temperature Sensors. *Int. J. Biol. Macromol.* **2025**, *290*, 138811. [CrossRef]
22. Wang, Y.; Zhang, Y.; Ren, P.; Yu, S.; Cui, P.; Nielsen, C.B.; Abrahams, I.; Briscoe, J.; Lu, Y. Versatile and Recyclable Double-Network PVA/Cellulose Hydrogels for Strain Sensors and Triboelectric Nanogenerators under Harsh Conditions. *Nano Energy* **2024**, *125*, 109599. [CrossRef]
23. Chen, S.; Guo, B.; Yu, J.; Yan, Z.; Liu, R.; Yu, C.; Zhao, Z.; Zhang, H.; Yao, F.; Li, J. A Polypyrrole-Dopamine/Poly(Vinyl Alcohol) Anisotropic Hydrogel for Strain Sensor and Bioelectrodes. *Chem. Eng. J.* **2024**, *486*, 150182. [CrossRef]
24. Hsiao, Y.-C.; Lee, L.-C.; Lin, Y.-T.; Hong, S.-H.; Wang, K.-C.; Tung, S.-H.; Liu, C.-L. Stretchable Polyvinyl Alcohol and Sodium Alginate Double Network Ionic Hydrogels for Low-Grade Heat Harvesting with Ultrahigh Thermopower. *Mater. Today Energy* **2023**, *37*, 101383. [CrossRef]
25. Li, M.; Wang, Y.; Wei, Q.; Zhang, J.; An, Y.; Chen, X.; Liu, Z.; Jiao, D. A Triple-Network PVA/Cellulose Nanofiber Composite Hydrogel with Excellent Strength, Transparency, Conductivity, and Antibacterial Properties. *J. Mater. Sci. Technol.* **2025**, *237*, 312–322. [CrossRef]
26. Peng, X.; Wang, W.; Yang, W.; Chen, J.; Peng, Q.; Wang, T.; Yang, D.; Wang, J.; Zhang, H.; Zeng, H. Stretchable, Compressible, and Conductive Hydrogel for Sensitive Wearable Soft Sensors. *J. Colloid Interface Sci.* **2022**, *618*, 111–120. [CrossRef]
27. Tavakoli, J.; Mirzaei, S.; Tang, Y. Cost-Effective Double-Layer Hydrogel Composites for Wound Dressing Applications. *Polymers* **2018**, *10*, 305. [CrossRef]
28. Pan, X.; Wang, Q.; He, P.; Liu, K.; Ni, Y.; Chen, L.; Ouyang, X.; Huang, L.; Wang, H.; Xu, S. A Bionic Tactile Plastic Hydrogel-Based Electronic Skin Constructed by a Nerve-like Nanonetwork Combining Stretchable, Compliant, and Self-Healing Properties. *Chem. Eng. J.* **2020**, *379*, 122271. [CrossRef]
29. Li, P.; Wang, H.; Ju, Z.; Jin, Z.; Ma, J.; Yang, L.; Zhao, X.; Xu, H.; Liu, Y. Ti3C2Tx MXene- and Sulfuric Acid-Treated Double-Network Hydrogel with Ultralow Conductive Filler Content for Stretchable Electromagnetic Interference Shielding. *ACS Nano* **2024**, *18*, 2906–2916. [CrossRef]
30. Pan, X.; Li, X.; Wang, R.; Wang, Z.; Ni, Y.; Wang, Q. Accelerated Hydrogel Strengthening: Synergy between Mechanical Training and Lignin Intake. *Nano Lett.* **2025**, *25*, 4526–4531. [CrossRef]
31. Shi, W.; Wang, Z.; Song, H.; Chang, Y.; Hou, W.; Li, Y.; Han, G. High-Sensitivity and Extreme Environment-Resistant Sensors Based on PEDOT:PSS@PVA Hydrogel Fibers for Physiological Monitoring. *ACS Appl. Mater. Interfaces* **2022**, *14*, 35114–35125. [CrossRef]
32. Wang, J.; Xu, P.; Ji, X.; Li, M.; Lu, W. Feature Selection in Machine Learning for Perovskite Materials Design and Discovery. *Materials* **2023**, *16*, 3134. [CrossRef] [PubMed]
33. Choi, W.; Advincula, R.C.; Wu, H.F.; Jiang, Y. Artificial Intelligence and Machine Learning in the Design and Additive Manufacturing of Responsive Composites. *MRS Commun.* **2023**, *13*, 714–724. [CrossRef]
34. Dean, J.; Scheffler, M.; Purcell, T.A.R.; Barabash, S.V.; Bhowmik, R.; Bazhirov, T. Interpretable Machine Learning for Materials Design. *J. Mater. Res.* **2023**, *38*, 4477–4496. [CrossRef]
35. Zhang, H.; Chen, W.; Iyer, A.; Apley, D.W.; Chen, W. Uncertainty-Aware Mixed-Variable Machine Learning for Materials Design. *Sci. Rep.* **2022**, *12*, 19760. [CrossRef]
36. Ling, N.; Wang, Y.; Song, S.; Liu, C.; Yang, F.; Qi, X.; Li, Y.; Zhang, J.; Wang, L. Experimentally Validated Screening Strategy for Alloys as Anode in Mg-Air Battery with Multi-Target Machine Learning Predictions. *Chem. Eng. J.* **2024**, *496*, 153824. [CrossRef]
37. Lai, J.-P.; Lin, Y.-L.; Lin, H.-C.; Shih, C.-Y.; Wang, Y.-P.; Pai, P.-F. Tree-Based Machine Learning Models with Optuna in Predicting Impedance Values for Circuit Analysis. *Micromachines* **2023**, *14*, 265. [CrossRef]
38. Hassanali, M.; Soltanaghaei, M.; Gandomani, T.J.; Boroujeni, F.Z. Software Development Effort Estimation Using Boosting Algorithms and Automatic Tuning of Hyperparameters with Optuna. *J. Softw. Evol. Process* **2024**, *36*, e2665. [CrossRef]
39. Xu, C.; Guan, S.; Dong, X.; Huang, H.; Qi, M. Ultratough and Freezing-Tolerant PVA–PAA-PANI Hybrid Hydrogel for Supercapacitors and Flexible Sensors. *ACS Sustain. Chem. Eng.* **2023**, *11*, 14886–14894. [CrossRef]
40. Yuan, H.; Han, S.; Wang, S.; Yang, P.; Li, S.; Mi, H.-Y.; Liu, C.; Shen, C. Plant-Derived Adhesive Hydrogel with High Stretchability and Conductivity for Wearable Electronics. *Sens. Actuators B Chem.* **2023**, *379*, 133195. [CrossRef]
41. Miao, L.; Wang, X.; Li, S.; Tu, Y.; Hu, J.; Huang, Z.; Lin, S.; Gui, X. An Ultra-Stretchable Polyvinyl Alcohol Hydrogel Based on Tannic Acid Modified Aramid Nanofibers for Use as a Strain Sensor. *Polymers* **2022**, *14*, 3532. [CrossRef] [PubMed]
42. Gu, Z.; Ma, R.; Chen, X.; Lin, Z.; Yang, Y.; Tan, B.; Sun, J.; Chen, T. Polyvinyl Alcohol Modified Plant Fiber Hydrogel Pressure and Strain Dual-Model Sensors for Biomedical Signal Detection. *Adv. Compos. Hybrid Mater.* **2025**, *8*, 214. [CrossRef]

43. Li, K.; Yang, X.; Dong, X.; Cao, H.; Zhuang, S.; Gu, X. Easy Regulation of Chitosan-Based Hydrogel Microstructure with Citric Acid as an Efficient Buffer. *Carbohydr. Polym.* **2023**, *300*, 120258. [CrossRef]
44. Zhang, S.; Guo, F.; Gao, X.; Yang, M.; Huang, X.; Zhang, D.; Li, X.; Zhang, Y.; Shang, Y.; Cao, A. High-Strength, Antiswelling Directional Layered PVA/MXene Hydrogel for Wearable Devices and Underwater Sensing. *Adv. Sci.* **2024**, *11*, 2405880. [CrossRef] [PubMed]
45. Ma, Y.; Gao, Y.; Liu, L.; Ren, X.; Gao, G. Skin-Contactable and Antifreezing Strain Sensors Based on Bilayer Hydrogels. *Chem. Mater.* **2020**, *32*, 8938–8946. [CrossRef]
46. Jiang, C.; Ding, X.; Xie, W.; Wu, D. Ultrastretchable Composite Organohydrogels with Dual Cross-Links Enabling Multimodal Sensing. *ACS Appl. Mater. Interfaces* **2022**, *14*, 55143–55154. [CrossRef]
47. Gu, J.; Huang, J.; Chen, G.; Hou, L.; Zhang, J.; Zhang, X.; Yang, X.; Guan, L.; Jiang, X.; Liu, H. Multifunctional Poly(Vinyl Alcohol) Nanocomposite Organohydrogel for Flexible Strain and Temperature Sensor. *ACS Appl. Mater. Interfaces* **2020**, *12*, 40815–40827. [CrossRef]
48. Kumar, A.; Behl, T.; Chadha, S. Synthesis of Physically Crosslinked PVA/Chitosan Loaded Silver Nanoparticles Hydrogels with Tunable Mechanical Properties and Antibacterial Effects. *Int. J. Biol. Macromol.* **2020**, *149*, 1262–1274. [CrossRef]
49. Li, Q.; Zhi, X.; Xia, Y.; Han, S.; Guo, W.; Li, M.; Wang, X. Ultrastretchable High-Conductivity MXene-Based Organohydrogels for Human Health Monitoring and Machine-Learning-Assisted Recognition. *ACS Appl. Mater. Interfaces* **2023**, *15*, 19435–19446. [CrossRef]
50. Gupta, B.; Bano, S.; Singh, R. High-Performance Silver Nanoparticles Embedded Conductive PVA Hydrogel for Stretchable Wearable Triboelectric Nanogenerators. *J. Power Sources* **2025**, *632*, 236271. [CrossRef]
51. Lei, T.; Pan, J.; Wang, N.; Xia, Z.; Zhang, Q.; Fan, J.; Tao, L.; Shou, W.; Gao, Y. Cold-Resistant, Highly Stretchable Ionic Conductive Hydrogels for Intelligent Motion Recognition in Winter Sports. *Mater. Horiz.* **2024**, *11*, 1234–1250. [CrossRef]
52. Hu, L.; Wang, Y.; Liu, Q.; Liu, M.; Yang, F.; Wang, C.; Pan, P.; Wang, L.; Chen, L.; Chen, J. Real-Time Monitoring Flexible Hydrogels Based on Dual Physically Cross-Linked Network for Promoting Wound Healing. *Chin. Chem. Lett.* **2023**, *34*, 108262. [CrossRef]
53. Luo, C.; Huang, M.; Sun, X.; Wei, N.; Shi, H.; Li, H.; Lin, M.; Sun, J. Super-Strong, Nonswellable, and Biocompatible Hydrogels Inspired by Human Tendons. *ACS Appl. Mater. Interfaces* **2022**, *14*, 2638–2649. [CrossRef] [PubMed]
54. Macarie, L.; Pekar, M.; Simulescu, V.; Plesu, N.; Iliescu, S.; Ilia, G.; Tara-Lunga-Mihali, M. Properties in Aqueous Solution of Homo- and Copolymers of Vinylphosphonic Acid Derivatives Obtained by UV-curing. *Macromol. Res.* **2017**, *25*, 214–221. [CrossRef]
55. Guo, Y.; Jiang, Y.; Zhang, Y.; Huang, L.; Tang, K. Highly Stretchable and Adhesive Hydrogel Dressing Possessing Rapid Hemostasis and Photothermal Antibacterial Properties for Joint Wound Healing. *Chem. Eng. J.* **2025**, *519*, 164834. [CrossRef]
56. Han, Y.; Liu, Y.; Liu, Y.; Jiang, D.; Wu, Z.; Jiang, B.; Yan, H.; Toktarbay, Z. High-Performance PVA-Based Hydrogels for Ultra-Sensitive and Durable Flexible Sensors. *Adv. Compos. Hybrid Mater.* **2025**, *8*, 154. [CrossRef]
57. Pu, L.; Yuan, Z.; Cai, Y.; Li, X.; Xue, Z.; Niu, Y.; Li, Y.; Ma, S.; Xu, W. Multiperformance PAM/PVA/CaCO3 Hydrogel for Flexible Sensing and Information Encryption. *ACS Appl. Mater. Interfaces* **2024**, *16*, 32762–32772. [CrossRef]
58. Li, Z.; Jiang, J.; Luo, J.; Meng, J.; Cheng, L.; Qin, H. A Robust and Conductive Hydrotalcite/Nanocellulose/PVA Hydrogel Constructed Based on the Hofmeister Effect. *Int. J. Biol. Macromol.* **2025**, *298*, 139994. [CrossRef]
59. Wang, T.; Xu, B.; Yu, T.; Yu, Y.; Fu, J.; Wang, Y.; Gao, X.; Xue, Z.; Li, R.; Chang, G. PVA/Chitosan-Based Multifunctional Hydrogels Constructed through Multi-Bonding Synergies and Their Application in Flexible Sensors. *Carbohydr. Polym.* **2025**, *350*, 123034. [CrossRef]
60. Liu, X.; Wu, Z.; Jiang, D.; Guo, N.; Wang, Y.; Ding, T.; Weng, L. A Highly Stretchable, Sensing Durability, Transparent, and Environmentally Stable Ion Conducting Hydrogel Strain Sensor Built by Interpenetrating Ca2+-SA and Glycerol-PVA Double Physically Cross-Linked Networks. *Adv. Compos. Hybrid Mater.* **2022**, *5*, 1712–1729. [CrossRef]
61. Prameswati, A.; Nurmaulia Entifar, S.A.; Han, J.W.; Wibowo, A.F.; Kim, J.H.; Sembiring, Y.S.B.; Park, J.; Lee, J.; Lee, A.-Y.; Song, M.H.; et al. Self-Healable Conductive Hydrogels with High Stretchability and Ultralow Hysteresis for Soft Electronics. *ACS Appl. Mater. Interfaces* **2023**, *15*, 24648–24657. [CrossRef]
62. Angel-Sánchez, K.D.; Treviño-Pacheco, A.V.; Perales-Martínez, I.A.; Martínez-Romero, O.; Olvera-Trejo, D.; Elías-Zúñiga, A. Antibacterial Crosslinker for Ternary PCL-Reinforced Hydrogels Based on Chitosan, Polyvinyl Alcohol, and Gelatin for Tissue Engineering. *Polymers* **2025**, *17*, 1520. [CrossRef] [PubMed]
63. Xu, Y.; Tan, C.; He, Y.; Luo, B.; Liu, M. Chitin Nanocrystals Stabilized Liquid Metal for Highly Stretchable and Anti-Freeze Hydrogels as Flexible Strain Sensor. *Carbohydr. Polym.* **2024**, *328*, 121728. [CrossRef] [PubMed]

gels

MDPI

Article

Interpretable Machine Learning Analysis of Design Factors in Hydrogel Supercapacitors

Liying Xu [1,*], Siqi Liu [2], Dandan Hu [2], Junhao Liu [2], Yuze Zhang [2], Ziqiang Li [2], Zichuan Su [3] and Daxin Liang [2,*]

[1] School of Food Engineering, Harbin University, Harbin 150086, China
[2] Key Laboratory of Bio-Based Material Science and Technology (Ministry of Education), Northeast Forestry University, Harbin 150040, China; siqiliu61@163.com (S.L.); 17868651902@nefu.edu.cn (D.H.); 2022210600@nefu.edu.cn (J.L.); zyz050611@163.com (Y.Z.); lzqqqq2001@163.com (Z.L.)
[3] College of Chemistry and Chemical Engineering, China University of Petroleum (East China), Qingdao 266580, China; z22060089@s.upc.edu.cn
* Correspondence: lyxutk@163.com (L.X.); daxin.liang@nefu.edu.cn (D.L.); Tel.: +86-18182806182 (L.X.)

Abstract: Understanding the relationships between design factors is crucial for the development of hydrogel supercapacitors, yet the relative importance and interdependencies of material properties and operating conditions remain unclear. This study employs interpretable machine learning to analyze the design factors that affect hydrogel supercapacitor performance, using 232 experimental samples from 41 recent studies. SHAP analysis was implemented to quantify parameter importance and reveal feature interactions among 16 key design parameters, including polymer types, electrolyte formulations, and operating conditions. Results show that synthetic vinyl polymers most strongly influence specific capacitance, while conductive polymers predominantly affect cycle stability. Ionic conductivity emerged as the most impactful parameter despite moderate feature importance, indicating complex nonlinear relationships. Critical interdependencies between polymer concentration and electrolyte formulation suggest that optimal design requires coordinated parameter selection rather than independent optimization. This interpretable framework provides quantitative insights into design factor hierarchies and parameter interdependencies, offering evidence-based guidelines for rational material selection in hydrogel supercapacitor development.

Keywords: hydrogel supercapacitor; machine learning; interpretable prediction

1. Introduction

Supercapacitors have emerged as critical energy storage devices that bridge the gap between conventional capacitors and batteries, offering unique advantages of high power density, rapid charge–discharge capabilities, and exceptional cycle stability [1]. With the growing demand for sustainable energy solutions and the proliferation of portable electronics, electric vehicles, and grid-scale energy storage systems, supercapacitors have found extensive applications ranging from consumer electronics and automotive systems to renewable energy integration and emergency power backup [2]. Their ability to deliver high power output within seconds while maintaining long-term reliability makes them indispensable components in modern energy infrastructure, particularly in applications requiring quick energy bursts and frequent cycling operations [3].

Hydrogel-based supercapacitors have attracted significant attention as promising candidates for flexible energy storage applications due to their unique combination of high ionic conductivity, mechanical flexibility, and environmental adaptability [4–6]. Recent

 https://doi.org/10.3390/gels11060464

studies have demonstrated that hydrogel electrolytes can achieve exceptional ionic conductivity owing to their three-dimensional polymer networks with high water content, facilitating efficient ion transport and enhanced electrochemical performance [7,8]. Li et al. developed flexible supercapacitors based on organohydrogel electrolytes that exhibit remarkable long-term anti-freezing and anti-drying properties, addressing key challenges in practical applications [9]. The structural design of hydrogel electrodes has also proven crucial for performance optimization, as evidenced by Huang et al., who employed multi-scale structural engineering strategies to develop high-performance MXene hydrogel supercapacitor electrodes with enhanced capacitive behavior [10]. The self-healing capability of hydrogel electrolytes represents another significant advantage, as reported by Zhao et al., who developed self-healing hydrogel electrolytes for flexible solid-state supercapacitors, enabling device recovery from mechanical damage and extending operational lifespan [11]. These inherent properties of hydrogels, including tunable mechanical strength, biocompatibility, and environmental stability, make them superior alternatives to conventional rigid electrolyte systems for next-generation flexible and wearable energy storage devices.

However, the development of high-performance hydrogel-based supercapacitors faces significant challenges in optimizing the complex interplay between polymer composition, electrolyte formulation, and operating conditions [12–15]. Traditional trial-and-error approaches for material design are time-consuming, resource-intensive [16], and often fail to capture the intricate relationships between multiple design parameters and performance outcomes [17]. The vast parameter space involving polymer types, electrolyte concentrations, ionic conductivity, current density, and voltage windows creates a multidimensional optimization problem that is difficult to navigate using conventional experimental methods [7,18,19]. However, interpretable machine learning (ML) approaches offer unprecedented opportunities to understand material design principles by identifying complex relationships in experimental data, quantifying parameter importance, and revealing the underlying factors governing optimal material combinations [20,21].

In this work, we developed a comprehensive interpretable machine learning framework to understand the design factors governing hydrogel-based supercapacitor performance and elucidate their relative importance. A systematic literature survey was conducted to collect 232 experimental data samples encompassing diverse material compositions and operating conditions to analyze parameter influences and material dependencies [9–13,15,22–56]. The dataset includes key parameters such as polymer types, electrolyte formulations, ionic conductivity, current density, voltage windows, and cycling conditions as input features to analyze their relationships with specific capacitance and cycle stability. Multiple machine learning algorithms were systematically evaluated to model these complex relationships, with SHAP analysis employed to quantify parameter importance and reveal the differential impacts of various polymer categories and material parameters on supercapacitor performance [57]. This interpretable approach provides mechanistic insights into design factor hierarchies and parameter interdependencies, offering evidence-based guidelines for understanding the critical factors governing hydrogel supercapacitor performance and rational material selection strategies. The overall workflow of this study is illustrated in Figure 1.

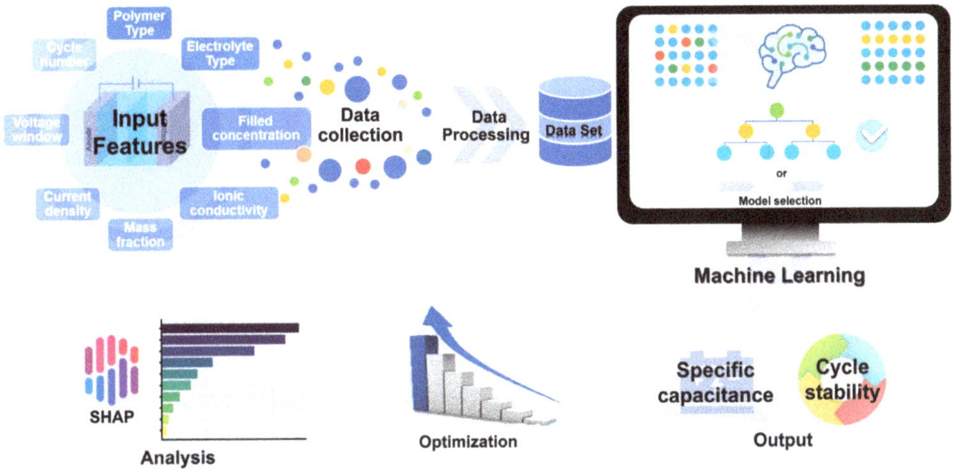

Figure 1. Workflow of interpretable machine learning framework for hydrogel supercapacitor performance prediction.

2. Results and Discussion

2.1. Analysis of Input Features Prior to Model Development

To understand the interrelationships among input parameters [17,58], Pearson correlation analysis was performed on all features before model development. The input parameters include: Polymer Concentration (PC), Alkali Metal (AM), Acid System (AS), Zinc System (ZS), Mixed System (MS), Organic Additive (OA), electrolyte Mass Fraction (MF), Ionic Conductivity (IC), Current Density (CD), Voltage Window (VW), Cycle Number (CN), Synthetic Vinyl (SV), Natural Polysaccharide (NP), Conductive Polymer (CP), Polybase (PB), and Polyacid (PA).

The correlation matrix in Figure 2 reveals generally weak to moderate correlations among most input features, with the majority of correlation coefficients ranging from −0.5 to 0.5. This indicates that the selected features provide relatively independent information for model training, which is beneficial for machine learning performance. Notable observations include moderate positive correlations between certain electrolyte system parameters (OA-IC: 0.39, MS-OA: 0.31) and some polymer type relationships. The electrolyte MF shows weak correlations with most other parameters, suggesting its independent contribution to supercapacitor performance [24,27,31,42].

Among polymer type features, some expected relationships are observed, such as the moderate negative correlation between SV and NP (−0.48), reflecting their distinct chemical nature. The relatively low correlation values across most feature pairs indicate minimal multicollinearity issues, supporting the robustness of subsequent machine learning model development. This correlation analysis confirms that the feature engineering approach effectively captures diverse and complementary aspects of hydrogel supercapacitor design parameters [2,59,60].

Among polymer type features, the most significant correlation is the negative relationship between SV and NP at −0.48, which reflects their fundamentally different chemical origins and application strategies. The SV category encompasses synthetic polymers including polyvinyl alcohol (PVA), polyacrylamide (PAM), and polyacrylonitrile (PAN), which are characterized by controlled molecular structures, predictable crosslinking behaviors, and consistent mechanical properties. In contrast, the NP category includes natural biopolymers such as sodium alginate (SA), cellulose, chitosan (CS), alginate, and starch, which offer excellent biocompatibility, strong crosslinking capabilities through inherent functional groups, and sustainable material advantages [41,43,46,49]. The Conductive Polymer (CP) category, comprising polyaniline (PANI), polypyrrole (PPY), polyethylenedioxythiophene (PEDOT), and polythiophene (PTH), shows relatively weak correlations with other polymer types, indicating its specialized role in enhancing electrical conductivity rather than serving as a primary structural matrix. Similarly, protein-based (PB) polymers, including gelatin, collagen, albumin, and lignin, demonstrate independent behavior, reflecting their unique position in biocompatible and sustainable supercapacitor designs [51–53]. This correlation pattern suggests that material selection follows targeted performance criteria: synthetic vinyl polymers for mechanical reliability, natural polysaccharides for biocompatibility and flexibility, conductive polymers for electrical enhancement, and protein-based materials for specialized biomedical applications.

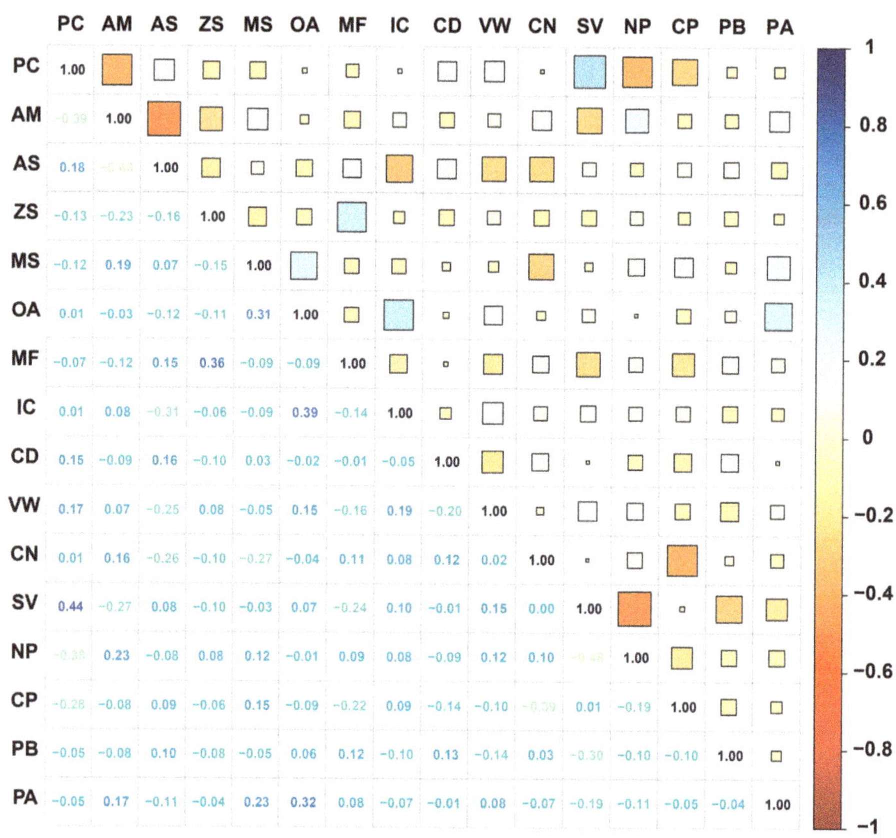

Figure 2. Pearson correlation matrix of input features.

2.2. Model Development

Six machine learning algorithms were systematically evaluated for predicting both specific capacitance and cycle stability of hydrogel supercapacitors. The performance of each model was assessed using training and testing datasets, with results presented as predicted vs. true value plots in Figure 3.

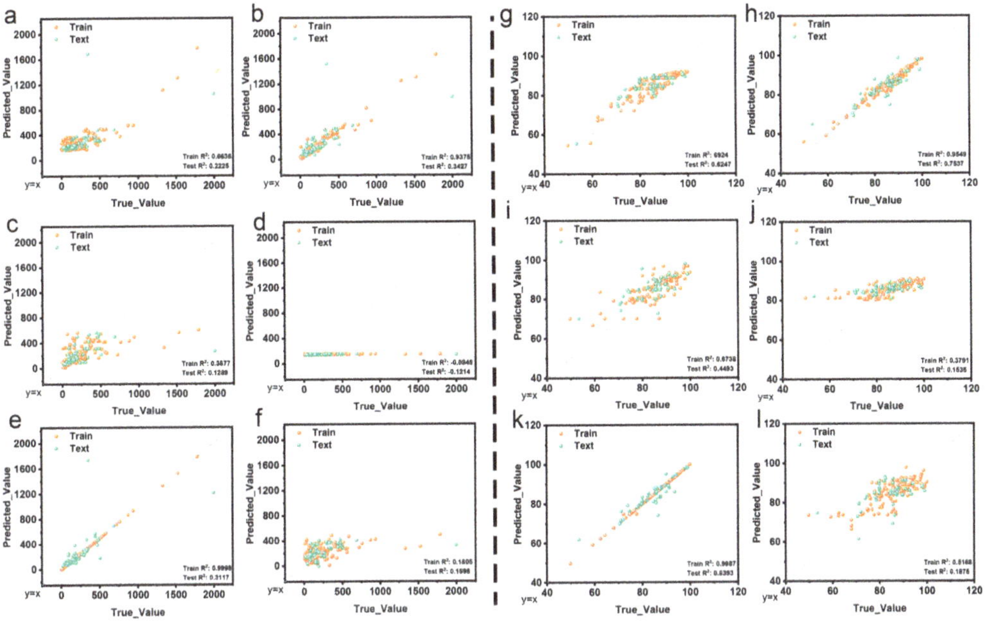

Figure 3. Machine learning model performance for specific capacitance (**a**–**f**) and cycle stability (**g**–**l**). (**a**,**g**) AdaBoost, (**b**,**h**) GBR, (**c**,**i**) KNN, (**d**,**j**) SVM, (**e**,**k**) XGBoost, (**f**,**l**) Lasso.

For specific capacitance prediction (Figure 3a–f), all models demonstrated poor overall performance with testing R^2 values below 0.35. Notably, XGBoost (Figure 3e) exhibited significant overfitting with an extremely high training R^2 of 0.9996 but a testing R^2 of only 0.3117, suggesting that while XGBoost has the potential to capture the underlying patterns in specific capacitance data, proper hyperparameter optimization is crucial to improve generalization performance. The other algorithms showed consistently poor performance across both training and testing datasets, with SVM (Figure 3d) particularly struggling to fit the data [61].

For cycle stability prediction (Figure 3g–l), the models generally achieved superior performance compared to specific capacitance prediction. GBR (Figure 3h) demonstrated exceptional performance with training R^2 of 0.9549 and testing R^2 of 0.7537, indicating both high accuracy and excellent generalization capability. XGBoost (Figure 3k) also showed strong performance with training R^2 of 0.9897 and testing R^2 of 0.3393. AdaBoost (Figure 3g), KNN (Figure 3i), and Lasso (Figure 3l) achieved reasonable performance with testing R^2 values ranging from 0.1675 to 0.6483. Notably, SVM (Figure 3j) again showed poor fitting with training R^2 of 0.3791 and testing R^2 of 0.1535.

The superior performance of tree-based ensemble methods (GBR and XGBoost) suggests that the relationships between input features and target properties involve complex nonlinear interactions that are effectively captured by these algorithms. The systematic model comparison through Taylor diagram analysis (Figure S1) confirmed the superior performance of optimized XGBoost for specific capacitance prediction and GBR for cycle stability prediction, while violin plot analysis (Figure S2) revealed distinct error distribution patterns that provide insights into model reliability and prediction consistency. The consistently poor performance of SVM indicates that the dataset characteristics may not be suitable for this particular algorithm's assumptions [62].

2.3. Hyperparameter Optimization

To address the overfitting issues identified in the initial model evaluation and improve predictive performance, systematic hyperparameter optimization was conducted for the most promising algorithms. The cross-validation results demonstrate that XGBoost optimization for cycle stability prediction achieved $R^2 = 0.857$ and GBR optimization for specific capacitance prediction achieved $R^2 = 0.821$, which are lower than the optimal pairings shown in Figure 4. These comparative results validate our target-specific model selection approach, confirming that XGBoost is most suitable for specific capacitance prediction while GBR excels in cycle stability prediction.

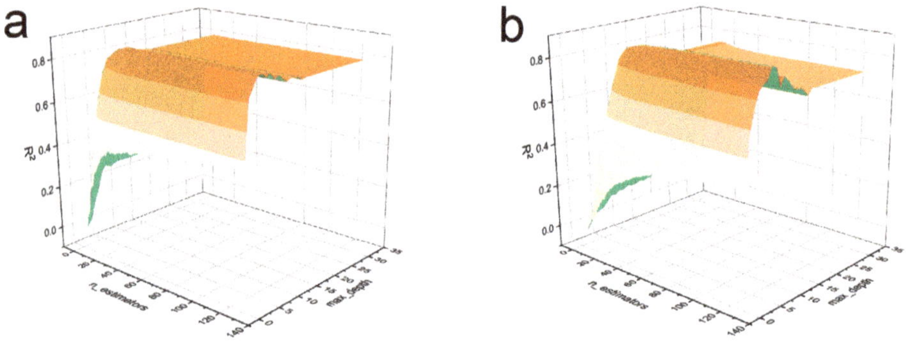

Figure 4. (**a**) XGBoost optimization for specific capacitance prediction and (**b**) GBR optimization for cycle stability prediction.

Figure 4a illustrates the hyperparameter optimization landscape for XGBoost in specific capacitance prediction, revealing a complex three-dimensional performance surface with multiple local optima and valleys. The optimization trajectory demonstrates the systematic exploration of the parameter space, starting from relatively low R^2 values and gradually ascending toward optimal regions. The surface topology indicates that XGBoost performance is highly sensitive to hyperparameter selection, with dramatic variations in R^2 across different parameter combinations. Notable features include steep gradients in certain regions, suggesting that small parameter changes can lead to significant performance differences [62,63].

Similarly, Figure 4b presents the hyperparameter optimization results for GBR in cycle stability prediction. The optimization landscape shows a smoother performance surface with a well-defined optimal region, indicating more stable behavior across different parameter combinations. The GBR optimization successfully identified parameter settings that maintain the model's strong fitting capability while enhancing generalization performance, building upon the already promising initial results for cycle stability prediction.

2.4. Feature Importance and Interpretability Analysis

The optimized XGBoost model for specific capacitance prediction shows substantially improved performance compared to the initial model (Figure 5a). The predicted vs. true value plot reveals excellent fitting with a training R^2 of 0.99 and testing R^2 of 0.8, representing a significant improvement in generalization capability while maintaining high training accuracy. The mean absolute error (MAE) of 2.44 and root mean square error (RMSE) of 4.05 indicate reasonable prediction precision across the range of specific capacitance values. The well-distributed data points along the diagonal line demonstrate that the hyperparameter optimization successfully addressed the overfitting issues identified in the preliminary analysis. Feature importance analysis (Figure 5b) reveals the relative contribution of different input parameters to specific capacitance prediction. SV emerges as the most influential feature with an importance score of approximately 0.15, suggesting that synthetic vinyl polymers like PVA, PAM, and PAN play a crucial role in determining capacitive performance. MF ranks as the second most important feature, indicating that electrolyte concentration significantly affects specific capacitance [34,37]. NP and CP also show substantial importance, reflecting the significant impact of polymer type selection on capacitive behavior [24,45,48]. Notably, CN and PC demonstrate moderate importance with feature importance scores of approximately 0.10 and 0.08, respectively, indicating that cycling protocols and polymer concentration have secondary but meaningful impacts on specific capacitance. The moderate influence of PC reflects the critical balance between mechanical integrity and ionic accessibility—optimal polymer concentrations must maintain structural stability while avoiding impedance of ion transport pathways [32,34]. In contrast, electrolyte system features (ZS, MS, PA) show relatively lower influence (importance scores < 0.06), suggesting that physical properties such as ionic conductivity and mass fraction are more critical than specific electrolyte chemistry for capacitive performance [39,41,43].

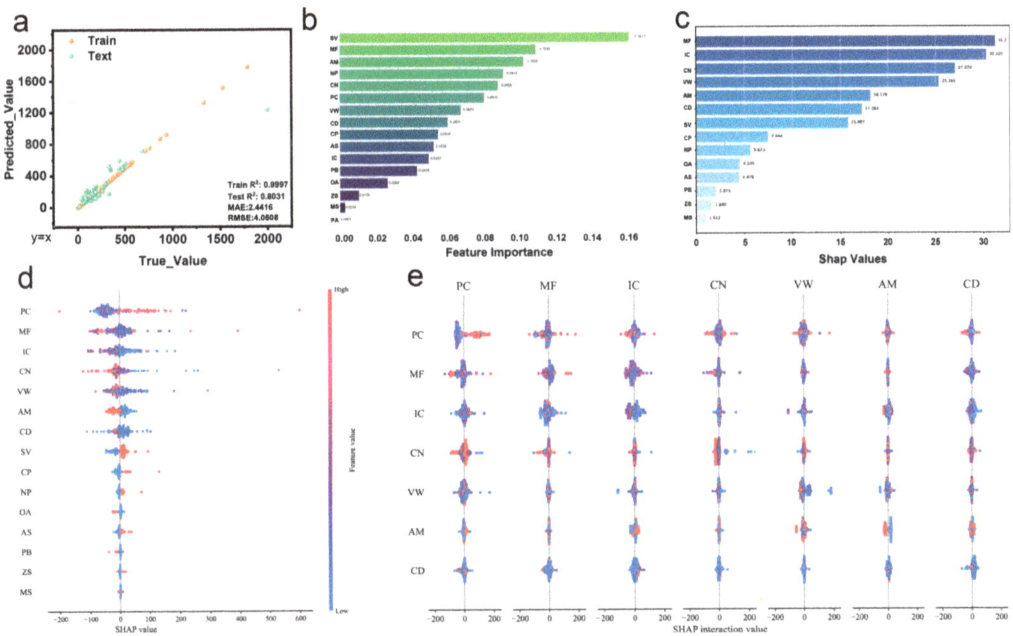

Figure 5. XGBoost analysis for specific capacitance prediction. (**a**) Optimized model performance, (**b**) feature importance ranking, (**c**) SHAP value summary, (**d**) SHAP value distribution for individual features, and (**e**) SHAP interaction analysis showing feature interdependencies.

SHAP summary analysis (Figure 5c) provides deeper insights into feature contributions by quantifying the impact magnitude of each parameter. IC shows the highest SHAP values at approximately 30, indicating its dominant influence on model predictions despite its moderate feature importance ranking. This apparent discrepancy suggests that while IC may not be frequently used for decision-making in the tree structure, its impact when activated is substantial. CN and PC display significant SHAP values around 27 and 24, respectively, confirming their important roles in specific capacitance determination [37,49]. The detailed SHAP value distribution (Figure 5d) illustrates how individual features contribute positively or negatively to predictions across different samples. PC, MF, IC, and CN exhibit the widest SHAP value ranges, with PC showing both substantial positive and negative contributions, indicating highly context-dependent effects on specific capacitance. PC exhibits the widest SHAP range, with both positive and negative contributions evenly distributed, indicating that polymer concentration effects are highly context-dependent and can either enhance or inhibit specific capacitance based on other system parameters [11,25]. MF shows a similarly broad range but with a slight bias toward positive contributions, suggesting that while electrolyte mass fraction generally benefits capacitive performance, optimal values depend on polymer matrix compatibility [55,56,62]. IC demonstrates predominantly positive SHAP values in the high-impact region, confirming its consistently beneficial role, while the color gradient shows that higher ionic conductivity values (red dots) correlate with stronger positive contributions [52,53]. Mid-range features, including VW, AM, CD, and SV, display moderate SHAP distributions, with SV showing primarily positive contributions that support its beneficial effect on capacitive performance. The remaining features demonstrate narrow SHAP distributions concentrated near zero, indicating stable but limited contributions to specific capacitance prediction. This hierarchical pattern confirms our feature importance ranking while revealing that high-impact features exhibit greater variability in their effects, whereas low-importance features provide consistent but modest contributions.

SHAP interaction analysis (Figure 5e) reveals the complex interdependencies between different input features. The interaction heatmap shows that PC exhibits significant interactions with multiple other features, particularly with MF and IC. These interactions suggest that the optimal polymer concentration depends on electrolyte formulation and conductivity levels, highlighting the importance of synergistic material design [11,12,23]. These interaction patterns provide valuable guidance for material optimization, suggesting that holistic design approaches considering parameter interdependencies are essential for achieving optimal specific capacitance performance.

The optimized GBR model achieves outstanding predictive performance for cycle stability (Figure 6a), with training R^2 of 0.9988 and testing R^2 of 0.8698, demonstrating excellent generalization capability that surpasses the specific capacitance prediction results. The MAE of 0.0392 and RMSE of 0.0688 indicate high prediction precision across the cycle stability range. Feature importance analysis (Figure 6b) reveals distinct patterns compared to specific capacitance prediction. CP emerges as the most critical feature with an importance score of approximately 0.23, indicating that conductive polymers play a dominant role in determining cycle stability. This finding suggests that the inherent electrical conductivity and structural stability of conductive polymers are crucial for maintaining performance during extended cycling. PC ranks second in importance, followed by VW and CN, indicating that polymer concentration, voltage window selection, and cycling protocols significantly influence long-term stability. Notably, MF shows moderate importance, while electrolyte system features demonstrate relatively lower influence on cycle stability.

SHAP summary analysis (Figure 6c) quantifies the impact magnitude of each feature on cycle stability predictions. CN shows the highest SHAP values at approximately 1.95, reflecting its fundamental role in defining cycling performance metrics. CP displays substantial SHAP values around 1.79, confirming its critical importance identified in the feature importance analysis [24,29,31]. MF, CD, and AS show moderate SHAP values, indicating their secondary but meaningful contributions to cycle stability. The relatively concentrated SHAP value distribution compared to specific capacitance analysis suggests more predictable and consistent feature impacts on cycle stability. The detailed SHAP value distribution (Figure 6d) illustrates the directional impacts of individual features on cycle stability predictions. CN and VW show the widest SHAP value ranges, indicating high variability in their contributions depending on specific experimental conditions. CP exhibits predominantly positive contributions, confirming that conductive polymer incorporation generally enhances cycle stability. PC shows balanced positive and negative contributions, suggesting that an optimal polymer concentration exists and deviations in either direction can negatively impact stability. The narrower SHAP distributions for electrolyte system features indicate more consistent and predictable impacts on cycle stability [37,38].

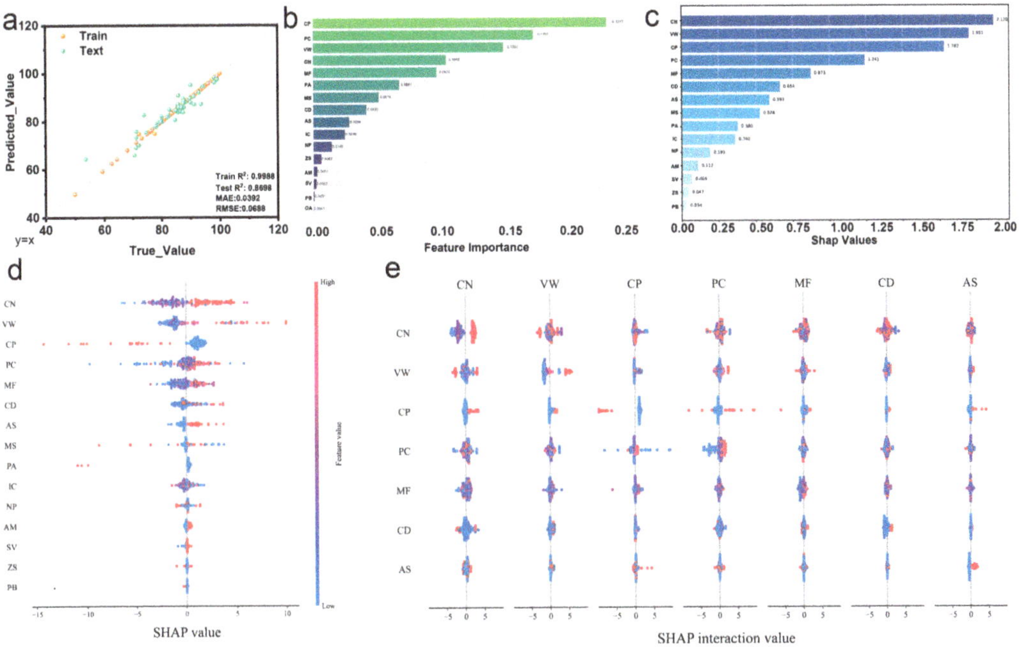

Figure 6. GBR analysis for cycle stability. (**a**) Optimized model performance, (**b**) feature importance ranking, (**c**) SHAP value summary, (**d**) SHAP value distribution for individual features, and (**e**) SHAP interaction analysis showing feature interdependencies.

SHAP interaction analysis (Figure 6e) reveals critical interdependencies between features affecting cycle stability. CN shows strong interactions with multiple parameters, particularly with VW, CP, and PC, indicating that optimal cycling protocols must consider material composition and operating conditions. The significant interaction between VW and CP suggests that voltage window selection should be tailored based on conductive polymer content to maximize cycle stability [15,40,42]. PC exhibits notable interactions with MF and CD, implying that polymer concentration optimization requires consideration of electrolyte formulation and current density conditions. These interaction patterns highlight the importance of integrated design approaches for achieving superior cycle stability, where material selection, concentration optimization, and operating parameter selection must be coordinated to prevent premature degradation and maintain long-term performance [46,47].

3. Conclusions

This study successfully developed an interpretable machine learning framework for understanding design factors in hydrogel-based supercapacitors using 232 experimental samples from the literature. The systematic evaluation of six ML algorithms enabled a comprehensive analysis of parameter–performance relationships, with hyperparameter optimization ensuring robust model behavior for reliable feature importance assessment.

SHAP analysis provided quantitative insights into the relative importance of design factors and their complex interdependencies.

SHAP interpretability analysis revealed key insights into design factor hierarchies and material dependencies. For specific capacitance, synthetic vinyl polymers (SV) emerged as the most influential material category, while ionic conductivity (IC) demonstrated the highest impact magnitude, emphasizing its critical role in capacitive performance. For cycle stability, conductive polymers (CP) proved most important, highlighting how PANI, PPY, PEDOT, and PTH significantly influence long-term performance maintenance [64]. SHAP interaction analysis identified crucial parameter interdependencies, particularly between polymer concentration (PC) and mass fraction (MF), revealing that coordinated parameter selection is essential for understanding optimal design strategies. The correlation analysis validated the feature engineering approach, with weak to moderate correlations ensuring minimal multicollinearity and enabling reliable importance assessment. The distinct material categories demonstrated appropriate independence, facilitating effective analysis of different polymer contributions to electrochemical performance.

This framework provides valuable insights for understanding hydrogel supercapacitor design principles by revealing parameter importance hierarchies and material dependencies through systematic analysis of literature data. The interpretable machine learning approach successfully identified critical design factors and their interdependencies. The comprehensive analysis offers evidence-based guidelines for rational material selection and demonstrates the potential of interpretable AI in elucidating complex material–property relationships.

However, a major challenge encountered during data collection was the lack of standardization in experimental protocols and reporting formats across different studies. Variations in testing conditions, measurement methods, and data presentation significantly hindered comprehensive analysis and limited the development of more sophisticated machine learning models. This heterogeneity represents a critical bottleneck for advancing data-driven materials research in the hydrogel supercapacitor field. To unlock the full potential of machine learning in materials discovery, we advocate for the research community to establish standardized experimental protocols, unified data reporting formats, and open-access databases with consistent metadata. Such initiatives would enable more robust predictive models, facilitate deeper mechanistic understanding, and accelerate the transition toward systematic data-driven materials design, ultimately realizing the transformative potential of artificial intelligence in hydrogel-based energy storage technologies [65,66].

4. Materials and Methods

4.1. Data Collection and Preprocessing

This study utilized Web of Science as the primary data source, employing "hydrogel supercapacitor", "hydrogel electrolyte", and "flexible supercapacitor" as search keywords to systematically collect experimental data from peer-reviewed literature published between 2020 and 2024. The literature survey employed systematic screening criteria including: (1) experimental studies with complete electrochemical performance data, (2) clear polymer and electrolyte specifications, and (3) quantitative reporting of both specific capacitance and cycle stability, resulting in 41 relevant studies that met all inclusion requirements. From these selected studies, 232 experimental data points were systematically extracted through manual data collection from tables, figures, and text, with each data point cross-verified against original sources to ensure accuracy. The extracted dataset encompasses material compositions (polymer types, concentrations), electrolyte formulations (ionic species, mass fractions), operating conditions (current density, voltage windows, cycle numbers), and electrochemical performance metrics (specific capacitance, cycle stability).

Rigorous data preprocessing procedures were implemented including: (1) removal of duplicate entries from identical experimental conditions, (2) treatment of missing values through literature cross-referencing, (3) elimination of statistical outliers using the interquartile range method (IQR > 1.5), and (4) min-max scaling of continuous variables to normalize value ranges. Categorical features including polymer types and electrolyte systems were systematically encoded using one-hot encoding to create binary numerical representations suitable for machine learning processing.

Polymer types were classified into six categories: Synthetic Vinyl (SV) including PVA, PAM, and PAN; Natural Polysaccharide (NP) encompassing SA, cellulose, chitosan, alginate, and starch; Conductive Polymer (CP) comprising PANI, PPY, PEDOT, and PTH; Polybase (PB), Polyacid (PA), and protein-based materials. Electrolyte systems were categorized as Alkali Metal (AM), Acid System (AS), Zinc System (ZS), Mixed System (MS), and Organic Additive (OA) based on chemical composition.

One-hot encoding was applied to categorical features to create binary numerical representations. The final dataset incorporated 16 input features: PC, AM, AS, ZS, MS, OA, MF, IC, CD, VW, CN, SV, NP, CP, PB, and PA, with specific capacitance and cycle stability as target variables.

4.2. ML Model Development

The dataset was split into training and testing sets using an 80:20 ratio to evaluate model generalization performance. Six machine learning algorithms were evaluated: AdaBoost (ADB), Gradient Boosting Regression (GBR), K-Nearest Neighbors (KNN), Support Vector Machine (SVM), XGBoost (version 2.0.2), and Lasso Regression. Tree-based ensemble models were emphasized for their effectiveness in handling categorical features and capturing complex feature interactions in material property prediction.

After model training, validation was performed using the test set, comparing predicted results with actual outcomes to verify the accuracy of different models. Model predictive performance was quantified using key statistical parameters, including root mean squared error (RMSE), mean absolute error (MAE), and coefficient of determination (R^2). RMSE measures the deviation between predicted and actual values, reflecting prediction accuracy. MAE serves as an intuitive indicator of model error, with values approaching zero indicating higher accuracy. R^2 indicates the degree of fit between model estimates and observed values, with values approaching 1 demonstrating superior performance. The RMSE, MAE, and R^2 were calculated using Equations (1)–(3):

$$RMSE = \sqrt{\frac{\sum_{i=1}^{n}\left(y_t - y_p\right)^2}{n}} \quad (1)$$

$$MAE = \frac{\sum_{i=1}^{n}\left|y_t - y_p\right|}{n} \quad (2)$$

$$R^2 = 1 - \frac{\sum_{i=1}^{n}\left(y_p - y_t\right)^2}{\sum_{i=1}^{n}\left(y_t - y_m\right)^2} \quad (3)$$

where yp represents the predicted output value, yt denotes the reported true output value, ym represents the mean of observed output values, n indicates the number of samples in the training or testing datasets. Hyperparameter optimization was conducted using Optuna for XGBoost and grid search for GBR to address overfitting and improve generalization capability. SHAP analysis was implemented to provide model interpretability, quantifying feature contributions and identifying parameter interactions affecting supercapacitor performance.

4.3. Model Optimization and Interpretation

Hyperparameter optimization was conducted for the most promising algorithms based on preliminary evaluation. XGBoost was optimized for specific capacitance prediction using Optuna framework, exploring learning rate (0.01–0.3), maximum depth (3–10), number of estimators (50–500), subsample ratio (0.6–1.0), and regularization parameters with 100 trials using Tree-structured Parzen Estimator algorithm.

GBR was optimized for cycle stability prediction using grid search, systematically evaluating number of estimators (50–300), learning rate (0.05–0.2), maximum depth (3–8), and minimum samples split (2–10) to identify optimal parameter combinations.

Model interpretability was achieved through SHAP analysis, providing feature importance ranking, summary plots visualizing feature impact magnitude and direction, and interaction analysis revealing interdependencies between input features. SHAP values quantify the marginal contribution of each feature to individual predictions, enabling mechanistic understanding of relationships between material parameters and electrochemical performance for rational design guidance.

Supplementary Materials: The following supporting information can be downloaded at: https://www.mdpi.com/article/10.3390/gels11060464/s1, Figure S1. (a) specific capacitance and (b) cycle stability prediction. Figure S2. Violin plots of prediction error distributions from different ML models for (a) specific capacitance and (b) cycle stability. Figure S3. XGBoost optimization for cycle stability prediction. Figure S4. GBR optimization for specific capacitance prediction.

Author Contributions: Conceptualization, L.X., D.L. and S.L.; methodology, L.X. and Z.S.; software, L.X. and S.L.; validation, L.X., S.L. and Z.L.; formal analysis, Z.L.; investigation, Y.Z.; resources, J.L.; data curation, S.L. and Y.Z.; writing—original draft preparation, S.L.; writing—review and editing, L.X., D.H. and Z.S.; visualization, D.H.; supervision, D.L. and L.X.; project administration, Z.S.; funding acquisition, L.X. All authors have read and agreed to the published version of the manuscript.

Funding: This work was supported by the National Natural Science Foundation of China (22378089), the Natural Science Foundation of Heilongjiang Province (LH2023B007).

Institutional Review Board Statement: Not applicable.

Informed Consent Statement: Not applicable.

Data Availability Statement: The data presented in this study are available on request from the corresponding authors.

Conflicts of Interest: The authors declare no conflicts of interest.

References

1. Zhu, Q.; Zhao, D.; Cheng, M.; Zhou, J.; Owusu, K.A.; Mai, L.; Yu, Y. A New View of Supercapacitors: Integrated Supercapacitors. *Adv. Energy Mater.* **2019**, *9*, 1901081. [CrossRef]
2. Pandey, D.; Kumar, K.S.; Thomas, J. Supercapacitor Electrode Energetics and Mechanism of Operation: Uncovering the Voltage Window. *Prog. Mater. Sci.* **2024**, *141*, 101219. [CrossRef]
3. Zhang, Y.-Z.; Wang, Y.; Cheng, T.; Yao, L.-Q.; Li, X.; Lai, W.-Y.; Huang, W. Printed Supercapacitors: Materials, Printing and Applications. *Chem. Soc. Rev.* **2019**, *48*, 3229–3264. [CrossRef]
4. Pacheco, M.; Lefort, B.; Pacheco, J.; Santana, A.; Valdivia, R. Hydrogel and Carbon Nanostructures Based Supercapacitor: Initial Steps toward Green Supercapatteries. *J. Energy Storage* **2023**, *72*, 108539. [CrossRef]
5. Mredha, T.I.; Jeon, I. Biomimetic Anisotropic Hydrogels: Advanced Fabrication Strategies, Extraordinary Functionalities, and Broad Applications. *Prog. Mater. Sci.* **2022**, *124*, 100870. [CrossRef]
6. Tadesse, M.G.; Lübben, J.F. Review on Hydrogel-Based Flexible Supercapacitors for Wearable Applications. *Gels* **2023**, *9*, 106. [CrossRef]

7. Pozo, B.; Garate, J.I.; Ferreiro, S.; Fernandez, I.; de Gorostiza, E.F. Supercapacitor Electro-Mathematical and Machine Learning Modelling for Low Power Applications. *Electronics* **2018**, *7*, 44. [CrossRef]
8. Ke, S.; Wang, Z.; Zhang, K.; Cheng, F.; Sun, J.; Wang, N.; Zhu, Y. Flexible Conductive Cellulose Network-Based Composite Hydrogel for Multifunctional Supercapacitors. *Polymers* **2020**, *12*, 1369. [CrossRef]
9. Li, X.; Lou, D.; Wang, H.; Sun, X.; Li, J.; Liu, Y.-N. Flexible Supercapacitor Based on Organohydrogel Electrolyte with Long-Term Anti-Freezing and Anti-Drying Property. *Adv. Funct. Mater.* **2020**, *30*, 2007291. [CrossRef]
10. Huang, X.; Huang, J.; Yang, D.; Wu, P. A Multi-Scale Structural Engineering Strategy for High-Performance MXene Hydrogel Supercapacitor Electrode. *Adv. Sci.* **2021**, *8*, 2101664. [CrossRef] [PubMed]
11. Zhao, J.; Gong, J.; Wang, G.; Zhu, K.; Ye, K.; Yan, J.; Cao, D. A Self-Healing Hydrogel Electrolyte for Flexible Solid-State Supercapacitors. *Chem. Eng. J.* **2020**, *401*, 125456. [CrossRef]
12. Zhang, Q.; Wang, H.; Chen, S.; Liu, X.; Liu, J.; Liu, X. Proton Hydrogel-Based Supercapacitors with Rapid Low-Temperature Self-Healing Properties. *ACS Appl. Mater. Interfaces* **2024**, *16*, 40980–40991. [CrossRef]
13. Zhang, X.; Zhao, J.; Xia, T.; Li, Q.; Ao, C.; Wang, Q.; Zhang, W.; Lu, C.; Deng, Y. Hollow Polypyrrole/Cellulose Hydrogels for High-Performance Flexible Supercapacitors. *Energy Storage Mater.* **2020**, *31*, 135–145. [CrossRef]
14. Gao, H.; Li, Y.; Xie, Y.; Liang, D.; Li, J.; Wang, Y.; Xiao, Z.; Wang, H.; Gan, W.; Pattelli, L.; et al. Optical Wood with Switchable Solar Transmittance for All-Round Thermal Management. *Compos. Part B Eng.* **2024**, *275*, 111287. [CrossRef]
15. Lu, C.; Chen, X. All-Temperature Flexible Supercapacitors Enabled by Antifreezing and Thermally Stable Hydrogel Electrolyte. *Nano Lett.* **2020**, *20*, 1907–1914. [CrossRef]
16. Yu, F.; Wang, W.; Shi, Y.; Li, Y.; Liang, D.; Du, M.; Liu, F. Design of High-Performance Wood-Derived N-Doped ECR Electrocatalysts Based on Marcus Theory. *Appl. Catal. B Environ. Energy* **2024**, *358*, 124340. [CrossRef]
17. Yan, B.; Shi, Y.; Liu, S.; Wang, W.; Ba, Z.; Li, H.; Wang, J.; Xiao, Z.; Peng, T.; Liang, D.; et al. Development and Optimization of Lithium-Ion Sieves through Machine Learning in Complex Brine Systems. *Sep. Purif. Technol.* **2025**, *374*, 133726. [CrossRef]
18. Yong, S.; Hillier, N.; Beeby, S.P. Phase-Inverted Copolymer Membrane for the Enhancement of Textile Supercapacitors. *Polymers* **2022**, *14*, 3399. [CrossRef]
19. Wang, W.; Cao, J.; Yu, J.; Tian, F.; Luo, X.; Hao, Y.; Huang, J.; Wang, F.; Zhou, W.; Xu, J.; et al. Flexible Supercapacitors Based on Stretchable Conducting Polymer Electrodes. *Polymers* **2023**, *15*, 1856. [CrossRef]
20. Yuan, C.; Shi, Y.; Ba, Z.; Liang, D.; Wang, J.; Liu, X.; Xu, Y.; Liu, J.; Xu, H. Machine Learning Models for Predicting Thermal Properties of Radiative Cooling Aerogels. *Gels* **2025**, *11*, 70. [CrossRef]
21. Ba, Z.; Yu, F.; Fan, H.; Wang, W.; Wang, Y.; Liang, D.; Xiao, Z.; Wang, H.; Xie, Y. Antileakage Performance of Schiff Base-Reinforced Thermal Energy Storage Wood for Indoor Temperature Control. *Ind. Crops Prod.* **2024**, *214*, 118561. [CrossRef]
22. Zou, Y.; Chen, C.; Sun, Y.; Gan, S.; Dong, L.; Zhao, J.; Rong, J. Flexible, All-Hydrogel Supercapacitor with Self-Healing Ability. *Chem. Eng. J.* **2021**, *418*, 128616. [CrossRef]
23. Li, G.; Zhang, X.; Sang, M.; Wang, X.; Zuo, D.; Xu, J.; Zhang, H. A Supramolecular Hydrogel Electrolyte for High-Performance Supercapacitors. *J. Energy Storage* **2021**, *33*, 101931. [CrossRef]
24. Tao, F.; Qin, L.; Wang, Z.; Pan, Q. Self-Healable and Cold-Resistant Supercapacitor Based on a Multifunctional Hydrogel Electrolyte. *ACS Appl. Mater. Interfaces* **2017**, *9*, 15541–15548. [CrossRef]
25. Cui, S.; Miao, W.; Wang, X.; Sun, K.; Peng, H.; Ma, G. Multifunctional Zincophilic Hydrogel Electrolyte with Abundant Hydrogen Bonds for Zinc-Ion Capacitors and Supercapacitors. *ACS Nano* **2024**, *18*, 12355–12366. [CrossRef]
26. Hu, Y.; Zhan, Y.; Xu, M.; Niu, F.; Chen, Y.; Yang, Q.; Xiong, C.; Shi, Z. One-Step Preparation of Flexible Nanocellulose-Based Composite Hydrogel Supercapacitors with High Specific Capacitance. *Compos. Sci. Technol.* **2022**, *230*, 109725. [CrossRef]
27. Li, R.; Jia, W.; Wen, J.; Hu, G.; Tang, T.; Li, X.; Jiang, L.; Li, M.; Huang, H.; Fang, G. MXene/Zwitterionic Hydrogel Oriented Anti-Freezing and High-Performance Zinc–Ion Hybrid Supercapacitor. *Adv. Funct. Mater.* **2024**, *34*, 2409207. [CrossRef]
28. Wei, J.; Zhou, J.; Su, S.; Jiang, J.; Feng, J.; Wang, Q. Water-Deactivated Polyelectrolyte Hydrogel Electrolytes for Flexible High-Voltage Supercapacitors. *ChemSusChem* **2018**, *11*, 3410–3415. [CrossRef]
29. Wang, D.; Yang, F.; Cong, L.; Feng, W.; Wang, C.; Chu, F.; Nan, J.; Chen, R. Lignin-Containing Hydrogel Matrices with Enhanced Adhesion and Toughness for All-Hydrogel Supercapacitors. *Chem. Eng. J.* **2022**, *450*, 138025. [CrossRef]
30. Shabeeba, A.; Rajan, L.; Sidheekha, M.P.; Thayyil, M.S.; Ismail, Y.A. Polypyrrole/Hydrogel Hybrid Films as Multi Sensing Supercapacitor Electrodes. *J. Energy Storage* **2022**, *55*, 105724. [CrossRef]
31. Chen, X.-X.; Ju, Y.-X.; Zhang, B.; Ge, X.-R.; Liu, E.-J.; Zhang, D.-Y.; Wang, J.; Yao, X.-H.; Zhao, W.-G.; Chen, T. High-Performance Supercapacitors Based on Coarse Nanofiber Bundle and Ordered Network Hydrogels. *Int. J. Biol. Macromol.* **2025**, *292*, 139208. [CrossRef]
32. Cui, S.; Lv, Y.; Jia, Y.; Hu, Q.; Hou, W.; Miao, W.; Shi, X.; Sun, K.; Peng, H.; Ma, G. A Flexible and Hydrophilic Hydrogel Film Based All-in-One Supercapacitor. *J. Energy Storage* **2023**, *68*, 107877. [CrossRef]
33. Ye, X.; Huang, H.; Chen, L.; Wang, Y.; Weng, M.; Zhang, L.; Luo, Z. All-Round Supercapacitors: A Rational Design of Sustainable Hydrogel Electrolyte Utilizing Calcium Salts. *Chem. Eng. J.* **2024**, *483*, 149158. [CrossRef]

34. Fang, L.; Cai, Z.; Ding, Z.; Chen, T.; Zhang, J.; Chen, F.; Shen, J.; Chen, F.; Li, R.; Zhou, X.; et al. Skin-Inspired Surface-Microstructured Tough Hydrogel Electrolytes for Stretchable Supercapacitors. *ACS Appl. Mater. Interfaces* **2019**, *11*, 21895–21903. [CrossRef] [PubMed]
35. Guo, G.; Sun, Y.; Fu, Q.; Ma, Y.; Zhou, Y.; Xiong, Z.; Liu, Y. Sol-Gel Synthesis of Ternary Conducting Polymer Hydrogel for Application in All-Solid-State Flexible Supercapacitor. *Int. J. Hydrog. Energy* **2019**, *44*, 6103–6115. [CrossRef]
36. Qi, Z.; Ren, R.; Hu, J.; Chen, Y.; Guo, Y.; Huang, Y.; Wei, J.; Zhang, H.; Pang, Q.; Zhang, X.; et al. Flexible Supercapacitor with Wide Electrochemical Stable Window Based on Hydrogel Electrolyte. *Small* **2024**, *20*, 2400369. [CrossRef]
37. Liu, S.; Zhong, Y.; Zhang, X.; Pi, M.; Wang, X.; Zhu, R.; Cui, W.; Ran, R. Highly Deformable, Conductive Double-Network Hydrogel Electrolytes for Durable and Flexible Supercapacitors. *ACS Appl. Mater. Interfaces* **2022**, *14*, 15641–15652. [CrossRef]
38. Zeng, J.; Dong, L.; Sha, W.; Wei, L.; Guo, X. Highly Stretchable, Compressible and Arbitrarily Deformable All-Hydrogel Soft Supercapacitors. *Chem. Eng. J.* **2020**, *383*, 123098. [CrossRef]
39. Chen, Y.; Liu, L.; Huang, Y.; Cao, H.; Liu, T.; Qi, Z.; Hu, J.; Guo, Y.; Sun, J.; Liang, M.; et al. Low-Salt Organohydrogel Electrolytes for Wide-Potential-Window Flexible All-Solid-State Supercapacitors. *Appl. Energy* **2024**, *363*, 123100. [CrossRef]
40. Yang, Y.; Zhang, D.; Liu, Y.; Shen, L.; Zhu, T.; Xu, X.; Zheng, J.; Gong, X. Solid-State Double-Network Hydrogel Redox Electrolytes for High-Performance Flexible Supercapacitors. *ACS Appl. Mater. Interfaces* **2021**, *13*, 34168–34177. [CrossRef]
41. Zhang, Y.; Sun, Y.; Nan, J.; Yang, F.; Wang, Z.; Li, Y.; Wang, C.; Chu, F.; Liu, Y.; Wang, C. In Situ Polymerization of Hydrogel Electrolyte on Electrodes Enabling the Flexible All-Hydrogel Supercapacitors with Low-Temperature Adaptability. *Small* **2024**, *20*, 2309900. [CrossRef] [PubMed]
42. Li, Z.; Gou, S.; Zhang, Z.; Yang, Y.; Wang, S.; Hu, Z.; Lu, X. Mechanically Stable All-Hydrogel Supercapacitor Achieved by Electrodes with Excellent Flexibility and High Capacitance Performance. *J. Energy Storage* **2024**, *84*, 110861. [CrossRef]
43. Lin, J.; Du, X. Self-Healable and Redox Active Hydrogel Obtained via Incorporation of Ferric Ion for Supercapacitor Applications. *Chem. Eng. J.* **2022**, *446*, 137244. [CrossRef]
44. Ma, J.; Zhu, J.; Zhou, S.; Zhao, C.; Liu, C.; Xin, Z.; Cai, J.; He, J.; Feng, P.; Guo, L.; et al. Multifunctional Starch-Based Conductive Hydrogels for Smart Sensors and Flexible Supercapacitors. *Int. J. Biol. Macromol.* **2025**, *302*, 140430. [CrossRef] [PubMed]
45. Qin, L.; Yang, G.; Li, D.; Ou, K.; Zheng, H.; Fu, Q.; Sun, Y. High Area Energy Density of All-Solid-State Supercapacitor Based on Double-Network Hydrogel with High Content of Graphene/PANI Fiber. *Chem. Eng. J.* **2022**, *430*, 133045. [CrossRef]
46. Cheng, T.; Liu, Z.-T.; Qu, J.; Meng, C.-F.; He, L.-J.; Li, L.; Yang, X.-L.; Cao, Y.-J.; Han, K.; Zhang, Y.-Z.; et al. High-Performance Organic–Inorganic Hybrid Conductive Hydrogels for Stretchable Elastic All-Hydrogel Supercapacitors and Flexible Self-Powered Integrated Systems. *Adv. Sci.* **2024**, *11*, 2403358. [CrossRef]
47. Guo, J.; Bao, D.; Zhang, S.; Zhang, X.; Ji, X.; Li, M.; Guan, F. Sodium Alginate-Based High Conductive, Ultra-Stretchable Hydrogel Fibers for Electrolytes of Flexible Solid-State Supercapacitors. *Int. J. Biol. Macromol.* **2024**, *276*, 133894. [CrossRef]
48. Fei, Y.; Jiang, Z.; Zhou, D.; Meng, F.; Wu, Y.; Xiong, Y.; Ye, Y.; Liu, T.; Fei, Z.; Kuang, T.; et al. Preparation a Highly Sensitive and Flexible Textile Supercapacitor Based on Lignin Hydrogel and Polyaniline@carbon Cloth Composites. *J. Energy Storage* **2023**, *73*, 108978. [CrossRef]
49. Xu, M.; Zhu, J.; Xie, J.; Mao, Y.; Hu, W. Dynamically Cross-Linked, Self-Healable, and Stretchable All-Hydrogel Supercapacitor with Extraordinary Energy Density and Real-Time Pressure Sensing. *Small* **2024**, *20*, 2305448. [CrossRef]
50. Zaidi, S.F.A.; Saeed, A.; Ho, V.-C.; Heo, J.H.; Cho, H.H.; Sarwar, N.; Lee, N.-E.; Mun, J.; Lee, J.H. Chitosan-Reinforced Gelatin Composite Hydrogel as a Tough, Anti-Freezing, and Flame-Retardant Gel Polymer Electrolyte for Flexible Supercapacitors. *Int. J. Biol. Macromol.* **2023**, *234*, 123725. [CrossRef]
51. Wanyan, H.; Li, Q.; Huang, H.; Li, J.; Huang, L.; Chen, L.; Wei, J.; Zhou, X.; Tang, Z.; Wu, H. Flexible High Electrochemical Active Hydrogel for Wearable Sensors and Supercapacitor Electrolytes. *Int. J. Biol. Macromol.* **2024**, *277*, 134356. [CrossRef] [PubMed]
52. Peng, K.; Wang, W.; Zhang, J.; Ma, Y.; Lin, L.; Gan, Q.; Chen, Y.; Feng, C. Preparation of Chitosan/Sodium Alginate Conductive Hydrogels with High Salt Contents and Their Application in Flexible Supercapacitors. *Carbohydr. Polym.* **2022**, *278*, 118927. [CrossRef] [PubMed]
53. Luo, N.; Wang, J.; Zhang, D.; Zhao, Y.; Wei, Y.; Liu, Y.; Zhang, Y.; Han, S.; Kong, X.; Huo, P. Inorganic Nanoparticle-Enhanced Double-Network Hydrogel Electrolytes for Supercapacitor with Superior Low-Temperature Adaptability. *Chem. Eng. J.* **2024**, *479*, 147741. [CrossRef]
54. Zhang, K.; Pang, Y.; Chen, C.; Wu, M.; Liu, Y.; Yu, S.; Li, L.; Ji, Z.; Pang, J. Stretchable and Conductive Cellulose Hydrogel Electrolytes for Flexible and Foldable Solid-State Supercapacitors. *Carbohydr. Polym.* **2022**, *293*, 119673. [CrossRef] [PubMed]
55. Zhang, R.; Zhang, Z.; Xu, P.; Xu, J.; Gao, Y.; Gao, G. Cellulose Nanofiber Hydrogel with High Conductivity Electrolytes for High Voltage Flexible Supercapacitors. *Carbohydr. Polym.* **2024**, *326*, 121654. [CrossRef]
56. Wang, R.; Lei, D.; Zhang, H.; Liang, J.; Shang, Z.; Wan, L.; Li, Z.; Si, C. An Interface-Integrated Hydrogel for All-in-One Flexible Supercapacitor with Excellent Wide-Temperature and Self-Healing Properties. *Compos. Part B Eng.* **2024**, *275*, 111345. [CrossRef]
57. Xu, B.; Yang, G. Interpretability Research of Deep Learning: A Literature Survey. *Inf. Fusion* **2025**, *115*, 102721. [CrossRef]

58. Ba, Z.; Yu, F.; Shi, Y.; Fan, H.; He, Z.; Wang, Y.; Liang, D.; Wang, H.; Xiao, Z.; Qiu, Z.; et al. Thermal Energy Storage Wood with Anti-Leakage and Fire-Retardant for Indoor Low-Grade Thermal Energy Utilization. *J. Energy Storage* **2025**, *126*, 117009. [CrossRef]
59. Li, W.; Gao, F.; Wang, X.; Zhang, N.; Ma, M. Strong and Robust Polyaniline-Based Supramolecular Hydrogels for Flexible Supercapacitors. *Angew. Chem. Int. Ed.* **2016**, *55*, 9196–9201. [CrossRef]
60. Zhu, T.; Ni, Y.; Biesold, G.M.; Cheng, Y.; Ge, M.; Li, H.; Huang, J.; Lin, Z.; Lai, Y. Recent Advances in Conductive Hydrogels: Classifications, Properties, and Applications. *Chem. Soc. Rev.* **2023**, *52*, 473–509. [CrossRef]
61. Sattari, M.T.; Falsafian, K.; Irvem, A.; S, S.; Qasem, S.N. Potential of Kernel and Tree-Based Machine-Learning Models for Estimating Missing Data of Rainfall. *Eng. Appl. Comput. Fluid Mech.* **2020**, *14*, 1078–1094. [CrossRef]
62. Asadi, N.; Lin, J.; de Vries, A.P. Runtime Optimizations for Tree-Based Machine Learning Models. *IEEE Trans. Knowl. Data Eng.* **2014**, *26*, 2281–2292. [CrossRef]
63. Duță, Ș.; Sultana, A.E. Optimizing Depression Classification Using Combined Datasets and Hyperparameter Tuning with Optuna. *Sensors* **2025**, *25*, 2083. [CrossRef] [PubMed]
64. Samukaite-Bubniene, U.; Valiūnienė, A.; Bucinskas, V.; Genys, P.; Ratautaite, V.; Ramanaviciene, A.; Aksun, E.; Tereshchenko, A.; Zeybek, B.; Ramanavicius, A. Towards Supercapacitors: Cyclic Voltammetry and Fast Fourier Transform Electrochemical Impedance Spectroscopy Based Evaluation of Polypyrrole Electrochemically Deposited on the Pencil Graphite Electrode. *Colloids Surf. A Physicochem. Eng. Asp.* **2021**, *610*, 125750. [CrossRef]
65. Zhu, C.; Bamidele, E.A.; Shen, X.; Zhu, G.; Li, B. Machine Learning Aided Design and Optimization of Thermal Metamaterials. *Chem. Rev.* **2024**, *124*, 4258–4331. [CrossRef]
66. Zheng, J.-J.; Li, Q.-Z.; Wang, Z.; Wang, X.; Zhao, Y.; Gao, X. Computer-Aided Nanodrug Discovery: Recent Progress and Future Prospects. *Chem. Soc. Rev.* **2024**, *53*, 9059–9132. [CrossRef]

MDPI AG
Grosspeteranlage 5
4052 Basel
Switzerland
Tel.: +41 61 683 77 34

Gels Editorial Office
E-mail: gels@mdpi.com
www.mdpi.com/journal/gels

www.ingramcontent.com/pod-product-compliance
Lightning Source LLC
Chambersburg PA
CBHW040838250926
51850CB00007BA/1346